AF361725

# Recent Numerical Advances in Fluid Mechanics

# Recent Numerical Advances in Fluid Mechanics

Special Issue Editor

**Omer San**

MDPI • Basel • Beijing • Wuhan • Barcelona • Belgrade • Manchester • Tokyo • Cluj • Tianjin

*Special Issue Editor*
Omer San
Oklahoma State University
USA

*Editorial Office*
MDPI
St. Alban-Anlage 66
4052 Basel, Switzerland

This is a reprint of articles from the Special Issue published online in the open access journal *Fluids* (ISSN 2311-5521) (available at: https://www.mdpi.com/journal/fluids/special_issues/numerical_fluid_mechanics).

For citation purposes, cite each article independently as indicated on the article page online and as indicated below:

LastName, A.A.; LastName, B.B.; LastName, C.C. Article Title. *Journal Name* **Year**, *Article Number*, Page Range.

**ISBN 978-3-03936-402-2 (Hbk)**
**ISBN 978-3-03936-403-9 (PDF)**

# Contents

# About the Special Issue Editor

**Omer San** served as Assistant Professor of Mechanical and Aerospace Engineering at OSU since 2015. He received his bachelor's in Aeronautical Engineering from Istanbul Technical University in 2005, his master's in Aerospace Engineering from Old Dominion University in 2007, and his Ph.D. in Engineering Mechanics from Virginia Tech in 2012. He worked as a postdoc at Virginia Tech from 2012 to 2014, and then from 2014 to 2015 at the University of Notre Dame, Indiana. His field of study is centered upon the development, analysis, and application of advanced computational methods in science and engineering with a particular emphasis on fluid dynamics across a variety of spatial and temporal scales.

*Editorial*

# Recent Numerical Advances in Fluid Mechanics

Omer San

School of Mechanical and Aerospace Engineering, Oklahoma State University, Stillwater, OK 74078, USA; osan@okstate.edu; Tel.: +1-405-744-2457; Fax: +1-405-744-7873

Received: 7 May 2020; Accepted: 11 May 2020; Published: 15 May 2020

---

In recent decades, the field of computational fluid dynamics has made significant advances in enabling advanced computing architectures to understand many phenomena in biological, geophysical, and engineering fluid flows. Almost all research areas in fluids use numerical methods at various complexities: from molecular to continuum descriptions; from laminar to turbulent regimes; from low-speed to hypersonic, from stencil-based computations to meshless approaches; from local basis functions to global expansions, as well as from 1st-order approximation to high order and spectral accuracy. Many successful efforts have been put forth in dynamic adaptation strategies, e.g., adaptive mesh refinement and multiresolution representation approaches. Furthermore, with recent advances in artificial intelligence and heterogeneous computing, broader fluids community has gained momentum to revisit and investigate such practices. This Special Issue, containing a collection of 13 papers, brings together researchers to address recent numerical advances in fluid mechanics.

Murea [1] presented an efficient semi-implicit monolithic method to simulate fluid-structure interaction problems within the Arbitrary Lagrangian Eulerian framework. The paper by Mohebbi et al. [2] exploits an accurate implementation of Kutta conditions for the airfoils with both finite angle and cusp trailing edges. Iqbal et al. [3] focused on the non-linear Schrödinger equation, and proposed a numerical method utilizing cubic B-spline Galerkin method. The paper by Karakouzian et al. [4] is devoted to the numerical modeling of hydraulic fracturing process in pressure tunnels. The authors performed a systematic finite element analysis to gauge guidelines for reducing the possibility of hydraulic fracturing in hydroelectric power plants. Cremades Rey et al. [5] concentrated their attention to the uncertainty quantification of Reynolds-averaged Navier-Stokes models using the open source package OpenFOAM. In their manuscript, Patel and Mathew [6] introduced a concept of adaptive explicit filtering for implicit large eddy simulations. The core idea is to utilize a second order accurate explicit filter in the presence of a discontinuity, and then gradually increase the order of explicit filter at the grid locations away from this discontinuity. The paper by Butcher and Spencer [7] takes up the concept of proper orthogonal decomposition (POD) for the identification of coherent structures. A new metric has been introduced for the selection of POD modes in reconstructing coherent or incoherent features. Idrissi et al. [8] dealt with an atmospheric dispersion study in highly realistic urban areas considering high level of details. They utilized large eddy simulations to explore the behaviour of the released pollutants in various topological settings. The article by Guilizzoni et al. [9] investigates the impact of multiple synchronized drops into a deep pool. In their paper, Bourantas et al. [10] put forth a strong form meshless point collocation method for solving the unsteady incompressible Navier-Stokes equations. Fakhari [11] considered to develop a wall model for large eddy simulations in simulating either the body fitted or immersed boundary problems. The paper by Ahmed and San [12] discusses a partitioning approach to break the Kolmogorov barrier in projection based reduced order modeling of transient flows. Finally, Xie et al. [13] introduced a machine learning approach to model the interaction between the resolved and unresolved modes in reduced order models.

**Conflicts of Interest:** The author declare no conflict of interest.

## References

1. Murea, C.M. Three-Dimensional Simulation of Fluid–Structure Interaction Problems Using Monolithic Semi-Implicit Algorithm. *Fluids* **2019**, *4*, 94. [CrossRef]
2. Mohebbi, F.; Evans, B.; Sellier, M. On the Kutta Condition in Compressible Flow over Isolated Airfoils. *Fluids* **2019**, *4*, 102. [CrossRef]
3. Iqbal, A.; Abd Hamid, N.N.; Ismail, M.; Izani, A. Soliton Solution of Schrödinger Equation Using Cubic B-Spline Galerkin Method. *Fluids* **2019**, *4*, 108. [CrossRef]
4. Karakouzian, M.; Nazari-Sharabian, M.; Karami, M. Effect of Overburden Height on Hydraulic Fracturing of Concrete-Lined Pressure Tunnels Excavated in Intact Rock: A Numerical Study. *Fluids* **2019**, *4*, 112. [CrossRef]
5. Cremades Rey, L.F.; Hinz, D.F.; Abkar, M. Reynolds Stress Perturbation for Epistemic Uncertainty Quantification of RANS Models Implemented in OpenFOAM. *Fluids* **2019**, *4*, 113. [CrossRef]
6. Patel, S.K.; Mathew, J. Shock Capturing in Large Eddy Simulations by Adaptive Filtering. *Fluids* **2019**, *4*, 132. [CrossRef]
7. Butcher, D.; Spencer, A. Cross-Correlation of POD Spatial Modes for the Separation of Stochastic Turbulence and Coherent Structures. *Fluids* **2019**, *4*, 134. [CrossRef]
8. Idrissi, M.S.; Ben Salah, N.; Chrigui, M. Numerical Modelling of Air Pollutant Dispersion in Complex Urban Areas: Investigation of City Parts from Downtowns Hanover and Frankfurt. *Fluids* **2019**, *4*, 137. [CrossRef]
9. Guilizzoni, M.; Santini, M.; Fest-Santini, S. Synchronized Multiple Drop Impacts into a Deep Pool. *Fluids* **2019**, *4*, 141. [CrossRef]
10. Bourantas, G.C.; Zwick, B.F.; Joldes, G.R.; Loukopoulos, V.C.; Tavner, A.C.; Wittek, A.; Miller, K. An Explicit Meshless Point Collocation Solver for Incompressible Navier-Stokes Equations. *Fluids* **2019**, *4*, 164. [CrossRef]
11. Fakhari, A. A New Wall Model for Large Eddy Simulation of Separated Flows. *Fluids* **2019**, *4*, 197. [CrossRef]
12. Ahmed, S.E.; San, O. Breaking the Kolmogorov Barrier in Model Reduction of Fluid Flows. *Fluids* **2020**, *5*, 26. [CrossRef]
13. Xie, X.; Webster, C.; Iliescu, T. Closure Learning for Nonlinear Model Reduction Using Deep Residual Neural Network. *Fluids* **2020**, *5*, 39. [CrossRef]

 *fluids*

*Article*

# Three-Dimensional Simulation of Fluid–Structure Interaction Problems Using Monolithic Semi-Implicit Algorithm

Cornel Marius Murea

Département de Mathématiques, IRIMAS, Université de Haute Alsace, 6, rue des Frères Lumière,
68093 Mulhouse CEDEX, France; cornel.murea@uha.fr

Received: 17 March 2019; Accepted: 13 May 2019; Published: 22 May 2019

**Abstract:** A monolithic semi-implicit method is presented for three-dimensional simulation of fluid–structure interaction problems. The updated Lagrangian framework is used for the structure modeled by linear elasticity equation and, for the fluid governed by the Navier–Stokes equations, we employ the Arbitrary Lagrangian Eulerian method. We use a global mesh for the fluid–structure domain where the fluid–structure interface is an interior boundary. The continuity of velocity at the interface is automatically satisfied by using globally continuous finite element for the velocity in the fluid–structure mesh. The method is fast because we solve only a linear system at each time step. Three-dimensional numerical tests are presented.

**Keywords:** fluid–structure interaction; monolithic method; Updated Lagrangian; Arbitrary Lagrangian Eulerian

---

## 1. Introduction

There exists a rich literature on solving numerically fluid–structure interaction problems. Some methods are based on partitioned procedures, the fluid and structure sub-problems are solved separately using iterative process: fixed point iterations [1–3], Newton-like methods [4–6] or optimization techniques [7–9]. Monolithic methods solve the fluid–structure interaction problem as a single system of equations, [10–13], or more recently [14–17].

In some methods such as the Arbitrary Lagrangian Eulerian (ALE) framework, the fluid equations are written over a moving mesh which follows the structure displacement (see [18,19]). Other methods use a fixed mesh for fluid domain: immersed boundary method [20], distributed Lagrange multiplier [21,22], penalization [23,24], extended finite element method (XFEM) [25,26], and Nitsche-XFEM [27]. Distributed Lagrange multiplier strategy with remeshing is used in [28] and a monolithic fictitious domain without Lagrange multiplier is employed in [29,30].

Most of these methods are implicit. For a long time, the explicit methods were considered not suitable because of the lack of stability, but these methods are successfully applied in [31,32]. A third class of methods are so-called semi-implicit methods, where the domain is computed explicitly while the fluid velocity and pressure as well as the structure displacement are computed implicitly, [33,34]. In [35], it is proved that a schema of this kind is unconditionally stable.

In this paper, a monolithic semi-implicit method is employed for three-dimensional simulation. For the structure modeled by the linear elasticity equations, we use the updated Lagrangian framework and, for the fluid governed by the Navier–Stokes equations, we employ the ALE method. A similar strategy is used in [36] for a bi-dimensional compressible neo-Hookean model or in [37] for a bi-dimensional linear elasticity model for the structure. As in [38], we employ a global mesh for the fluid–structure domain where the fluid–structure interface is an interior boundary. Using globally continuous finite element for the velocity in the fluid–structure mesh, the continuity of velocity at the interface is automatically satisfied. The method is fast because we solve only a linear system at each time step. Three-dimensional numerical tests are presented.

In [14–17], a global mesh obtained from the deformed structure mesh and a fluid mesh generated at each time step, compatible at the interface with the structure mesh are used. Remeshing the fluid domain improves the quality of the mesh in the case of large deformation. The non-linear structure equation written in the Eulerian coordinates is obtained by using Cayley–Hamilton theorem. The fluid equations are solved by the characteristics method. The weak formulation of the fluid–structure interaction problem is written in the Eulerian domain, which is unknown, and a fixed-point algorithm solves the global non-linear problem at each time step.

In [30], it is assumed that the structure is viscoelastic with the same viscosity as the fluid. Based on fictitious domain without Lagrange multiplier, the fluid is solved in a fixed mesh of the fluid–structure domain. The weak formulation contains integrals over the unknown Eulerian domain of the structure. At each time step, a fixed-point algorithm is employed.

In [36], by using the Updated Lagrangian framework for a compressible Neo-Hookean structure, the weak formulation is written in the known configuration obtained at the precedent time step. By linearization around this configuration, at each time step, only a linear system is solved for the fluid–structure coupled equations and consequently the computing time is reduced. In the present paper, we follow this approach for three-dimensional simulations using linear elastic model for the structure.

If at each time step of the monolithic implicit methods, the fixed-point algorithm does not converge quickly or the computational time by fixed-point iteration is very expensive, thus the monolithic semi-implicit methods, which have similar stability properties and a reduced computational time, could be a good alternative.

## 2. Problem Statement

The initial fluid domain $\Omega_0^F$ is a right circular cylinder of bases $\Sigma_1$, $\Sigma_3$ and lateral surface $\Gamma_0$. We denote by $\Omega_0^S$ the initial structure domain and we assume that it is a right annular cylinder of bases $\Sigma_5$, $\Sigma_7$, interior lateral surface $\Gamma_0$ and exterior lateral surface $\Gamma_0^N$ (see Figure 1). We suppose that the initial structure domain is undeformed (stress-free). The boundary $\Gamma_0$ is common of both domains and it represents the initial position of the fluid–structure interface.

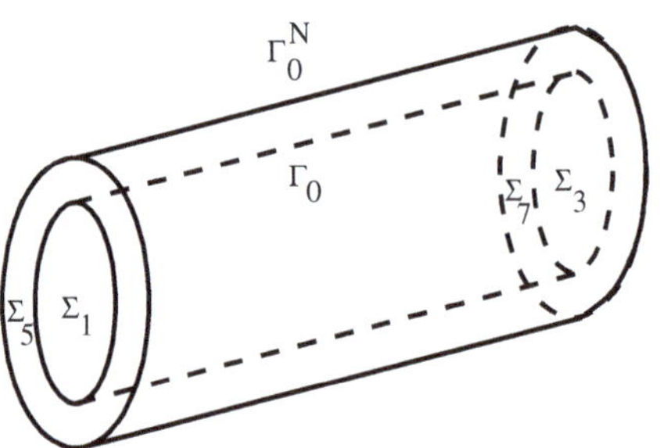


**Figure 1.** Initial geometrical configuration.

At the time instant $t$, the fluid occupies the domain $\Omega_t^F$ bounded by the moving interface $\Gamma_t$ and by the fixed boundaries $\Sigma_1$, $\Sigma_3$, while the structure occupies the domain $\Omega_t^S$ bounded by the moving lateral surfaces $\Gamma_t$, $\Gamma_t^N$ and by the fixed boundaries $\Sigma_5$, $\Sigma_7$.

We denote by $\mathbf{U}^S : \Omega_0^S \times [0, T] \to \mathbb{R}^3$ the displacement of the structure. A particle of the structure whose initial position was the point $\mathbf{X}$ will occupy the position $\mathbf{x} = \mathbf{X} + \mathbf{U}^S(\mathbf{X}, t)$ in the deformed domain $\Omega_t^S$. At the time instant $t$, the interface $\Gamma_t$ is the image of $\Gamma_0$ via the map $\mathbf{X} \to \mathbf{X} + \mathbf{U}^S(\mathbf{X}, t)$. The same relationship exists between $\Gamma_t^N$ and $\Gamma_0^N$. On $\Gamma_0^D = \Sigma_5 \cup \Sigma_7$, we impose zero displacements.

We set $\nabla_{\mathbf{X}}\mathbf{U}^S$ the gradient of the displacement $\mathbf{U}^S = \left(U_1^S, U_2^S, U_3^S\right)^T$ with respect to the Lagrangian coordinates $\mathbf{X} = (X_1, X_2, X_3)^T$. We denote by $\mathbf{F}(\mathbf{X},t) = \mathbf{I} + \nabla_{\mathbf{X}}\mathbf{U}^S(\mathbf{X},t)$ the gradient of the deformation, where $\mathbf{I}$ is the unity matrix and we set $J(\mathbf{X},t) = \det \mathbf{F}(\mathbf{X},t)$. The second Piola–Kirchhoff stress tensor is denoted by $\mathbf{\Sigma}$.

We assume that the fluid is governed by the Navier–Stokes equations. For each time instant $t \in [0,T]$, we denote the fluid velocity by $\mathbf{v}^F(t) = \left(v_1^F(t), v_2^F(t), v_3^F(t)\right)^T : \Omega_t^F \to \mathbb{R}^3$ and the fluid pressure by $p^F(t) : \Omega_t^F \to \mathbb{R}$. Let $\epsilon\left(\mathbf{v}^F\right) = \frac{1}{2}\left(\nabla\mathbf{v}^F + \left(\nabla\mathbf{v}^F\right)^T\right)$ be the fluid rate of strain tensor and let $\sigma^F = -p^F\mathbf{I} + 2\mu^S\epsilon\left(\mathbf{v}^F\right)$ be the fluid stress tensor. To simplify the notation, we write $\nabla\mathbf{v}^F$ in place of $\nabla_{\mathbf{x}}\mathbf{v}^F$, when the gradients are computed with respect to the Eulerian coordinates $\mathbf{x}$.

The problem is to find the structure displacement $\mathbf{U}^S$, the fluid velocity $\mathbf{v}^F$ and the fluid pressure $p^F$ such that:

$$\rho_0^S(\mathbf{X})\frac{\partial^2\mathbf{U}^S}{\partial t^2}(\mathbf{X},t) - \nabla_{\mathbf{X}}\cdot(\mathbf{F\Sigma})(\mathbf{X},t) \;=\; \rho_0^S(\mathbf{X})\,\mathbf{g}, \quad \text{in } \Omega_0^S \times (0,T) \tag{1}$$

$$\mathbf{U}^S(\mathbf{X},t) \;=\; 0, \quad \text{on } \Gamma_0^D \times (0,T) \tag{2}$$

$$(\mathbf{F\Sigma})(\mathbf{X},t)\,\mathbf{N}^S(\mathbf{X}) \;=\; 0, \quad \text{on } \Gamma_0^N \times (0,T) \tag{3}$$

$$\rho^F\left(\frac{\partial\mathbf{v}^F}{\partial t} + (\mathbf{v}^F\cdot\nabla)\mathbf{v}^F\right) - 2\mu^F\nabla\cdot\epsilon\left(\mathbf{v}^F\right) + \nabla p^F \;=\; \rho^F\mathbf{g}, \; \forall t \in (0,T), \forall \mathbf{x} \in \Omega_t^F \tag{4}$$

$$\nabla\cdot\mathbf{v}^F \;=\; 0, \; \forall t \in (0,T), \forall \mathbf{x} \in \Omega_t^F \tag{5}$$

$$\sigma^F\mathbf{n}^F \;=\; \mathbf{h}_{in}, \quad \text{on } \Sigma_1 \times (0,T) \tag{6}$$

$$\sigma^F\mathbf{n}^F \;=\; \mathbf{h}_{out}, \quad \text{on } \Sigma_3 \times (0,T) \tag{7}$$

$$\mathbf{v}^F\left(\mathbf{X} + \mathbf{U}^S(\mathbf{X},t), t\right) \;=\; \frac{\partial\mathbf{U}^S}{\partial t}(\mathbf{X},t), \text{ on } \Gamma_0 \times (0,T) \tag{8}$$

$$\left(\sigma^F\mathbf{n}^F\right)_{\left(\mathbf{X}+\mathbf{U}^S(\mathbf{X},t),t\right)} \;=\; -(\mathbf{F\Sigma})(\mathbf{X},t)\,\mathbf{N}^S(\mathbf{X}), \text{ on } \Gamma_0 \times (0,T) \tag{9}$$

$$\mathbf{U}^S(\mathbf{X},0) \;=\; \mathbf{U}^{S,0}(\mathbf{X}), \text{ in } \Omega_0^S \tag{10}$$

$$\frac{\partial\mathbf{U}^S}{\partial t}(\mathbf{X},0) \;=\; \mathbf{V}^{S,0}(\mathbf{X}), \text{ in } \Omega_0^S \tag{11}$$

$$\mathbf{v}^F(\mathbf{X},0) \;=\; \mathbf{v}^{F,0}(\mathbf{X}), \text{ in } \Omega_0^F \tag{12}$$

where $\rho_0^S : \Omega_0^S \to \mathbb{R}$ is the initial mass density of the structure, $\mathbf{g}$ is the acceleration of gravity vector and it is assumed to be constant, $\mathbf{N}^S$ is the unit outer normal vector along the boundary $\partial\Omega_0^S$, $\mathbf{U}^{S,0}$ and $\mathbf{V}^{S,0}$ are the initial displacement and velocity of the structure, $\rho^F > 0$ and $\mu^F > 0$ are constants representing the mass density and the viscosity of the fluid, $\mathbf{h}_{in}$ and $\mathbf{h}_{out}$ are prescribed boundary stress, $\mathbf{n}^F$ is the unit outer normal vector along the boundary $\partial\Omega_t^F$, and $\mathbf{v}^{F,0}$ is the initial velocity of the fluid.

## 3. Updated Lagrangian Framework for the Structure Approximation

Introducing $\mathbf{V}^S$, the velocity of the structure in the Lagrangian coordinates, Equation (1) can be rewritten as

$$\rho_0^S(\mathbf{X}) \frac{\partial \mathbf{V}^S}{\partial t}(\mathbf{X}, t) - \nabla_{\mathbf{X}} \cdot (\mathbf{F}\boldsymbol{\Sigma})(\mathbf{X}, t) = \rho_0^S(\mathbf{X})\mathbf{g}, \quad \text{in } \Omega_0^S \times (0, T) \tag{13}$$

$$\frac{\partial \mathbf{U}^S}{\partial t}(\mathbf{X}, t) = \mathbf{V}^S(\mathbf{X}, t), \quad \text{in } \Omega_0^S \times (0, T). \tag{14}$$

Let $N \in \mathbb{N}^*$ be the number of time steps and $\Delta t = T/N$ the time step. We set $t_n = n\Delta t$ for $n = 0, 1, \ldots, N$. Let $\mathbf{V}^{S,n}(\mathbf{X})$ and $\mathbf{U}^{S,n}(\mathbf{X})$ be approximations of $\mathbf{V}^S(\mathbf{X}, t_n)$ and $\mathbf{U}^S(\mathbf{X}, t_n)$. In the sequel,

$$\mathbf{F}^n = \mathbf{I} + \nabla_{\mathbf{X}}\mathbf{U}^{S,n}, \quad \boldsymbol{\Sigma}^n = \boldsymbol{\Sigma}(\mathbf{F}^n), \; n \geq 0.$$

Using the implicit Euler scheme, we approach the system in Equations (13) and (14) by

$$\rho_0^S(\mathbf{X}) \frac{\mathbf{V}^{S,n+1}(\mathbf{X}) - \mathbf{V}^{S,n}(\mathbf{X})}{\Delta t} - \nabla_{\mathbf{X}} \cdot \left(\mathbf{F}^{n+1}\boldsymbol{\Sigma}^{n+1}\right)(\mathbf{X}) = \rho_0^S(\mathbf{X})\mathbf{g}, \; \text{in } \Omega_0^S \tag{15}$$

$$\frac{\mathbf{U}^{S,n+1}(\mathbf{X}) - \mathbf{U}^{S,n}(\mathbf{X})}{\Delta t} = \mathbf{V}^{S,n+1}(\mathbf{X}), \; \text{in } \Omega_0^S \tag{16}$$

Using Equation (16), we get $\mathbf{F}^{n+1} = \mathbf{F}^n + \Delta t \nabla_{\mathbf{X}}\mathbf{V}^{S,n+1}$ and, consequently, $\mathbf{F}^{n+1}$ and $\boldsymbol{\Sigma}^{n+1}$ depend on the velocity $\mathbf{V}^{S,n+1}$. We have eliminated the unknown displacement and we have now an equation of unknown $\mathbf{V}^{S,n+1}$.

We can put Equation (15) in a weak form: find $\mathbf{V}^{S,n+1} : \Omega_0^S \to \mathbb{R}^3$, $\mathbf{V}^{S,n+1} = 0$ on $\Gamma_0^D$, such that

$$\int_{\Omega_0^S} \rho_0^S \frac{\mathbf{V}^{S,n+1} - \mathbf{V}^{S,n}}{\Delta t} \cdot \mathbf{W}^S \, d\mathbf{X} + \int_{\Omega_0^S} \mathbf{F}^{n+1}\boldsymbol{\Sigma}^{n+1} : \nabla_{\mathbf{X}}\mathbf{W}^S \, d\mathbf{X}$$

$$= \int_{\Omega_0^S} \rho_0^S \mathbf{g} \cdot \mathbf{W}^S \, d\mathbf{X} + \int_{\Gamma_0} \mathbf{F}^{n+1}\boldsymbol{\Sigma}^{n+1}\mathbf{N}^S \cdot \mathbf{W}^S \, dS \tag{17}$$

for all $\mathbf{W}^S : \Omega_0^S \to \mathbb{R}^3$, $\mathbf{W}^S = 0$ on $\Gamma_0^D$. For instance, we have assumed that the forces $\mathbf{F}^{n+1}\boldsymbol{\Sigma}^{n+1}\mathbf{N}^S$ on the interface $\Gamma_0$ are known.

The above formulation is also called the total Lagrangian framework, since the equations are written in the undeformed domain $\Omega_0^S$. Now, we present the updated Lagrangian framework, where the equations are written in the domain $\Omega_n^S$. We follow a similar method as in [36], where the structure is a bi-dimensional compressible neo-Hookean material, or in [37], where the bi-dimensional linear elasticity model is used.

We set $\Omega_n^S$ the image of $\Omega_0^S$ via the map $\mathbf{X} \to \mathbf{X} + \mathbf{U}^{S,n}(\mathbf{X})$ and we denote by $\widehat{\Omega}^S = \Omega_n^S$ the computational domain for the structure. The map from $\Omega_0^S$ to $\Omega_{n+1}^S$ defined by $\mathbf{X} \to \mathbf{x} = \mathbf{X} + \mathbf{U}^{S,n+1}(\mathbf{X})$ is the composition of the map from $\Omega_0^S$ to $\widehat{\Omega}^S$ given by $\mathbf{X} \to \widehat{\mathbf{x}} = \mathbf{X} + \mathbf{U}^{S,n}(\mathbf{X})$ with the map from $\widehat{\Omega}^S$ to $\Omega_{n+1}^S$ given by

$$\widehat{\mathbf{x}} \to \mathbf{x} = \widehat{\mathbf{x}} + \mathbf{U}^{S,n+1}(\mathbf{X}) - \mathbf{U}^{S,n}(\mathbf{X}) = \widehat{\mathbf{x}} + \widehat{\mathbf{u}}(\widehat{\mathbf{x}}).$$

Using the notations $\widehat{\mathbf{F}} = \mathbf{I} + \nabla_{\widehat{\mathbf{x}}}\widehat{\mathbf{u}}$ and $\widehat{J} = \det\widehat{\mathbf{F}}$, $J^n = \det\mathbf{F}^n$, we get

$$\mathbf{F}^{n+1}(\mathbf{X}) = \widehat{\mathbf{F}}(\widehat{\mathbf{x}})\mathbf{F}^n(\mathbf{X}), \quad J^{n+1}(\mathbf{X}) = \widehat{J}(\widehat{\mathbf{x}})J^n(\mathbf{X}). \tag{18}$$

We have the relation between the Cauchy stress tensor of the structure $\sigma^S$ and the second Piola–Kirchhoff stress tensor $\boldsymbol{\Sigma}$,

$$\sigma^S(\mathbf{x}, t) = \left(\frac{1}{J}\mathbf{F}\boldsymbol{\Sigma}\mathbf{F}^T\right)(\mathbf{X}, t), \quad \mathbf{x} = \mathbf{X} + \mathbf{U}^S(\mathbf{X}, t).$$

The mass conservation assumption gives $\rho^S(\mathbf{x}, t) = \frac{\rho_0^S(\mathbf{X})}{J(\mathbf{X}, t)}$, where $\rho^S(\mathbf{x}, t)$ is the mass density of the structure in the Eulerian framework.

For the semi-discrete scheme, we use the notations

$$\sigma^{S,n+1}\left(\mathbf{x}\right) = \left(\frac{1}{J^{n+1}}\mathbf{F}^{n+1}\mathbf{\Sigma}^{n+1}\left(\mathbf{F}^{n+1}\right)^{T}\right)(\mathbf{X}), \quad \mathbf{x} = \mathbf{X} + \mathbf{U}^{S,n+1}\left(\mathbf{X}\right)$$

and $\rho^{S,n}\left(\widehat{\mathbf{x}}\right) = \frac{\rho_0^S(\mathbf{X})}{J^n(\mathbf{X})}, \quad \widehat{\mathbf{x}} = \mathbf{X} + \mathbf{U}^{S,n}\left(\mathbf{X}\right).$

Let us introduce $\widehat{\mathbf{v}}^{S,n+1} : \widehat{\Omega}^S \to \mathbb{R}^3$ and $\mathbf{v}^{S,n} : \widehat{\Omega}^S \to \mathbb{R}^3$ defined by $\widehat{\mathbf{v}}^{S,n+1}\left(\widehat{\mathbf{x}}\right) = \mathbf{V}^{S,n+1}\left(\mathbf{X}\right)$ and $\mathbf{v}^{S,n}\left(\widehat{\mathbf{x}}\right) = \mathbf{V}^{S,n}\left(\mathbf{X}\right).$ In addition, for $\mathbf{W}^S : \Omega_0^S \to \mathbb{R}^3$, we define $\widehat{\mathbf{w}}^S : \widehat{\Omega}^S \to \mathbb{R}^3$ and $\mathbf{w}^S : \Omega_{n+1}^S \to \mathbb{R}^3$ by $\widehat{\mathbf{w}}^S\left(\widehat{\mathbf{x}}\right) = \mathbf{w}^S\left(\mathbf{x}\right) = \mathbf{W}^S\left(\mathbf{X}\right).$

Now, we rewrite Equation (17) over the domain $\widehat{\Omega}^S$. For the first term of Equation (17), we get

$$\int_{\Omega_0^S} \rho_0^S \frac{\mathbf{V}^{S,n+1} - \mathbf{V}^{S,n}}{\Delta t} \cdot \mathbf{W}^S\, d\mathbf{X} = \int_{\widehat{\Omega}^S} \rho^{S,n} \frac{\widehat{\mathbf{v}}^{S,n+1} - \mathbf{v}^{S,n}}{\Delta t} \cdot \widehat{\mathbf{w}}^S\, d\widehat{\mathbf{x}}$$

and similarly

$$\int_{\Omega_0^S} \rho_0^S \mathbf{g} \cdot \mathbf{W}^S\, d\mathbf{X} = \int_{\widehat{\Omega}^S} \rho^{S,n} \mathbf{g} \cdot \widehat{\mathbf{w}}^S\, d\widehat{\mathbf{x}}.$$

Using the identity $\left(\nabla\mathbf{w}^S\left(\mathbf{x}\right)\right)\mathbf{F}^{n+1}\left(\mathbf{X}\right) = \nabla_{\mathbf{X}}\mathbf{W}^S\left(\mathbf{X}\right)$ and the definition of $\sigma^{S,n+1}$, we get

$$\int_{\Omega_0^S} \mathbf{F}^{n+1}\mathbf{\Sigma}^{n+1} : \nabla_{\mathbf{X}}\mathbf{W}^S\, d\mathbf{X} = \int_{\Omega_{n+1}^S} \sigma^{S,n+1} : \nabla\mathbf{w}^S\, d\mathbf{x}.$$

Details about this kind of transformation can be found in [39], Chapter 1.2.
To write the above integral over the domain $\widehat{\Omega}^S$, let us introduce the tensor

$$\widehat{\mathbf{\Sigma}}\left(\widehat{\mathbf{x}}\right) = \widehat{J}\left(\widehat{\mathbf{x}}\right)\widehat{\mathbf{F}}^{-1}\left(\widehat{\mathbf{x}}\right)\sigma^{S,n+1}\left(\mathbf{x}\right)\widehat{\mathbf{F}}^{-T}\left(\widehat{\mathbf{x}}\right). \tag{19}$$

Since $\left(\nabla\mathbf{w}^S\left(\mathbf{x}\right)\right)\widehat{\mathbf{F}}\left(\widehat{\mathbf{x}}\right) = \nabla_{\widehat{\mathbf{x}}}\widehat{\mathbf{w}}^S\left(\widehat{\mathbf{x}}\right)$ (see [39], Chapter 1.2) and taking into account Equation (19), we get

$$\int_{\Omega_{n+1}^S} \sigma^{S,n+1} : \nabla\mathbf{w}^S\, d\mathbf{x} = \int_{\widehat{\Omega}^S} \widehat{\mathbf{F}}\widehat{\mathbf{\Sigma}} : \nabla_{\widehat{\mathbf{x}}}\widehat{\mathbf{w}}^S\, d\widehat{\mathbf{x}}.$$

Now, it is possible to present the updated Lagrangian version of Equation (17). Knowing $\mathbf{U}^{S,n} : \Omega_0^S \to \mathbb{R}^3$, $\widehat{\Omega}^S = \Omega_n^S$ and $\mathbf{v}^{S,n} : \widehat{\Omega}^S \to \mathbb{R}^3$, we try to find $\widehat{\mathbf{v}}^{S,n+1} : \widehat{\Omega}^S \to \mathbb{R}^3$, $\widehat{\mathbf{v}}^{S,n+1} = 0$ on $\Gamma_0^D$ such that

$$\int_{\widehat{\Omega}^S} \rho^{S,n} \frac{\widehat{\mathbf{v}}^{S,n+1} - \mathbf{v}^{S,n}}{\Delta t} \cdot \widehat{\mathbf{w}}^S\, d\widehat{\mathbf{x}} + \int_{\widehat{\Omega}^S} \widehat{\mathbf{F}}\widehat{\mathbf{\Sigma}} : \nabla_{\widehat{\mathbf{x}}}\widehat{\mathbf{w}}^S\, d\widehat{\mathbf{x}}$$
$$= \int_{\widehat{\Omega}^S} \rho^{S,n} \mathbf{g} \cdot \widehat{\mathbf{w}}^S\, d\widehat{\mathbf{x}} + \int_{\Gamma_0} \mathbf{F}^{n+1}\mathbf{\Sigma}^{n+1}\mathbf{N}^S \cdot \mathbf{W}^S\, dS \tag{20}$$

for all $\widehat{\mathbf{w}}^S : \widehat{\Omega}^S \to \mathbb{R}^3$, $\widehat{\mathbf{w}}^S = 0$ on $\Gamma_0^D$. We recall that the forces $\mathbf{F}^{n+1}\mathbf{\Sigma}^{n+1}\mathbf{N}^S$ on the interface $\Gamma_0$ are assumed known.

Using the identity $\widehat{\mathbf{u}}\left(\widehat{\mathbf{x}}\right) = \mathbf{U}^{S,n+1}\left(\mathbf{X}\right) - \mathbf{U}^{S,n}\left(\mathbf{X}\right) = \Delta t\, \mathbf{V}^{S,n+1}\left(\mathbf{X}\right) = \Delta t\, \widehat{\mathbf{v}}^{S,n+1}\left(\widehat{\mathbf{x}}\right)$, we obtain

$$\widehat{\mathbf{F}} = \mathbf{I} + \Delta t \nabla_{\widehat{\mathbf{x}}}\widehat{\mathbf{v}}^{S,n+1}. \tag{21}$$

Using Equations (18) and (19), it follows that

$$\begin{aligned}
\widehat{\mathbf{\Sigma}} &= \widehat{J}\widehat{\mathbf{F}}^{-1}\sigma^{S,n+1}\widehat{\mathbf{F}}^{-T} = \widehat{J}\widehat{\mathbf{F}}^{-1}\frac{1}{J^{n+1}}\mathbf{F}^{n+1}\mathbf{\Sigma}^{n+1}\left(\mathbf{F}^{n+1}\right)^{T}\widehat{\mathbf{F}}^{-T} \\
&= \frac{1}{J^n}\mathbf{F}^n\mathbf{\Sigma}^{n+1}\left(\mathbf{F}^n\right)^{T}.
\end{aligned} \tag{22}$$

For the linear elastic material, we have

$$\mathbf{\Sigma}(\mathbf{U}) = \lambda^S (\nabla_{\mathbf{X}} \cdot \mathbf{U}) + \mu^S \left( \nabla_{\mathbf{X}} \mathbf{U} + (\nabla_{\mathbf{X}} \mathbf{U})^T \right)$$

where $\lambda^S$ and $\mu^S$ are the Lamé coefficients. We have

$$\mathbf{\Sigma}^{n+1} = \mathbf{\Sigma}(\mathbf{U}^{S,n+1}) = \mathbf{\Sigma}(\mathbf{U}^{S,n}) + (\Delta t)\mathbf{\Sigma}(\mathbf{V}^{S,n+1}) = \mathbf{\Sigma}^n + (\Delta t)\mathbf{\Sigma}(\mathbf{V}^{S,n+1}).$$

From Equations (21) and (22) and the above equality, we get

$$
\begin{aligned}
\widehat{\mathbf{F}\mathbf{\Sigma}} \;=\; & \frac{1}{J^n}\mathbf{F}^n\mathbf{\Sigma}^n\left(\mathbf{F}^n\right)^T + \Delta t \nabla_{\widehat{\mathbf{x}}}\widehat{\mathbf{v}}^{S,n+1}\frac{1}{J^n}\mathbf{F}^n\mathbf{\Sigma}^n\left(\mathbf{F}^n\right)^T \\
& + \frac{\Delta t}{J^n}\mathbf{F}^n\mathbf{\Sigma}(\mathbf{V}^{S,n+1})\left(\mathbf{F}^n\right)^T + \frac{(\Delta t)^2}{J^n}\nabla_{\widehat{\mathbf{x}}}\widehat{\mathbf{v}}^{S,n+1}\mathbf{F}^n\mathbf{\Sigma}(\mathbf{V}^{S,n+1})\left(\mathbf{F}^n\right)^T \\
\;=\; & \sigma^{S,n} + \Delta t \nabla_{\widehat{\mathbf{x}}}\widehat{\mathbf{v}}^{S,n+1}\sigma^{S,n} \\
& + \frac{\Delta t}{J^n}\mathbf{F}^n\mathbf{\Sigma}(\mathbf{V}^{S,n+1})\left(\mathbf{F}^n\right)^T + \frac{(\Delta t)^2}{J^n}\nabla_{\widehat{\mathbf{x}}}\widehat{\mathbf{v}}^{S,n+1}\mathbf{F}^n\mathbf{\Sigma}(\mathbf{V}^{S,n+1})\left(\mathbf{F}^n\right)^T .
\end{aligned}
$$

We introduce $\mathbf{\Sigma}_{\widehat{\mathbf{x}}}(\widehat{\mathbf{u}}) = \lambda^S(\nabla_{\widehat{\mathbf{x}}} \cdot \widehat{\mathbf{u}}) + \mu^S\left(\nabla_{\widehat{\mathbf{x}}}\widehat{\mathbf{u}} + (\nabla_{\widehat{\mathbf{x}}}\widehat{\mathbf{u}})^T\right)$ and $\widehat{\mathbf{u}}^{S,n}(\widehat{\mathbf{x}}) = \mathbf{U}^{S,n}(\mathbf{X})$. For small deformations, we have $\mathbf{F}^n \approx \mathbf{I}$, $J^n \approx 1$, then $\mathbf{\Sigma}(\mathbf{V}^{S,n+1})$ could be approached by $\mathbf{\Sigma}_{\widehat{\mathbf{x}}}(\widehat{\mathbf{v}}^{S,n+1})$ and $\sigma^{S,n}$ by $\mathbf{\Sigma}_{\widehat{\mathbf{x}}}(\widehat{\mathbf{u}}^{S,n})$.

Finally, we can approach the map $\widehat{\mathbf{v}}^{S,n+1} \to \widehat{\mathbf{F}\mathbf{\Sigma}}$, by the simplified linear application

$$\widehat{\mathbf{L}}\left(\widehat{\mathbf{v}}^{S,n+1}\right) \;=\; \mathbf{\Sigma}_{\widehat{\mathbf{x}}}(\widehat{\mathbf{u}}^{S,n}) + (\Delta t)\mathbf{\Sigma}_{\widehat{\mathbf{x}}}(\widehat{\mathbf{v}}^{S,n+1}). \tag{23}$$

The linearized updated Lagrangian weak formulation of the structure is: knowing $\mathbf{U}^{S,n} : \Omega_0^S \to \mathbb{R}^3$, $\widehat{\Omega}^S = \Omega_n^S$ and $\mathbf{v}^{S,n} : \widehat{\Omega}^S \to \mathbb{R}^3$, find $\widehat{\mathbf{v}}^{S,n+1} : \widehat{\Omega}^S \to \mathbb{R}^3$, $\widehat{\mathbf{v}}^{S,n+1} = 0$ on $\Gamma_0^D$ such that

$$
\begin{aligned}
\int_{\widehat{\Omega}^S} \rho^{S,n}\frac{\widehat{\mathbf{v}}^{S,n+1} - \mathbf{v}^{S,n}}{\Delta t} \cdot \widehat{\mathbf{w}}^S \, d\widehat{\mathbf{x}} + \int_{\widehat{\Omega}^S} \widehat{\mathbf{L}}\left(\widehat{\mathbf{v}}^{S,n+1}\right) : \nabla_{\widehat{\mathbf{x}}}\widehat{\mathbf{w}}^S \, d\widehat{\mathbf{x}} \\
= \int_{\widehat{\Omega}^S} \rho^{S,n}\mathbf{g} \cdot \widehat{\mathbf{w}}^S \, d\widehat{\mathbf{x}} + \int_{\Gamma_0} \mathbf{F}^{n+1}\mathbf{\Sigma}^{n+1}\mathbf{N}^S \cdot \mathbf{W}^S \, dS
\end{aligned}
\tag{24}
$$

for all $\widehat{\mathbf{w}}^S : \widehat{\Omega}^S \to \mathbb{R}^3$, $\widehat{\mathbf{w}}^S = 0$ on $\Gamma_0^D$.

**Remark 1.** *We can use a non-linear model for the structure such as St. Venant Kirchhoff, neo-Hookean, etc. By linearization of $\widehat{\mathbf{F}\mathbf{\Sigma}}$ (around the deformed state at the precedent time instant), we obtain in place of Equation (23)*

$$\widehat{\mathbf{L}}\left(\widehat{\mathbf{v}}^{S,n+1}\right) = n\ell(\widehat{\mathbf{u}}^{S,n}) + (\Delta t)\ell(\widehat{\mathbf{v}}^{S,n+1})$$

*where $n\ell$ is a non-linear operator and $\ell$ a linear one. Since $n\ell(\widehat{\mathbf{u}}^{S,n})$ is a known term, we can transfer it to the right-hand side, then the problem to solve is linear, similar to Equation (24).*

## 4. Arbitrary Lagrangian Eulerian (ALE) Framework for Approximation of Fluid Equations

The Arbitrary Eulerian Lagrangian (ALE) framework is a successful method to solve fluid equations in moving domain (see [19]). Let $\widehat{\Omega}^F$ be a reference fluid domain and let $\mathcal{A}_t$, $t \in [0, T]$ be a family of transformations such that: $\mathcal{A}_t(\widehat{\mathbf{x}}) = \widehat{\mathbf{x}}$ for all $\widehat{\mathbf{x}} \in \Sigma_1 \cup \Sigma_3$ and $\mathcal{A}_t(\widehat{\Omega}^F) = \Omega_t^F$, where $\widehat{\mathbf{x}} = (\widehat{x}_1, \widehat{x}_2, \widehat{x}_3)^T \in \widehat{\Omega}^F$ are the ALE coordinates and $\mathbf{x} = (x_1, x_2, x_3)^T = \mathcal{A}_t(\widehat{\mathbf{x}})$ the Eulerian coordinates.

Let $\mathbf{v}^F$ be the fluid velocity in the Eulerian coordinates. We denote by $\widehat{\mathbf{v}}^F : \widehat{\Omega}^F \to \mathbb{R}^3$ the corresponding function in the ALE coordinates, which is defined by $\widehat{\mathbf{v}}^F(\widehat{\mathbf{x}}, t) = \mathbf{v}^F(\mathcal{A}_t(\widehat{\mathbf{x}}), t) = \mathbf{v}^F(\mathbf{x}, t)$.

We denote the mesh velocity by $\boldsymbol{\vartheta}(\mathbf{x}, t) = \frac{\partial \mathcal{A}_t}{\partial t}(\widehat{\mathbf{x}}) = \frac{\partial \mathcal{A}_t}{\partial t}(\mathcal{A}_t^{-1}(\mathbf{x}))$ and the ALE time derivative of the fluid velocity by $\frac{\partial \mathbf{v}^F}{\partial t}\big|_{\widehat{\mathbf{x}}}(\mathbf{x}, t) = \frac{\partial \widehat{\mathbf{v}}^F}{\partial t}(\widehat{\mathbf{x}}, t)$.

The Navier–Stokes equations in the ALE framework give:

$$\rho^F \left( \frac{\partial \mathbf{v}^F}{\partial t}\bigg|_{\widehat{\mathbf{x}}} + \left( \left( \mathbf{v}^F - \boldsymbol{\vartheta} \right) \cdot \nabla \right) \mathbf{v}^F \right) - 2\mu^F \nabla \cdot \epsilon \left( \mathbf{v}^F \right) + \nabla p^F \;=\; \rho^F \mathbf{g}, \text{ in } \Omega_t^F \times (0, T)$$

$$\nabla \cdot \mathbf{v}^F \;=\; 0, \text{ in } \Omega_t^F \times (0, T).$$

We denote by $\mathbf{v}^{F,n}$, $p^{F,n}$, and $\boldsymbol{\vartheta}^n$ the approximations of $\mathbf{v}^F(\cdot, t_n)$, $p^F(\cdot, t_n)$, and $\boldsymbol{\vartheta}(\cdot, t_n)$, respectively, all defined in $\Omega_n^F$. Here, we set $\widehat{\Omega}^F = \Omega_n^F$. The time advancing scheme for fluid equations is: find $\widehat{\mathbf{v}}^{F,n+1} : \Omega_n^F \to \mathbb{R}^3$ and $\widehat{p}^{F,n+1} : \Omega_n^F \to \mathbb{R}$ such that

$$\rho^F \left( \frac{\widehat{\mathbf{v}}^{F,n+1} - \mathbf{v}^{F,n}}{\Delta t} + \left( \left( \mathbf{v}^{F,n} - \boldsymbol{\vartheta}^n \right) \cdot \nabla_{\widehat{\mathbf{x}}} \right) \widehat{\mathbf{v}}^{F,n+1} \right)$$

$$-2\mu^F \nabla_{\widehat{\mathbf{x}}} \cdot \epsilon \left( \widehat{\mathbf{v}}^{F,n+1} \right) + \nabla_{\widehat{\mathbf{x}}} \widehat{p}^{F,n+1} \;=\; \rho^F \mathbf{g}, \text{ in } \Omega_n^F \tag{25}$$

$$\nabla_{\widehat{\mathbf{x}}} \cdot \widehat{\mathbf{v}}^{F,n+1} \;=\; 0, \text{ in } \Omega_n^F \tag{26}$$

$$\sigma^F(\widehat{\mathbf{v}}^{F,n+1}, \widehat{p}^{F,n+1})\mathbf{n}^F \;=\; \mathbf{h}_{in}^{n+1}, \text{ on } \Sigma_1 \tag{27}$$

$$\sigma^F(\widehat{\mathbf{v}}^{F,n+1}, \widehat{p}^{F,n+1})\mathbf{n}^F \;=\; \mathbf{h}_{out}^{n+1}, \text{ on } \Sigma_3 \tag{28}$$

The above time discretization scheme is based on the backward Euler scheme and a linearization of the convective term.

We multiply Equation (25) by a test function $\widehat{\mathbf{w}}^F : \Omega_n^F \to \mathbb{R}^3$ and Equation (26) by a test function $\widehat{q} : \Omega_n^F \to \mathbb{R}$. After integrating them over the domain $\Omega_n^F$ and using the Green's formula and the corresponding boundary conditions, we get the below discrete weak form. Find $\widehat{\mathbf{v}}^{F,n+1} : \Omega_n^F \to \mathbb{R}^3$ and $\widehat{p}^{F,n+1} : \Omega_n^F \to \mathbb{R}$ such that:

$$\int_{\Omega_n^F} \rho^F \frac{\widehat{\mathbf{v}}^{F,n+1}}{\Delta t} \cdot \widehat{\mathbf{w}}^F \, d\widehat{\mathbf{x}} + \int_{\Omega_n^F} \rho^F \left( \left( \left( \mathbf{v}^{F,n} - \boldsymbol{\vartheta}^n \right) \cdot \nabla_{\widehat{\mathbf{x}}} \right) \widehat{\mathbf{v}}^{F,n+1} \right) \cdot \widehat{\mathbf{w}}^F \, d\widehat{\mathbf{x}}$$

$$- \int_{\Omega_n^F} \left( \nabla_{\widehat{\mathbf{x}}} \cdot \widehat{\mathbf{w}}^F \right) \widehat{p}^{F,n+1} d\widehat{\mathbf{x}} + \int_{\Omega_n^F} 2\mu^F \epsilon \left( \widehat{\mathbf{v}}^{F,n+1} \right) : \epsilon \left( \widehat{\mathbf{w}}^F \right) d\widehat{\mathbf{x}}$$

$$= \mathcal{L}_F(\widehat{\mathbf{w}}^F) + \int_{\Gamma_n} \left( \sigma^F(\widehat{\mathbf{v}}^{F,n+1}, \widehat{p}^{F,n+1})\mathbf{n}^F \right) \cdot \widehat{\mathbf{w}}^F \, ds, \tag{29}$$

$$\int_{\Omega_n^F} (\nabla_{\widehat{\mathbf{x}}} \cdot \widehat{\mathbf{v}}^{F,n+1}) \widehat{q} \, d\widehat{\mathbf{x}} = 0, \tag{30}$$

for all $\widehat{\mathbf{w}}^F : \Omega_n^F \to \mathbb{R}^3$ and for all $\widehat{q} : \Omega_n^F \to \mathbb{R}$, where

$$\mathcal{L}_F(\widehat{\mathbf{w}}^F) = \int_{\Omega_n^F} \rho^F \frac{\widehat{\mathbf{v}}^{F,n}}{\Delta t} \cdot \widehat{\mathbf{w}}^F \, d\widehat{\mathbf{x}} + \int_{\Omega_n^F} \rho^F \mathbf{g} \cdot \widehat{\mathbf{w}}^F + \int_{\Sigma_1} \mathbf{h}_{in}^{n+1} \cdot \widehat{\mathbf{w}}^F + \int_{\Sigma_3} \mathbf{h}_{out}^{n+1} \cdot \widehat{\mathbf{w}}^F.$$

We have assumed, for instance, that the forces $\sigma^F(\widehat{\mathbf{v}}^{F,n+1}, \widehat{p}^{F,n+1})\mathbf{n}^F$ on the interface $\Gamma_n$ are known. The mesh velocity $\widehat{\boldsymbol{\vartheta}}^{n+1} : \Omega_n^F \to \mathbb{R}^3$ can be computed from

$$\begin{cases} \Delta_{\widehat{\mathbf{x}}} \widehat{\boldsymbol{\vartheta}}^{n+1} &=\; 0, \quad \text{in } \Omega_n^F \\ \widehat{\boldsymbol{\vartheta}}^{n+1} &=\; 0, \quad \text{on } \Sigma_1 \cup \Sigma_3 \\ \widehat{\boldsymbol{\vartheta}}^{n+1} &=\; \widehat{\mathbf{v}}^{F,n+1}, \quad \text{on } \Gamma_n. \end{cases}$$

For all $n = 0, \cdots, N-1$, we denote by $\mathcal{A}_{t_{n+1}}$ the map from $\overline{\Omega}_n^F$ to $\mathbb{R}^3$ defined by $\mathcal{A}_{t_{n+1}}(\widehat{x}_1, \widehat{x}_2, \widehat{x}_3) = (\widehat{x}_1 + \Delta t \vartheta_1^{n+1}, \widehat{x}_2 + \Delta t \vartheta_2^{n+1}, \widehat{x}_3 + \Delta t \vartheta_3^{n+1})$. We set $\Omega_{n+1}^F = \mathcal{A}_{t_{n+1}}(\Omega_n^F)$, $\Gamma_{n+1} = \mathcal{A}_{t_{n+1}}(\Gamma_n)$ and we remark that $\widehat{\mathbf{x}} = \mathcal{A}_{t_{n+1}}(\widehat{\mathbf{x}})$, for all $\widehat{\mathbf{x}} \in \Sigma_1 \cup \Sigma_3$.

We define the fluid velocity $\mathbf{v}^{F,n+1} : \Omega^F_{n+1} \to \mathbb{R}^3$, the fluid pressure $p^{F,n+1} : \Omega^F_{n+1} \to \mathbb{R}$ and the mesh velocity $\boldsymbol{\vartheta}^{n+1} : \Omega^F_{n+1} \to \mathbb{R}^3$ by:

$$\mathbf{v}^{F,n+1}(\mathbf{x}) = \widehat{\mathbf{v}}^{F,n+1}(\widehat{\mathbf{x}}), \quad p^{F,n+1}(\mathbf{x}) = \widehat{p}^{F,n+1}(\widehat{\mathbf{x}}), \quad \boldsymbol{\vartheta}^{n+1}(\mathbf{x}) = \widehat{\boldsymbol{\vartheta}}^{n+1}(\widehat{\mathbf{x}}),$$

for all $\widehat{\mathbf{x}} \in \Omega^F_n$, $\mathbf{x} = \mathcal{A}_{t_{n+1}}(\widehat{\mathbf{x}}) \in \Omega^F_{n+1}$.

## 5. Monolithic Formulation for the Fluid–Structure Equations

Let $\Omega_n = \Omega^F_n \cup \Gamma_n \cup \Omega^S_n$ be the global fluid–structure domain at time instant $n$ and let us introduce the global velocity and test function

$$\widehat{\mathbf{v}}^{n+1} : \Omega_n \to \mathbb{R}^3, \quad \widehat{\mathbf{w}} : \Omega_n \to \mathbb{R}^3$$

$$\widehat{\mathbf{v}}^{n+1} = \begin{cases} \widehat{\mathbf{v}}^{F,n+1} \text{ in } \Omega^F_n \\ \widehat{\mathbf{v}}^{S,n+1} \text{ in } \Omega^S_n \end{cases} \qquad \widehat{\mathbf{w}} = \begin{cases} \widehat{\mathbf{w}}^F \text{ in } \Omega^F_n \\ \widehat{\mathbf{w}}^S \text{ in } \Omega^S_n. \end{cases}$$

At each time step, we solve the linear coupled problem: find $\widehat{\mathbf{v}}^{n+1} \in \left( H^1(\Omega_n) \right)^2$, $\widehat{\mathbf{v}}^{n+1} = 0$ on $\Gamma^D_0$ and $\widehat{p}^{F,n+1} \in L^2(\Omega^F_n)$, such that:

$$\int_{\Omega^F_n} \rho^F \frac{\widehat{\mathbf{v}}^{n+1}}{\Delta t} \cdot \widehat{\mathbf{w}} d\widehat{\mathbf{x}} + \int_{\Omega^F_n} \rho^F \left( ((\mathbf{v}^n - \boldsymbol{\vartheta}^n) \cdot \nabla_{\widehat{\mathbf{x}}}) \widehat{\mathbf{v}}^{n+1} \right) \cdot \widehat{\mathbf{w}} d\widehat{\mathbf{x}}$$

$$- \int_{\Omega^F_n} (\nabla_{\widehat{\mathbf{x}}} \cdot \widehat{\mathbf{w}}) \widehat{p}^{F,n+1} d\widehat{\mathbf{x}} + \int_{\Omega^F_n} 2\mu^F \epsilon\left( \widehat{\mathbf{v}}^{n+1} \right) : \epsilon\left( \widehat{\mathbf{w}} \right) d\widehat{\mathbf{x}}$$

$$+ \int_{\Omega^S_n} \rho^{S,n} \frac{\widehat{\mathbf{v}}^{n+1}}{\Delta t} \cdot \widehat{\mathbf{w}} d\widehat{\mathbf{x}} + \int_{\Omega^S_n} \widehat{\mathbf{L}}\left( \widehat{\mathbf{v}}^{n+1} \right) : \nabla_{\widehat{\mathbf{x}}} \widehat{\mathbf{w}} d\widehat{\mathbf{x}}$$

$$= \mathcal{L}_F(\widehat{\mathbf{w}}) + \int_{\Omega^S_n} \rho^{S,n} \frac{\mathbf{v}^n}{\Delta t} \cdot \widehat{\mathbf{w}} d\widehat{\mathbf{x}} + \int_{\Omega^S_n} \rho^{S,n} \mathbf{g} \cdot \widehat{\mathbf{w}} d\widehat{\mathbf{x}}, \tag{31}$$

$$\int_{\Omega^F_n} (\nabla_{\widehat{\mathbf{x}}} \cdot \widehat{\mathbf{v}}^{n+1}) \widehat{q} d\widehat{\mathbf{x}} = 0, \tag{32}$$

for all $\widehat{\mathbf{w}} \in \left( H^1(\Omega_n) \right)^3$, $\widehat{\mathbf{w}} = 0$ on $\Gamma^D_0$ and for all $\widehat{q} \in L^2(\Omega^F_n)$.

From the regularity $\widehat{\mathbf{v}}^{n+1} \in \left( H^1(\Omega_n) \right)^2$, the traces of $\widehat{\mathbf{v}}^{F,n+1}$ and $\widehat{\mathbf{v}}^{S,n+1}$ on $\Gamma_n$ are well defined and $\widehat{\mathbf{v}}^{F,n+1}_{|\Gamma_n} = \widehat{\mathbf{v}}^{S,n+1}_{|\Gamma_n}$ which is a discrete form of the continuity of the velocity at the interface (8). Similarly, we get $\widehat{\mathbf{w}}^{F,n+1}_{|\Gamma_n} = \widehat{\mathbf{w}}^{S,n+1}_{|\Gamma_n}$.

Equation (31) is obtained by adding Equations (24) and (29). By enforcement of the hypothesis of continuity of stress (Equation (9)) at the discrete level, the expression

$$\int_{\Gamma_0} \mathbf{F}^{n+1} \boldsymbol{\Sigma}^{n+1} \mathbf{N}^S \cdot \mathbf{W}^S dS + \int_{\Gamma_n} \left( \sigma^F(\widehat{\mathbf{v}}^{F,n+1}, \widehat{p}^{F,n+1}) \mathbf{n}^F \right) \cdot \widehat{\mathbf{w}}^F ds$$

does not appear anymore in Equation (31).

### Algorithm for fluid–structure interaction
**Time advancing scheme from $n$ to $n+1$**

We assume that we know $\Omega_n = \Omega^F_n \cup \Gamma_n \cup \Omega^S_n$, $\mathbf{v}^n : \Omega_n \to \mathbb{R}^3$, $\boldsymbol{\vartheta}^n : \Omega^F_n \to \mathbb{R}^3$.

**Step 1**: Solve the **linear** system in Equations (31) and (32) and get the velocity $\widehat{\mathbf{v}}^{n+1}$ and the pressure $\widehat{p}^{F,n+1}$.

**Step 2**: Compute the fluid mesh velocity $\widehat{\boldsymbol{\vartheta}}^{n+1} : \Omega_n^F \to \mathbb{R}^3$

$$
\begin{cases}
\Delta_{\widehat{x}}\widehat{\boldsymbol{\vartheta}}^{n+1} & = \quad 0, \quad \text{in } \Omega_n^F \\
\widehat{\boldsymbol{\vartheta}}^{n+1} & = \quad 0, \quad \text{on } \Sigma_1 \cup \Sigma_3 \\
\widehat{\boldsymbol{\vartheta}}^{n+1} & = \quad \widehat{\mathbf{v}}^{n+1}, \quad \text{on } \Gamma_n.
\end{cases}
\tag{33}
$$

To improve the quality of the mesh, we can replace in Equation (33) the Laplacian by a linear elasticity operator.

**Step 3**: Define the map $\mathbb{T}_n : \overline{\Omega}_n \to \mathbb{R}^3$ by:

$$
\mathbb{T}_n(\widehat{\mathbf{x}}) = \widehat{\mathbf{x}} + (\Delta t)\widehat{\boldsymbol{\vartheta}}^{n+1}(\widehat{\mathbf{x}})\chi_{\Omega_n^F}(\widehat{\mathbf{x}}) + (\Delta t)\widehat{\mathbf{v}}^{n+1}(\widehat{\mathbf{x}})\chi_{\Omega_n^S}(\widehat{\mathbf{x}})
$$

where $\chi_{\Omega_n^F}$ and $\chi_{\Omega_n^S}$ are the characteristic functions of fluid and structure domains.

**Step 4**: Set $\Omega_{n+1}^F = \mathbb{T}_n(\Omega_n^F), \Omega_{n+1}^S = \mathbb{T}_n(\Omega_n^S)$, and $\Gamma_{n+1} = \mathbb{T}_n(\Gamma_n)$; consequently, $\Omega_{n+1} = \mathbb{T}_n(\Omega_n)$. Define $\mathbf{v}^{n+1} : \Omega_{n+1} \to \mathbb{R}^3$ by

$$
\mathbf{v}^{n+1}(\mathbf{x}) = \widehat{\mathbf{v}}^{n+1}(\widehat{\mathbf{x}}), \ \forall \widehat{\mathbf{x}} \in \Omega_n, \ \mathbf{x} = \mathbb{T}_n(\widehat{\mathbf{x}})
$$

and $p^{F,n+1} : \Omega_{n+1}^F \to \mathbb{R}, \boldsymbol{\vartheta}^{n+1} : \Omega_{n+1}^F \to \mathbb{R}^3$ by:

$$
p^{F,n+1}(\mathbf{x}) = \widehat{p}^{F,n+1}(\widehat{\mathbf{x}}), \ \boldsymbol{\vartheta}^{n+1}(\mathbf{x}) = \widehat{\boldsymbol{\vartheta}}^{n+1}(\widehat{\mathbf{x}}), \ \forall \widehat{\mathbf{x}} \in \Omega_n^F, \ \mathbf{x} = \mathbb{T}_n(\widehat{\mathbf{x}}).
$$

**Remark 2.** *(i) The domain is computed explicitly while the velocity and the pressure are computed implicitly. This kind of schema is also called semi-implicit. The monolithic system in Equations (31) and (32) is linear.*

*(ii) Using globally continuous finite element for the velocity $\widehat{\mathbf{v}}^{n+1} \in \left(H^1\left(\Omega_n\right)\right)^2$ defined all over the fluid–structure global mesh, the continuity of the velocity at the interface holds, automatically.*

*(iii) The vertices in the structure mesh are moved using the structure velocity, thus the structure mesh is of updated Lagrangian type.*

**Remark 3.** *In [35], for linear elastic model for the structure and Navier–Stokes equations for the fluid, a semi-implicit monolithic algorithm is introduced. The unconditional stability in time is established. In a forthcoming paper, a proof of the unconditional stability of an algorithm will be presented for a non-linear model of the structure. This stable algorithm is similar to the one presented in this paper, only a stabilization term has been added, as in [35].*

**Remark 4.** *In this paper, the derivative of fluid velocity as well as the derivative of structure velocity are approached by the implicit Euler scheme. It is possible to use different time discretization schemes, for example Newmark for the structure and implicit Euler for the fluid. We have to pay attention to the time advancing algorithm for the interface. One solution is to solve the fluid–structure coupled equations written in the domain at the precedent time instant to find the fluid–structure velocity and the fluid pressure. Then, the structure including the interface is advanced by the Newmark scheme, and finally the fluid mesh velocity is computed using the new position and velocity of the interface.*

## 6. Numerical Experiments

The numerical tests were produced using *FreeFem++* (see [40]).

### 6.1. Straight Cylinder

We tested the benchmark studied in [3,4] concerning blood flow in artery. The geometrical configuration is represented in Figure 1. The fluid occupies initially the straight cylinder of length

$L = 5$ cm and radius $R = 0.5$ cm. The disk $\Sigma_1$ is in the plane $x_1Ox_2$ and the axis of the cylinder is $Ox_3$. The viscosity of the fluid is $\mu^F = 0.03 \frac{\text{g}}{\text{cm}\cdot\text{s}}$ and its density is $\rho^F = 1 \frac{\text{g}}{\text{cm}^3}$.

The fluid is surrounded by a structure of constant thickness $h^S = 0.1$ cm. The others physical parameters of the structure are: the Young's modulus is $E = 3 \times 10^6 \frac{\text{g}}{\text{cm}\cdot\text{s}^2}$, the Poisson ratio is $\nu^S = 0.3$, and the density is $\rho_0^S = 1.2 \frac{\text{g}}{\text{cm}^3}$. The Lamé parameters are computed by the formulas $\lambda^S = \frac{\nu^S E}{(1-2\nu^S)(1+\nu^S)}$ and $\mu^S = \frac{E}{2(1+\nu^S)}$.

For the volume force in fluid and structure, we put $\mathbf{g} = (0,0,0)^T$. The prescribed boundary stress at the inlet $\Sigma_1$ is $\mathbf{h}_{in} = \left(0,\, 0,\, 1.3332 \times 10^4\right) \frac{\text{g}}{\text{cm}\cdot\text{s}^2}$ for $t \leq 0.005$ s and $\mathbf{h}_{out} = (0,\, 0,\, 0) \frac{\text{g}}{\text{cm}\cdot\text{s}^2}$ at the outlet $\Sigma_3$. The structure is clamped at both ends, $\Sigma_5$ and $\Sigma_7$. The remaining boundary conditions are Equations (3), (8) and (9). Initially, the fluid and the structure are at rest.

Using *FreeFem++*, it is possible to construct a global fluid–structure mesh with an "interior boundary" that is the fluid–structure interface. For the finite element approximation of the fluid–structure velocity, we used the finite element $\mathbb{P}_1 + bubble$ and we employed for the pressure the finite element $\mathbb{P}_1$. The parameters of meshes used for the numerical tests are presented in Table 1.

**Table 1.** The number of vertices, tetrahedra and degrees of freedom (DOF) of fluid–structure linear system for each mesh.

|        | Vertices | Tetrahedra | DOF       |
|--------|----------|------------|-----------|
| Mesh 1 | 768      | 3510       | 13,602    |
| Mesh 2 | 2314     | 11,400     | 43,456    |
| Mesh 3 | 8976     | 47,100     | 177,204   |
| Mesh 4 | 63,731   | 356,400    | 1,324,124 |

The time step was set to $\Delta t = 0.0005$ s and the number of time steps to $N = 40$. The radial displacement of the interface was measured at three points $A(R,0,L/4)$, $B(R,0,L/2)$, and $C(R,0,3L/4)$. We observed that the displacements were small, less than 0.012 cm (see Figure 2, left). Similar results are observed in [41] using non-conforming meshes. In addition, we measured the radial displacement along the line $(R,0,x_3)$, $x_3 \in [0,L]$ on the interface (see Figure 2, right.)

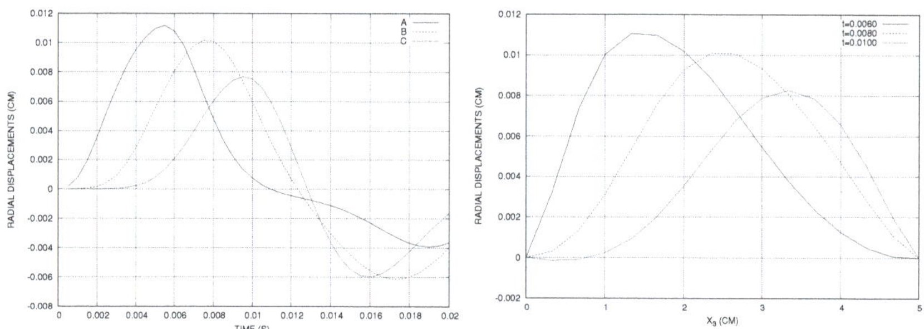

**Figure 2.** Constant stress at the inlet: the time history of the displacement of three points on the interface (**left**); and the radial displacement at time instants $t = 0.0060$, $t = 0.0080$, $t = 0.0100$ (**right**).

To obtain visible displacement, we also tested with the sin-like stress at the inlet:

$$\mathbf{h}_{in}(\mathbf{x},t) = \begin{cases} \left(0,\, 0,\, 5 \times 1.3332 \times 10^4 \times \frac{(1-\cos(2\pi t/0.001))}{2}\right), & \mathbf{x} \in \Sigma_1, 0 \leq t \leq 0.001 \\ (0,\, 0,\, 0), & \mathbf{x} \in \Sigma_1, 0.001 \leq t \leq T \end{cases}$$

where $T = 0.02$ s. The other parameters were the same. The displacements at three points are plotted in Figure 3 and the pressure at three time instants is plotted in Figure 4.

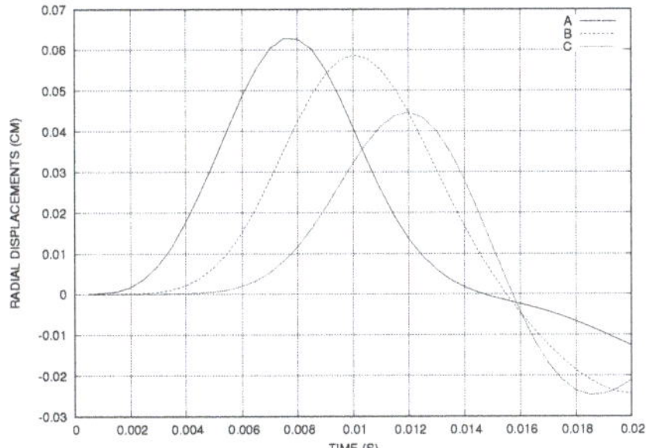

**Figure 3.** The time history of the displacement of three points on the interface using sin-like stress at the inlet.

At each time step, we had to solve a sparse non-symmetric linear system for the fluid–structure velocity and pressure and a sparse symmetric positive definite linear system for the mesh displacement. The linear systems were solved using MUMPS (MUltifrontal Massively Parallel sparse direct Solver), implemented in *FreeFem++*. This method was very efficient; the total CPU time for the first three meshes were: 44 s (1.1 s/iteration) for Mesh 1, 173 s (4.3 s/iteration) for Mesh 2 and 1083 s (27 s/iteration) for Mesh 3, on a computer with a processor of $4 \times 3.30$ GHz frequency and 16 Go RAM. For Mesh 4, the CPU time was about 10 min/iteration on a noed Intel Sandy-Bridge $16 \times 3.30$ GHz and 64 Go RAM.

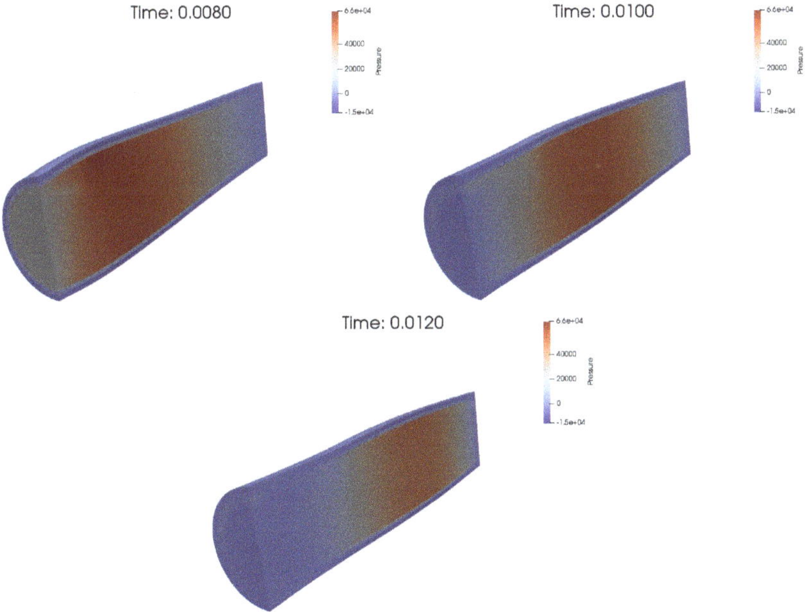

**Figure 4.** The pressure obtained with Mesh 4 using sin-like stress at the inlet.

In [4], the mean number of iterations per time step is: 33.9 for the fixed-point with Aitken acceleration and 8.9 iterations for a quasi-Newton algorithm. At each iteration of quasi-Newton algorithm, a linear system is solved. In [41,42], the average number of Newton iterations per time step is 3.

## 6.2. Cane Cylinder

We considered the fluid–structure interaction in a cane-like geometry inspired from [4,43]. The parameters of the fluid domain were: $R = 0.5$ cm, $R_1 = 1$ cm, $L_1 = 1$ cm, $L_2 = 5$ cm (see Figure 5, left). The fluid was surrounded by a structure of constant thickness $h^S = 0.1$ cm.

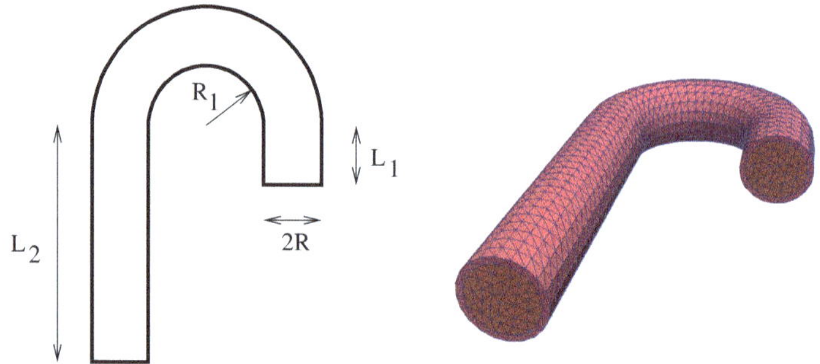


**Figure 5.** The parameters of the fluid domain (**left**); and a global mesh for fluid–structure domain (**right**).

The time step was $\Delta t = 0.0005$ s, but the final time was $T = 0.04$ s. The other parameters were the same as in the case of the straight cylinder. The details of meshes used for the numerical tests are presented in Table 2.

**Table 2.** The number of vertices, tetrahedra and degrees of freedom (DOF) of fluid–structure linear system for each mesh.

|        | Vertices | Tetrahedra | DOF     |
| ------ | -------- | ---------- | ------- |
| Mesh 1 | 2079     | 9828       | 37,800  |
| Mesh 2 | 3632     | 17,784     | 67,880  |
| Mesh 3 | 5802     | 29,106     | 110,526 |

We measured the displacement of the interface ar three points: $U_1^S$ at $A(R_1 + 2R, 0, 0)$, $U_3^S$ at $B(0, 0, R_1 + 2R)$ and $U_1^S$ at $C(-R_1 - 2R, 0, 0)$ (see Figure 6). The pressure at three time instants is plotted in Figure 7. We observed that, for the sin-like stress at the inlet, the structure displacements were greater than 0.23 (cm) and a non-linear model for the structure could be more appropriate. We chose a sin-like stress at the inlet with maximal value five times greater than the constant case to obtain visible deformations. Even though we used the same sin-like stress at the inlet, the structure displacements were larger than in the straight cylinder case because of the shape as well as because the cane was longer. Recall that the structure was fixed at both ends.

The total CPU time for the three meshes were: 255 s (3.18 s/iteration) for Mesh 1, 502 s (6.2 s/iteration) for Mesh 2 and 915 s (11.2 s/iteration) for Mesh 3, on a computer with a processor of $4 \times 3.30$ GHz frequency and 16 Go RAM.

In [4], the mean number of iterations per time step is 8.9 for quasi-Newton algorithm. The fixed-point algorithm with Aitken acceleration failed.

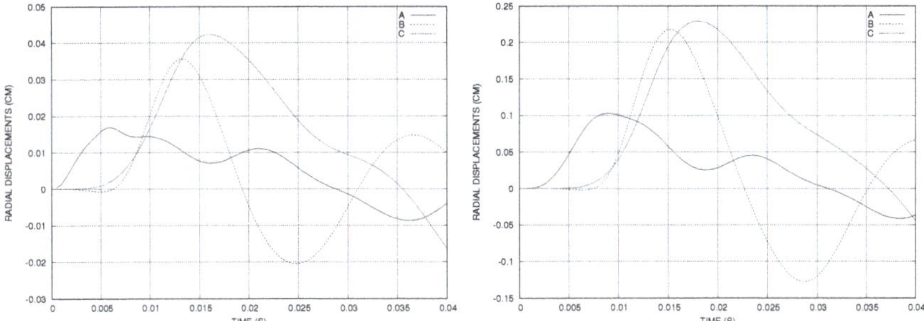

**Figure 6.** The time history of the displacement of three points on the interface using constant (**left**) and sin-like (**right**) stress at the inlet.

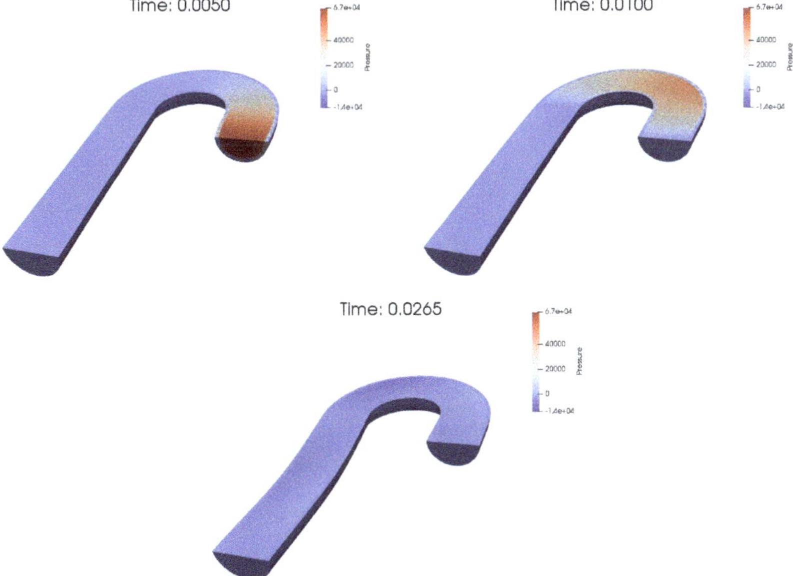

**Figure 7.** The computed pressure obtained with sin-like stress at the inlet (Mesh 2).

## 6.3. Artery Stenosis

Now, we considered a fluid–structure interaction in artery stenosis inspired from the paper [42]. The inlet and outlet surfaces $\Sigma_1$, $\Sigma_3$ were disks of radius $R$, normal to the axis $Ox_1$ of centers $(x_{min}, 0, 0)$ and $(x_{max}, 0, 0)$, respectively. The lateral surface $\Gamma_0$ of the initial fluid domain was composed by the bottom half straight cylinder surface

$$\left\{ (x_1, x_2, x_3) \in \mathbb{R}^3; \; x_{min} < x_1 < x_{max}, \; \sqrt{x_2^2 + x_3^2} = R, \; x_3 < 0 \right\}$$

and the stenosis surface (see Figure 8), obtained from the top half straight cylinder surface

$$\left\{ (x_1, x_2, x_3) \in \mathbb{R}^3; \; x_{min} < x_1 < x_{max}, \; \sqrt{x_2^2 + x_3^2} = R, \; x_3 \geq 0 \right\}$$

via the map

$$(x_1, x_2, x_3) \rightarrow \left( x_1, x_2, x_3 - (h^S_{max} - h^S_{min}) \frac{(\ell^2 - x_1^2)}{\ell^2} \frac{(R^2 - x_2^2)}{R^2} \right), \ |x_1| \leq \ell, |x_2| \leq R.$$

The initial boundary of the structure domain was composed by: the interior lateral surface $\Gamma_0$, which is the fluid–structure interface; the exterior lateral surface

$$\Gamma_0^N = \left\{ (x_1, x_2, x_3) \in \mathbb{R}^3; \ x_{min} < x_1 < x_{max}, \ \sqrt{x_2^2 + x_3^2} = R + h^S_{min} \right\};$$

and two annular surfaces $\Sigma_5, \Sigma_7$, of radii $R$ and $R + h^S_{min}$, normal to the axis $Ox_1$, of centers $(x_{min}, 0, 0)$ and $(x_{max}, 0, 0)$, respectively.

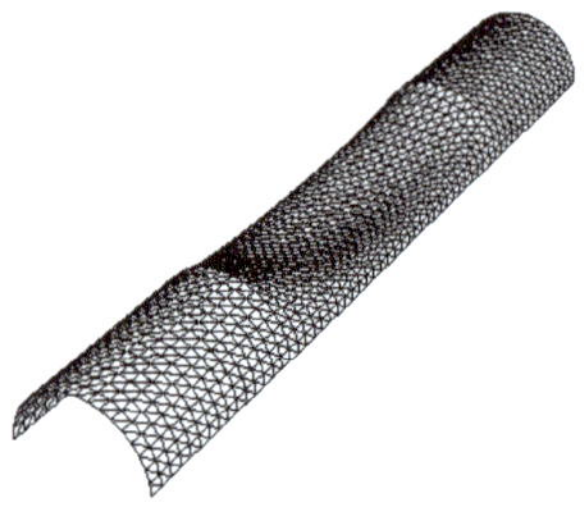

**Figure 8.** The stenosis interface.

The numerical values were: $R = 0.5$ cm, $x_{min} = -3$, $x_{max} = 3$, $\ell = 1.5$ cm, $h^S_{min} = 0.1$ cm, and $h^S_{max} = 0.5$ cm. The details of meshes used for the numerical tests are presented in Table 3.

**Table 3.** The number of vertices, tetrahedra and degrees of freedom (DOF) of fluid–structure linear system for each mesh.

|        | Vertices | Tetrahedra | DOF       |
| ------ | -------- | ---------- | --------- |
| Mesh 1 | 5756     | 28,395     | 108,209   |
| Mesh 2 | 20,970   | 95,202     | 369,486   |
| Mesh 3 | 48,245   | 220,266    | 805,533   |
| Mesh 4 | 92,228   | 425,872    | 1,554,300 |

We used the same sin-like stress at the inlet similar to the precedent experiments

$$\mathbf{h}_{in}(\mathbf{x}, t) = \begin{cases} \left(0, \ 0, \ mag \times 1.3332 \times 10^4 \times \frac{(1 - \cos(2\pi t/0.001))}{2}\right), & \mathbf{x} \in \Sigma_1, 0 \leq t \leq 0.001 \\ (0, \ 0, \ 0), & \mathbf{x} \in \Sigma_1, 0.001 \leq t \leq T \end{cases}$$

where $T = 0.02$ s and $mag > 0$ is a parameter.

For *Mesh 1*, the time step was set to $\Delta t = 5 \times 10^{-4}$ s, the number of time steps to $N = 40$ and $mag = 5$. The total CPU time was 323 s (8.07 s/iteration), on a computer with a processor of $4 \times 3.30$ GHz frequency and 16 Go RAM. The radial displacement of the interface was measured at three points: $A(-\ell, 0, R)$, $B(0, 0, R - (h^S_{max} - h^S_{min}))$, and $C(\ell, 0, R)$ (see Figure 9). We observed that the maximal displacement of point A was more than 0.08 cm, which was more important than in the case of straight cylinder (see Figure 3). The artery deformation was more important in the uphill zone of the stenosis than in the case of healthy artery.

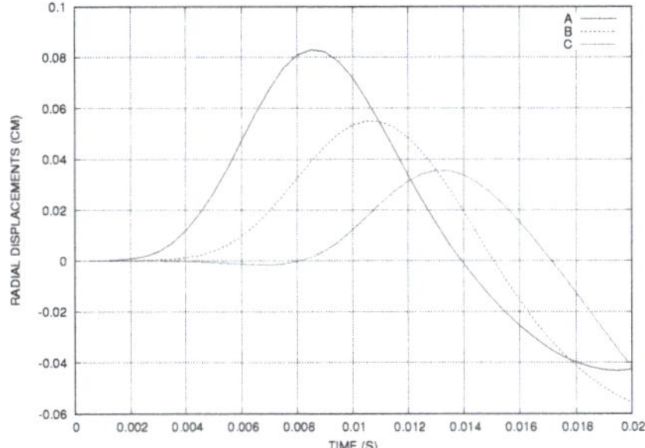

**Figure 9.** The time history of the displacement of three points on the interface using sin-like stress at the inlet (Mesh 1).

For *Mesh 2*, the time step was set to $\Delta t = 10^{-4}$ s, the number of time steps to $N = 200$ and *mag* = 2. The total CPU time was 15,123 s (75.6 s/iteration), on a noed Intel Sandy-Bridge 16 × 3.30 GHz and 64 Go RAM.

For *Mesh 3*, the time step was set to $\Delta t = 10^{-4}$ s, the number of time steps to $N = 200$ and *mag* = 1. The average CPU time was 4 min 55 s by iteration, on a noed Intel Sandy-Bridge 16 × 3.30 GHz and 64 Go RAM. The pressure at three time instants is plotted in Figure 10.

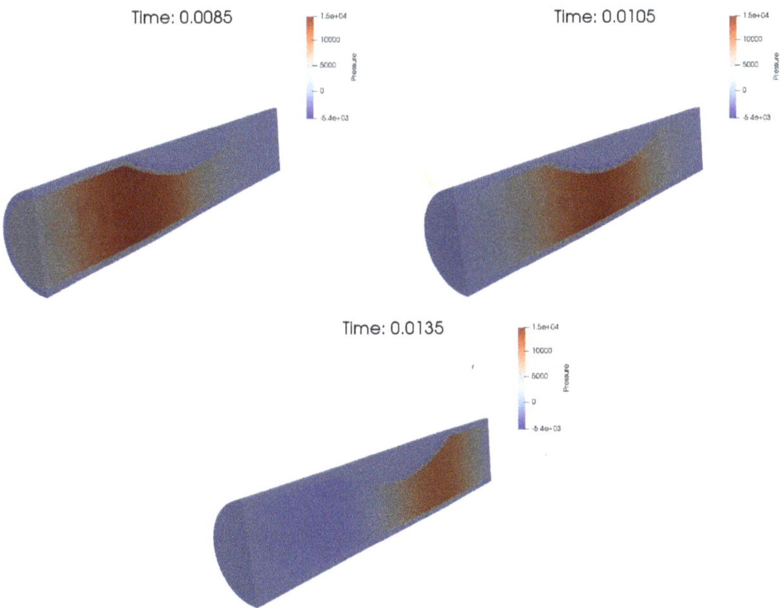


**Figure 10.** The computed pressure obtained with *Mesh 3*.

For *Mesh 4*, the time step was set to $\Delta t = 5 \times 10^{-5}$ s, the number of time steps to $N = 400$ and *mag* = 1. The average CPU time was 18 min 8 s by iteration, on a noed Intel Sandy-Bridge 16 × 3.30 GHz and 64 Go RAM.

For finer mesh, we were forced to use smaller time steps. This phenomenon was observed for neither the three-dimensional tests presented above nor for the two-dimensional benchmark flow around a flexible thin structure attached to a fixed cylinder (see [37]). The semi-implicit algorithms had good stability properties (see Remark 3). We suppose that the source of the problem is this particular surface mesh of the stenosis zone obtained from the mesh of the top half straight cylinder surface by vertical projection on the stenosis surface.

## 7. Conclusions

We have presented a monolithic semi-implicit method for three-dimensional fluid–structure interaction problems. At each time step, we solve only a linear system to find the fluid–structure velocity and the fluid pressure, thus the method is fast. We use a global mesh for the fluid–structure domain where the fluid–structure interface is an interior boundary. Using globally continuous finite element for the velocity in the fluid–structure mesh, the continuity of velocity at the interface is automatically satisfied.

**Funding:** This research received no external funding.

**Acknowledgments:** I gratefully thank Computing Center, University of Strasbourg for giving me access to their computing resources.

**Conflicts of Interest:** The author declares no conflict of interest.

## References

1. Le Tallec, P.; Mouro, J. Fluid structure interaction with large structural displacements. *Comput. Methods Appl. Mech. Eng.* **2001**, *190*, 3039–3067. [CrossRef]
2. Nobile, F. Numerical Approximation of Fluid-Structure Interaction Problems with Application to Haemodynamics. Ph.D. Thesis, EPFL, Lausanne, Switzerland, 2001.
3. Formaggia, L.; Gerbeau, J.-F.; Nobile, F.; Quarteroni, A. On the coupling of 3D and 1D Navier-Stokes equations for flow problems in compliant vessels. *Comput. Methods Appl. Mech. Eng.* **2001**, *191*, 561–582. [CrossRef]
4. Gerbeau, J.-F.; Vidrascu, M. A quasi-Newton algorithm based on a reduced model for fluid-structure interaction problems in blood flows. *M2AN Math. Model. Numer. Anal.* **2003**, *37*, 631–647. [CrossRef]
5. Fernàndez, M.A.; Moubachir, M. A Newton method using exact jacobians for solving fluid-structure coupling. *Comput. Struct.* **2005**, *83*, 127–142. [CrossRef]
6. Dettmer, W.; Perić, D. A computational framework for fluid-structure interaction: Finite element formulation and applications. *Comput. Methods Appl. Mech. Eng.* **2006**, *195*, 5754–5779. [CrossRef]
7. Murea, C.M. Numerical simulation of a pulsatile flow through a flexible channel. *ESAIM Math. Model. Numer. Anal.* **2006**, *40*, 1101–1125. [CrossRef]
8. Mbaye, I.; Murea, C.M. Numerical procedure with analytic derivative for unsteady fluid-structure interaction. *Commun. Numer. Meth. Eng.* **2008**, *24*, 1257–1275. [CrossRef]
9. Kuberry, P.; Lee, H. A decoupling algorithm for fluid-structure interaction problems based on optimization. *Comput. Methods Appl. Mech. Eng.* **2013**, *267*, 594–605. [CrossRef]
10. Hubner, B.; Walhorn, E.; Dinkler, D. A monolithic approach to fluid-structure interaction using space-time finite elements. *Comput. Meth. Appl. Mech. Eng.* **2004**, *193*, 2087–2104. [CrossRef]
11. Hron, J.; Turek, S. A monolithic FEM/multigrid solver for an ALE formulation of fluid-structure interaction with application in biomechanics. In *Fluid-Structure Interaction*; Lecture Notes in Computational Science and Engineering 53; Springer: Berlin, Germany, 2006; pp. 146–170.
12. Dunne, T. An Eulerian approach to fluid-structure interaction and goal-oriented mesh adaptation. *Int. J. Numer. Methods Fluids* **2006**, *51*, 1017–1039. [CrossRef]
13. Heil, M.; Hazed, A.L.; Boyle, J. Solvers for large-displacement fluid-structure interaction problems: Segregated versus monolithic approaches. *Cumput. Mech.* **2008**, *193*, 91–101. [CrossRef]
14. Pironneau, O. Numerical study of a monolithic fluid-structure formulation. In *Variational Analysis and Aerospace Engineering*; Springer Optimization and Its Applications 116; Springer: Cham, Seitzerland, 2016; pp. 401–420.
15. Hecht, F.; Pironneau, O. An energy stable monolithic Eulerian fluid-structure finite element method. *Int. J. Numer. Methods Fluids* **2017**, *85*, 430–446. [CrossRef]

16. Chiang, C.-Y.; Pironneau, O.; Sheu, T.W.H.; Thiriet, M. Numerical Study of a 3D Eulerian Monolithic Formulation for Incompressible Fluid-Structures Systems. *Fluids* **2017**, *2*, 34. [CrossRef]
17. Pironneau, O. An energy stable monolithic Eulerian fluid-structure numerical scheme. *Chin. Ann. Math. Ser. B* **2018**, *39*, 213–232. [CrossRef]
18. Donea, J. Arbitrary Lagrangian-Eulerian finite element methods. In *Computational Methods for Transient Analysis*; Belytschko, T., Hughes, T.J.R., Eds.; North-Holland: Amsterdam, The Netherlands, 1983; pp. 474–516.
19. Quarteroni, A.; Formaggia, L. Mathematical modelling and numerical simulation of the cardiovascular system. In *Handbook of Numerical Analysis, Vol. XII*; Ciarlet, P.G., Ed.; North-Holland: Amsterdam, The Netherlands, 2004; pp. 3–127.
20. Peskin, C.S. The immersed boundary method. *Acta Numer.* **2002**, *11*, 479–517. [CrossRef]
21. Glowinski, R.; Hesla, T.I.; Joseph, D.; Pan, T.; Périaux, J. A distributed Lagrange multiplier/fictitious domain method for the simulation of flow around moving rigid bodies: Application to particulate flow. *Comput. Meth. Appl. Mech. Eng.* **2000**, *184*, 241–267. [CrossRef]
22. Boffi, D.; Gastaldi, L. A fictitious domain approach with Lagrange multiplier for fluid-structure interactions. *Numer. Math.* **2017**, *135*, 711–732. [CrossRef]
23. Bost, C.; Cottet, G.-H.; Maitre, E. Convergence analysis of a penalization method for the three-dimensional motion of a rigid body in an incompressible viscous fluid. *SIAM J. Numer. Anal.* **2010**, *48*, 1313–1337. [CrossRef]
24. Yakhlef, O.; Murea, C.M. Numerical procedure for fluid-structure interaction with the structure displacements limited by a rigid obstacle. *Appl. Comput. Mech.* **2017**, *11*, 91–104. [CrossRef]
25. Court, S.; Fournié, M.; Lozinski, A. A fictitious domain approach for fluid-structure interactions based on the extended finite element method. In Proceedings of the SMAI Congress 2013 6th French Biennale of Applied and Industrial Mathematics, Seignosse, France, 27–31 May 2013; pp. 308–317.
26. Court, S.; Fournié, M. A fictitious domain finite element method for simulations of fluid-structure interactions: The Navier-Stokes equations coupled with a moving solid. *J. Fluids Struct.* **2015**, *55*, 398–408. [CrossRef]
27. Alauzet, F.; Fabrèges, B.; Fernández, M.A.; Landajuela, M. Nitsche-XFEM for the coupling of an incompressible fluid with immersed thin-walled structures. *Comput. Methods Appl. Mech. Eng.* **2016**, *301*, 300–335. [CrossRef]
28. Boffi, D.; Hecht, F.; Pironneau, O. Distributed Lagrange multiplier for fluid-structure interactions. In *Numerical Methods for PDEs*; Di Pietro, D., Ern, A., Formaggia, L., Eds.; SEMA SIMAI Springer Series; Springer: Berlin/Heidelberg, Germany, 2018; pp. 129–145.
29. Wang, Y.; Jimack, P.; Walkley, M. A one-field monolithic fictitious domain method for fluid-structure, interactions. *Comput. Methods Appl. Mech. Eng.* **2017**, *317*, 1146–1168. [CrossRef]
30. Wang, Y.; Jimack, P.; Walkley, M. Energy analysis for the one-field fictitious domain method for fluid-structure interactions. *Appl. Numer. Math.* **2019**, *140*, 165–182. [CrossRef]
31. Fernàndez, M.A.; Mullaert, J.; Vidrascu, M. Explicit Robin-Neumann schemes for the coupling of incompressible fluids with thin-walled structures. *Comput. Methods Appl. Mech. Eng.* **2013**, *267*, 566–593. [CrossRef]
32. Dettmer, W.; Perić, D. A new staggered scheme for fluid-structure interaction. *Int. J. Numer. Meth. Eng.* **2013**, *93*, 1–22. [CrossRef]
33. Fernàndez, M.A.; Gerbeau, J.-F.; Grandmont, C. A projection semi-implicit scheme for the coupling of an elastic structure with an incompressible fluid. *Int. J. Numer. Methods Eng.* **2007**, *69*, 794–821. [CrossRef]
34. Murea, C.M.; Sy, S. A fast method for solving fluid-structure interaction problem numerically. *Int. J. Numer. Meth. Fluids* **2009**, *60*, 1149–1172. [CrossRef]
35. Sy, S.; Murea, C.M. A stable time advancing scheme for solving fluid-structure interaction problem at small structural displacements. *Comput. Meth. Appl. Mech. Eng.* **2008**, *198*, 210–222. [CrossRef]
36. Murea, C.M.; Sy, S. Updated Lagrangian/Arbitrary Lagrangian Eulerian framework for interaction between a compressible Neo-Hookean structure and an incompressible fluid. *Int. J. Numer. Meth. Eng.* **2017**, *103*, 1067–1084. [CrossRef]
37. Murea, C.M. Monolithic algorithm for dynamic fluid-structure interaction problem. In *Mathematical Modelling in Solid Mechanics*; Dell Isola, F., Sofonea, M., Steigmann, D., Eds.; Advanced Structured Materials 69; Springer: Singapore, 2017; pp. 135–146.
38. Sy, S.; Murea, C.M. Algorithm for solving fluid-structure interaction problem on a global moving mesh. *Coupled Syst. Mech.* **2012**, *1*, 99–113. [CrossRef]

39. Ciarlet, P.G. *Élasticité Tridimensionnelle*; Masson: Paris, France, 1986.
40. Hecht, F. New development in FreeFem++. *J. Numer. Math.* **2012**, *20*, 251–265. [CrossRef]
41. Deparis, S.; Forti, D.; Quarteroni, A. A Fluid-Structure Interaction Algorithm Using Radial Basis Function Interpolation Between Non-Conforming Interfaces. In *Advances in Computational Fluid-Structure Interaction and Flow Simulation*; Bazilevs, Y., Takizawa, K., Eds.; Springer: Berlin/Heidelberg, Germany, 2016; pp. 439–450.
42. Forti, D.; Bukac, M.; Quaini, A.; Canic, S.; Deparis, S. A monolithic approach to fluid-composite structure interaction. *J. Sci. Comput.* **2017**, *72*, 396–421. [CrossRef]
43. Lefieux, A.; Auricchio, F.; Conti, M.; Morganti, S.; Reali, A.; Trimarchi, S.; Veneziani, A. Computational Study of Aortic Hemodynamics: From Simplified to Patient-Specific Geometries. In *Advances in Computational Fluid-Structure Interaction and Flow Simulation*; Bazilevs, Y., Takizawa, K., Eds.; Modeling and Simulation in Science, Engineering and Technology; Springer: Berlin/Heidelberg, Germany, 2016; pp. 397–407.

Article

# On the Kutta Condition in Compressible Flow over Isolated Airfoils


Farzad Mohebbi [1,*], Ben Evans [1] and Mathieu Sellier [2]

[1] Zienkiewicz Centre for Computational Engineering, College of Engineering, Bay Campus, Swansea University, Swansea SA1 8EN, UK; b.j.evans@swansea.ac.uk

[2] Department of Mechanical Engineering, University of Canterbury, Private Bag 4800, Christchurch 8140, New Zealand; mathieu.sellier@canterbury.ac.nz

* Correspondence: farzadmohebbi@yahoo.com


Received: 10 April 2019; Accepted: 27 May 2019; Published: 1 June 2019


**Abstract:** This paper presents a novel and accurate method to implement the Kutta condition in solving subsonic (subcritical) inviscid isentropic compressible flow over isolated airfoils using the stream function equation. The proposed method relies on body-fitted grid generation and solving the stream function equation for compressible flows in computational domain using finite-difference method. An expression is derived for implementing the Kutta condition for the airfoils with both finite angles and cusped trailing edges. A comparison of the results obtained from the proposed numerical method and the results from experimental and other numerical methods reveals that they are in excellent agreement, which confirms the accuracy and correctness of the proposed method.



**Keywords:** computational aerodynamics; Kutta condition; compressible flow; stream function


---

## 1. Introduction

Nowadays, computational fluid dynamics complements its experimental and theoretical counterpart. Benefiting from high-speed digital computers, the use of sophisticated numerical methods has made possible the numerical solution of fluid flow problems which were heretofore intractable. Compressible flows over airfoils and wings play a vital role in computational aerodynamics, and demand advanced computational techniques. Since introducing the source and vortex panel methods in the late 1960s [1], they have become the standard tools to numerically solve low-speed flows over bodies of arbitrary shape, and have had extensive applications in flow modeling [2–6]. The panel methods used for the simulation of flows over an airfoil are concerned with the vortex panel strength and circulation quantities; the evaluation of such quantities allows one to calculate the velocity distribution over the airfoil surface, and hence, to determine the pressure coefficients. The Kutta condition (a viscous boundary condition based on physical observation used with inviscid theoretical model) states that the flow leaves the sharp trailing edge of an airfoil smoothly [7]. Various methods have been proposed to impose the Kutta condition [8–10]. In panel methods, the Kutta condition is incorporated in the numerical formulation by requiring that the strength of vortex sheet is zero at the airfoil trailing edge. Moreover, the use of panel methods along with the compressibility corrections such as the Prandtl-Glauert method [11] allow one to consider the compressible flows over bodies. The panel methods have been extensively investigated in the aerodynamics literature, so these will not be discussed further here. However, these methods often have trouble with accuracy at the trailing edge of airfoils with zero angle cusped trailing edges [12]. Moreover, the compressibility corrections do not give accurate results for the compressible flows over airfoils of any shape at any angle of attack. For example, the Prandtl-Glauert method is based on the linearized perturbation velocity potential equation, and hence, it is limited to thin airfoils at small angles of attack [4].

In this paper, we propose a novel method to numerically solve the steady irrotational compressible flow over an airfoil which is exempt from considerations of quantities such as the vortex panel strength and circulation. This method takes advantage of an O-grid, generated by the elliptic grid generation technique, over the flow field and approximates the flow field quantities such as stream function, density, velocity, pressure, speed of sound, and Mach number at the nodal points. An accurate Kutta condition scheme is proposed and implemented into the computational loop by an exact derived expression for the stream function. The exact expression is general and encompasses both the finite-angle and cusped trailing edges. Finally, the results obtained from the proposed numerical method, and the results from experimental and other numerical methods, are compared to reveal the accuracy of the proposed method (The material given in this article are implemented in the code FOILcom which is freely available at: DOI:10.13140/RG.2.2.36459.64801/1).

It is worth emphasizing that the stream function equation considered in this study is completely equivalent to the full potential equation. Here, the stream function equation is solved in non-conservative form and can be employed to obtain accurate results for subsonic (subcritical) inviscid isentropic compressible flow over isolated airfoils. The conservative stream function equation, along with upwinding schemes such as artificial compressibility [13], can be used to solve transonic flows over airfoils which is not considered in this study.

## 2. Governing Equations

The stream function equation for two-dimensional, irrotational, steady, and isentropic flow of a compressible fluid in non-conservative form is as follows [14–16]

$$(c^2 - u^2)\psi_{xx} + (c^2 - v^2)\psi_{yy} - 2uv\psi_{xy} = 0 \tag{1}$$

where $\psi$ is the stream function, $u$ and $v$ are the components of the velocity vector $\mathbf{V}$, i.e., $\mathbf{V} = u\mathbf{i} + v\mathbf{j}$ ($\mathbf{i}$ and $\mathbf{j}$ are the unit vectors in $x$ and $y$ directions, respectively). For a two-dimensional compressible flow,

$$u = \frac{\rho_0}{\rho}\psi_y \tag{2}$$

$$v = -\frac{\rho_0}{\rho}\psi_x \tag{3}$$

$c$ is the local sound speed [4]

$$c^2 = c_0^2 - \frac{\gamma - 1}{2}V^2 = c_0^2 - \frac{\gamma - 1}{2}(u^2 + v^2) \tag{4}$$

$c_0$ is the stagnation speed of sound, $\rho$ is the density, $\rho_0$ is the stagnation density, and $\gamma = c_p/c_v$ ($c_v$ and $c_p$ are the specific heats at constant volume and constant pressure, respectively) is the ratio of specific heats of gas (for air at standard conditions, $\gamma = 1.4$). Dividing both sides of Equation (1) by $c^2$ and then substituting the expressions in Equations (2) and (3) into Equation (1) gives the stream function equation as

$$[1 - \frac{1}{c^2}\left(\frac{\rho_0}{\rho}\right)^2\psi_y^2]\psi_{xx} + [1 - \frac{1}{c^2}\left(\frac{\rho_0}{\rho}\right)^2\psi_x^2]\psi_{yy} + \frac{2}{c^2}\left(\frac{\rho_0}{\rho}\right)^2\psi_x\psi_y\psi_{xy} = 0 \tag{5}$$

by substituting Equation (4) into Equation (5), we get

$$[1 - \frac{1}{c_0^2 - \frac{\gamma-1}{2}\left(\frac{\rho_0}{\rho}\right)^2(\psi_x^2+\psi_y^2)}\left(\frac{\rho_0}{\rho}\right)^2\psi_y^2]\psi_{xx} + [1 - \frac{1}{c_0^2 - \frac{\gamma-1}{2}\left(\frac{\rho_0}{\rho}\right)^2(\psi_x^2+\psi_y^2)}\left(\frac{\rho_0}{\rho}\right)^2\psi_x^2]\psi_{yy} +$$
$$\frac{2}{c_0^2 - \frac{\gamma-1}{2}\left(\frac{\rho_0}{\rho}\right)^2(\psi_x^2+\psi_y^2)}\left(\frac{\rho_0}{\rho}\right)^2\psi_x\psi_y\psi_{xy} = 0 \tag{6}$$

and from the local isentropic stagnation properties of an ideal gas, we have

$$\frac{p_0}{\rho} = (1 + \frac{\gamma-1}{2}M^2)^{\frac{1}{\gamma-1}} = (1 + \frac{\gamma-1}{2}\frac{u^2+v^2}{c^2})^{\frac{1}{\gamma-1}} \tag{7}$$

$M = \frac{V}{c}$ is the local Mach number. Equations (6) and (7) should be solved simultaneously to obtain the local values of $\psi$ and $\rho$ in the flow field.

### 2.1. Transformation

The solution of the governing PDE is based on transformation of the physical domain $(x, y)$ and the governing equations into the regular computational domain $(\xi, \eta)$ (see Figure 1). Therefore, the derivatives such as $\psi_x$, $\psi_y$, $\psi_{xx}$, $\psi_{yy}$, and $\psi_{xy}$ in the stream function equation, Equation (1), should be transformed from the physical domain $(x, y)$ to the computational domain $(\xi, \eta)$ [17–19]. This transformation can be stated as

$$\xi \equiv \xi(x, y) \tag{8}$$

$$\eta \equiv \eta(x, y) \tag{9}$$

and the inverse transformation is given as below.

$$x \equiv x(\xi, \eta) \tag{10}$$

$$y \equiv y(\xi, \eta) \tag{11}$$

Since the stream function equation involves first and second derivatives, relationships are needed to transform such derivatives from the $(x, y)$ system to the $(\xi, \eta)$ one. In order to do this, the Jacobian of the transformation is needed, which is given below

$$2\text{D} : J = J(\frac{x, y}{\xi, \eta}) = \begin{vmatrix} x_\xi & y_\xi \\ x_\eta & y_\eta \end{vmatrix} = x_\xi y_\eta - x_\eta y_\xi \neq 0 \tag{12}$$

As will be shown, the transformation relations involve the Jacobian in denominator. Hence, it cannot be zero. Since we deal with the stream function equation, it is necessary to find relationships for the transformation of the first and second derivatives of the variable $\psi$ with respect to the variables $x$ and $y$. By using the chain rule, it can be concluded that

$$\frac{\partial \psi}{\partial x} = \frac{\partial \psi}{\partial \xi}\frac{\partial \xi}{\partial x} + \frac{\partial \psi}{\partial \eta}\frac{\partial \eta}{\partial x} = \frac{\partial \psi}{\partial \xi}\xi_x + \frac{\partial \psi}{\partial \eta}\eta_x \tag{13}$$

$$\frac{\partial \psi}{\partial y} = \frac{\partial \psi}{\partial \xi}\frac{\partial \xi}{\partial y} + \frac{\partial \psi}{\partial \eta}\frac{\partial \eta}{\partial y} = \frac{\partial \psi}{\partial \xi}\xi_y + \frac{\partial \psi}{\partial \eta}\eta_y \tag{14}$$

By interchanging $x$ and $\xi$, and $y$ and $\eta$, the following relationships can also be derived

$$\frac{\partial \psi}{\partial \xi} = \frac{\partial \psi}{\partial x}\frac{\partial x}{\partial \xi} + \frac{\partial \psi}{\partial y}\frac{\partial y}{\partial \xi} = \frac{\partial \psi}{\partial x}x_\xi + \frac{\partial \psi}{\partial y}y_\xi \tag{15}$$

$$\frac{\partial \psi}{\partial \eta} = \frac{\partial \psi}{\partial x}\frac{\partial x}{\partial \eta} + \frac{\partial \psi}{\partial y}\frac{\partial y}{\partial \eta} = \frac{\partial \psi}{\partial x}x_\eta + \frac{\partial \psi}{\partial y}y_\eta \tag{16}$$

By solving Equations (15) and (16) for $\frac{\partial \psi}{\partial x}$ and $\frac{\partial \psi}{\partial y}$, we finally obtain

$$\frac{\partial \psi}{\partial x} = \frac{1}{J}(y_\eta \frac{\partial \psi}{\partial \xi} - y_\xi \frac{\partial \psi}{\partial \eta}) \tag{17}$$

$$\frac{\partial \psi}{\partial y} = \frac{1}{J}\left(-x_\eta \frac{\partial \psi}{\partial \xi} + x_\xi \frac{\partial \psi}{\partial \eta}\right) \tag{18}$$

where $J = x_\xi y_\eta - x_\eta y_\xi$ is Jacobian of the transformation. By comparing Equations (13) and (17), and (14) and (18), it can be shown that

$$\xi_x = \frac{1}{J}y_\eta, \, \xi_y = -\frac{1}{J}x_\eta \tag{19}$$

$$\eta_x = -\frac{1}{J}y_\xi, \, \eta_y = \frac{1}{J}x_\xi \tag{20}$$

To transform terms in the stream function equation (Equation (6)), the second order derivatives are needed. Therefore,

$$\psi_{xx} = (\psi_x)_x = \left(\frac{1}{J}(y_\eta\psi_\xi - y_\xi\psi_\eta)\right)_x \tag{21}$$

$$\psi_{yy} = (\psi_y)_y = \left(\frac{1}{J}(-x_\eta\psi_\xi + x_\xi\psi_\eta)\right)_y \tag{22}$$

$$\psi_{xy} = (\psi_x)_y = \left(\frac{1}{J}(y_\eta\psi_\xi - y_\xi\psi_\eta)\right)_y \tag{23}$$

which result in the following expressions for the second order derivatives

$$\psi_{xx} = \frac{(y_\eta^2\psi_{\xi\xi} - 2y_\xi y_\eta\psi_{\xi\eta} + y_\xi^2\psi_{\eta\eta})}{J^2} + \frac{(y_\eta^2 y_{\xi\xi} - 2y_\xi y_\eta y_{\xi\eta} + y_\xi^2 y_{\eta\eta})(x_\eta\psi_\xi - x_\xi\psi_\eta)}{J^3} + \frac{(y_\eta^2 x_{\xi\xi} - 2y_\xi y_\eta x_{\xi\eta} + y_\xi^2 x_{\eta\eta})(y_\xi\psi_\eta - y_\eta\psi_\xi)}{J^3} \tag{24}$$

$$\psi_{yy} = \frac{(x_\eta^2\psi_{\xi\xi} - 2x_\xi x_\eta\psi_{\xi\eta} + x_\xi^2\psi_{\eta\eta})}{J^2} + \frac{(x_\eta^2 y_{\xi\xi} - 2x_\xi x_\eta y_{\xi\eta} + x_\xi^2 y_{\eta\eta})(x_\eta\psi_\xi - x_\xi\psi_\eta)}{J^3} + \frac{(x_\eta^2 x_{\xi\xi} - 2x_\xi x_\eta x_{\xi\eta} + x_\xi^2 x_{\eta\eta})(y_\xi\psi_\eta - y_\eta\psi_\xi)}{J^3} \tag{25}$$

$$\psi_{xy} = \frac{1}{J}\left[-x_\eta\left(\frac{y_{\xi\eta}\psi_\xi + y_\eta\psi_{\xi\xi} - y_{\xi\xi}\psi_\eta - y_\xi\psi_{\xi\eta}}{J} - \frac{(y_\eta\psi_\xi - y_\xi\psi_\eta)(x_{\xi\xi}y_\eta + x_\xi y_{\xi\eta} - x_{\xi\eta}y_\xi - x_\eta y_{\xi\xi})}{J^2}\right) + x_\xi\left(\frac{y_{\eta\eta}\psi_\xi + y_\eta\psi_{\xi\eta} - y_{\xi\eta}\psi_\eta - y_\xi\psi_{\eta\eta}}{J} - \frac{(y_\eta\psi_\xi - y_\xi\psi_\eta)(x_{\xi\eta}y_\eta + x_\xi y_{\eta\eta} - x_{\eta\eta}y_\xi - x_\eta y_{\xi\eta})}{J^2}\right)\right] \tag{26}$$

The finite-difference method can be used to discretize the above expressions in the regular computational domain $(\xi, \eta)$. The actual values of $\xi$ and $\eta$ in the computational domain are immaterial, because they do not appear in the final expressions. Thus, without a loss of generality, we can select the coordinates of the node $A$ in the computational domain as $\xi = \eta = 1$ and the mesh size as $\Delta\xi = \Delta\eta = 1$ [19]. Therefore, we have [17]

$$(f_\xi)_{i,j} = \frac{1}{2}(f_{i+1,j} - f_{i-1,j}) \tag{27}$$

$$(f_\eta)_{i,j} = \frac{1}{2}(f_{i,j+1} - f_{i,j-1}) \tag{28}$$

$$(f_{\xi\xi})_{i,j} = (f_{i+1,j} - 2f_{i,j} + f_{i-1,j}) \tag{29}$$

$$(f_{\eta\eta})_{i,j} = (f_{i,j+1} - 2f_{i,j} + f_{i,j-1}) \tag{30}$$

$$(f_{\xi\eta})_{i,j} = \frac{1}{4}(f_{i+1,j+1} - f_{i-1,j+1} - f_{i+1,j-1} + f_{i-1,j-1}) \tag{31}$$

where $f \equiv x, y, \psi$.

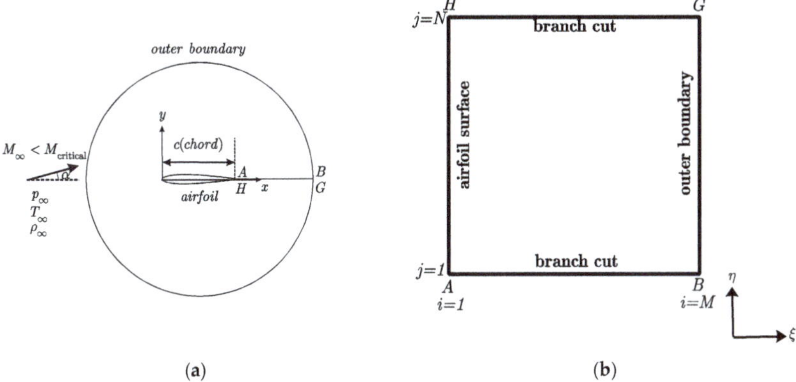

**Figure 1.** The physical and the computational domains. (**a**) Physical domain. (**b**) Computational domain.

### 2.2. Boundary Conditions

*Conditions at infinity*: Far away from the airfoil surface (toward infinity), in all directions, the flow approaches the free stream conditions. The known free stream conditions are the velocity $V_\infty$, the pressure $p_\infty$, the density $\rho_\infty$, and the temperature $T_\infty$. Thus, the free stream Mach number is

$$M_\infty = \frac{V_\infty}{c_\infty} < M_{\text{critical}} \tag{32}$$

and the speed of sound $c_\infty$ is

$$c_\infty = \sqrt{\gamma R T_\infty} \tag{33}$$

where $R$ is the specific gas constant, which is a different value for different gases. For air at standard conditions, $R_{\text{air}} = 287\text{J}/(\text{kg.K})$, $\gamma_{\text{air}} = 1.4$, $\rho_\infty = 1.23\text{kg/m}^3$, and $p_\infty = 1.01 \times 10^5 \text{ N/m}^2$. The critical Mach number $M_{\text{critical}}$ is that free stream Mach number at which sonic flow is first achieved on the airfoil surface.

*Condition on the airfoil surface*: The relevant boundary condition at the airfoil surface for the inviscid flow is the no-penetration boundary condition. Thus, the velocity vector must be tangential to the surface. This wall boundary condition can be expressed by

$$\frac{\partial \psi}{\partial s} = 0 \text{ or } \psi = \text{constant} \tag{34}$$

where $s$ is tangent to the surface.

### 2.3. Grid Generation

Here, an O-grid is initially generated around the airfoil using elliptic grid generation method [17] and then the stream function equation is solved in the computational domain. The size of mesh is $M \times N$ where $M$ is the number of nodes on the branch cut (a horizontal line connecting the trailing edge to outer boundary) and $N$ is the number of nodes on the airfoil surface (and hence, the outer boundary), as shown in Figure 1. The physical domain before and after meshing using an O-grid of size $110 \times 101$ is shown in Figure 2 (using a NACA 0012 airfoil). Moreover, the magnified view of different parts of domain is shown in Figure 3 to highlight the orthogonality of the gridlines at airfoil surface and outer boundary.

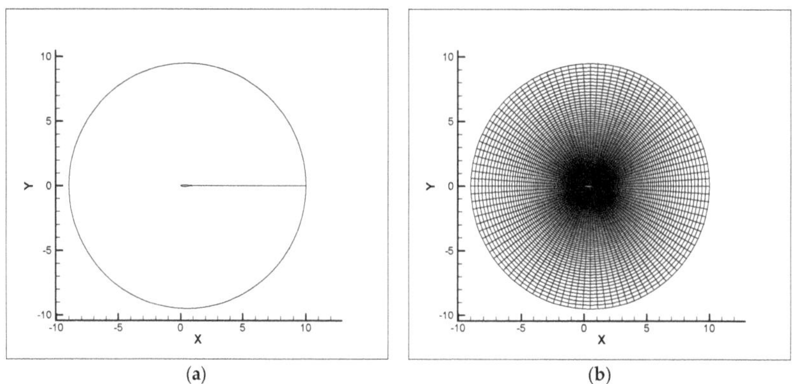

**Figure 2.** The physical domain before and after meshing. (**a**) Physical domain before meshing. (**b**) Physical domain after meshing.

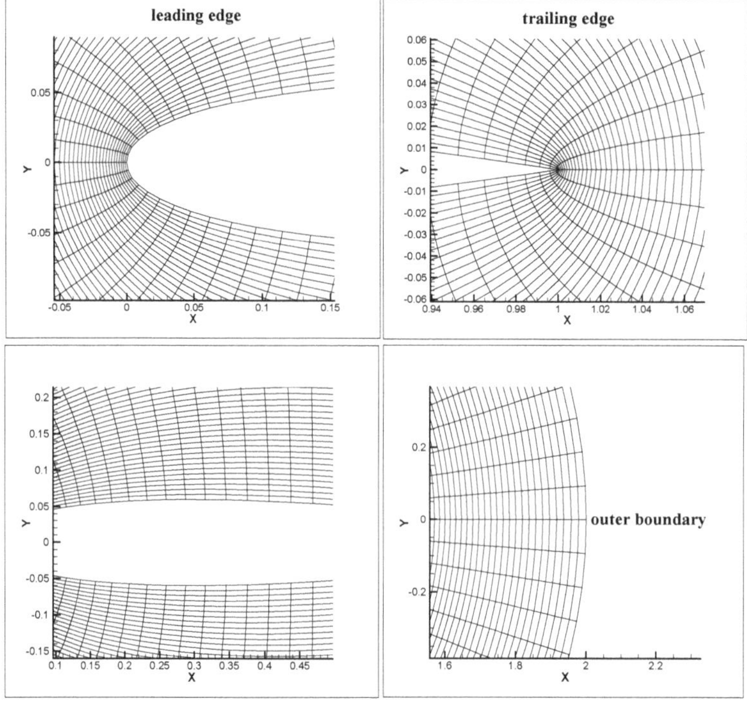

**Figure 3.** Magnified view of different parts of domain.

### 2.4. Kutta Condition

In inviscid flows, because the flow cannot penetrate the surface, the velocity vector must be tangential to the surface. In other words, the component of velocity normal to the surface must be zero and only the tangential velocity component must be considered. The unit tangent vector on the airfoil surface can be expressed as

$$\tau^{(\xi)} = \mathbf{n}^{(\xi)} \times \mathbf{k} \tag{35}$$

$$\tau^{(\eta)} = \mathbf{n}^{(\eta)} \times \mathbf{k} \tag{36}$$

where $\mathbf{n}^{(\xi)}$ and $\mathbf{n}^{(\eta)}$ are the outward-pointing unit normal vector to a airfoil surface in $\xi$ and $\eta$ directions, respectively, and $\mathbf{k}$ is the unit vector in $z$ direction.

At the airfoil surface, corresponding to surface $S_3$ in Figure 4, we have

$$\mathbf{n}^{(\xi)} = -\frac{\nabla\xi}{|\nabla\xi|} = -\frac{\xi_x\mathbf{i} + \xi_y\mathbf{j}}{\sqrt{\xi_x^2 + \xi_y^2}} \tag{37}$$

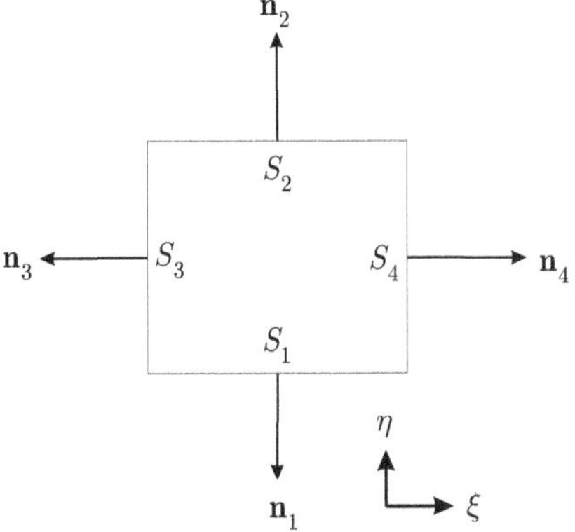

**Figure 4.** The outward—pointing unit normal vectors to $\xi$ = constant and $\eta$ = constant lines.

From the transformation relationships (Equation (19)), we have

$$\mathbf{n}^{(\xi)} = -\frac{\frac{1}{J}y_\eta\mathbf{i} + (-\frac{1}{J}x_\eta\mathbf{j})}{\sqrt{\left(\frac{y_\eta}{J}\right)^2 + \left(-\frac{x_\eta}{J}\right)^2}} = \frac{-y_\eta\mathbf{i} + x_\eta\mathbf{j}}{\sqrt{\alpha}} \tag{38}$$

where $\alpha = x_\eta^2 + y_\eta^2$. Using Equation (35), we get

$$\tau_{S_3}^{(\xi)} = \mathbf{n}_{S_3}^{(\xi)} \times \mathbf{k} = \frac{1}{\sqrt{\alpha}}(-y_\eta\mathbf{i} + x_\eta\mathbf{j}) \times \mathbf{k} = \frac{1}{\sqrt{\alpha}}(-y_\eta(-\mathbf{j}) + x_\eta\mathbf{i}) = \frac{1}{\sqrt{\alpha}}(x_\eta\mathbf{i} + y_\eta\mathbf{j}) \tag{39}$$

The velocity component tangential to the airfoil surface ($S_3$) is

$$V_t = \mathbf{V}.\tau_{S_3}^{(\xi)} = (u\mathbf{i} + v\mathbf{j}).\frac{1}{\sqrt{\alpha}}(x_\eta\mathbf{i} + y_\eta\mathbf{j}) \tag{40}$$

For compressible flows, the velocity components of $u$ and $v$ can be expressed in terms of stream function $\psi$ as

$$u = \frac{\rho_0}{\rho}\psi_y = \frac{\rho_0}{\rho}\frac{1}{J}(-x_\eta\psi_\xi + x_\xi\psi_\eta) \tag{41}$$

and

$$v = -\frac{\rho_0}{\rho}\psi_x = -\frac{\rho_0}{\rho}\frac{1}{J}(y_\eta\psi_\xi - y_\xi\psi_\eta) \tag{42}$$

where $\rho_0$ is a constant. On the airfoil surface, $\psi_s = \psi_\eta = 0$ where $s$ is the distance measured along the airfoil surface because the airfoil contour is a streamline of the flow. Thus Equations (41) and (42) become

$$u = \frac{\rho_0}{\rho}\frac{1}{J}(-x_\eta\psi_\xi) \tag{43}$$

and

$$v = -\frac{\rho_0}{\rho}\frac{1}{J}(y_\eta \psi_\xi) \tag{44}$$

By substituting Equations (43) and (44) into Equation (40), we get

$$V_t = \left(-\frac{\rho_0}{\rho}\frac{1}{J}x_\eta\psi_\xi\mathbf{i} - \frac{\rho_0}{\rho}\frac{1}{J}y_\eta\psi_\xi\mathbf{j}\right)\cdot\frac{1}{\sqrt{\alpha}}(x_\eta\mathbf{i} + y_\eta\mathbf{j}) = -\frac{\rho_0}{\rho}\frac{1}{J}\frac{1}{\sqrt{\alpha}}x_\eta^2\psi_\xi - \frac{\rho_0}{\rho}\frac{1}{J}\frac{1}{\sqrt{\alpha}}y_\eta^2\psi_\xi =$$
$$-\frac{\rho_0}{\rho}\frac{1}{J}\frac{1}{\sqrt{\alpha}}(x_\eta^2 + y_\eta^2)\psi_\xi = -\frac{\rho_0}{\rho}\frac{1}{J}\frac{1}{\sqrt{\alpha}}(\alpha)\psi_\xi = -\frac{\rho_0}{\rho}\frac{\sqrt{\alpha}}{J}\psi_\xi \tag{45}$$

Based on the Kutta condition, the velocity at the trailing edge of the airfoil must be the same when approached from the upstream direction along the upper and lower airfoil surfaces [20] (see Figure 5). Thus,

$$V_t|_{1,1} = V_t|_{1,N} = C \begin{cases} C = 0 \text{ for finite angle trailing edge} \\ C \neq 0 \text{ for cusped trailing edge} \end{cases} \tag{46}$$

For the finite angle trailing edge, we have

$$V_t|_{1,1} = 0$$

$$-\frac{\rho_0}{\rho}\frac{\sqrt{\alpha}}{J}\psi_\xi\bigg|_{1,1} = 0$$

hence

$$\psi_\xi|_{1,1} = 0 \, \psi_{2,1} - \psi_{1,1} = 0$$

$$\psi_{2,1} = \psi_{1,1} \tag{47}$$

(The above expression is based on forward finite-difference coefficient with first-order accuracy. The forward finite-difference coefficient with second-order accuracy can be used if more accurate Kutta condition implementation is needed.)

Therefore, the wall boundary condition for the finite angle trailing edge becomes

$$\psi_{1,j} = \psi_{2,1}(j = 1,\ldots,N) \tag{48}$$

because the stream function on the airfoil surface is constant.

For the cusped trailing edge, we have

$$V_t|_{1,1} = V_t|_{1,N} - \frac{\rho_0}{\rho}\frac{\sqrt{\alpha}}{J}\psi_\xi\bigg|_{1,1} = -\frac{\rho_0}{\rho}\frac{\sqrt{\alpha}}{J}\psi_\xi\bigg|_{1,N} \tag{49}$$

The points $(1,1)$ and $(1,N)$ as well as the points $(2,1)$ and $(2,N)$ are the same points (see Figure 5). Therefore, $\psi_{1,1} = \psi_{1,N}$ and $\psi_{2,1} = \psi_{2,N}$

$$\psi_\xi|_{1,1} = \psi_{2,1} - \psi_{1,1} = \psi_{2,N} - \psi_{1,N} = \psi_\xi|_{1,N}$$

On the other hand, we know that

$$-\frac{\rho_0}{\rho}\frac{\sqrt{\alpha}}{J}\bigg|_{1,1} \neq -\frac{\rho_0}{\rho}\frac{\sqrt{\alpha}}{J}\bigg|_{1,N} \tag{50}$$

Thus, the only possibility to satisfy Equation (49) is that

$$\psi_\xi|_{1,1} = \psi_\xi|_{1,N} = 0$$

which this relation again results in the same expression as for the finite angle trailing edge case. So, the general expression to implement the Kutta condition based on the stream function for the 2D compressible flow is

$$\psi_{1,1} = \psi_{2,1} \tag{51}$$

which is the same as one for the incompressible flow [21]

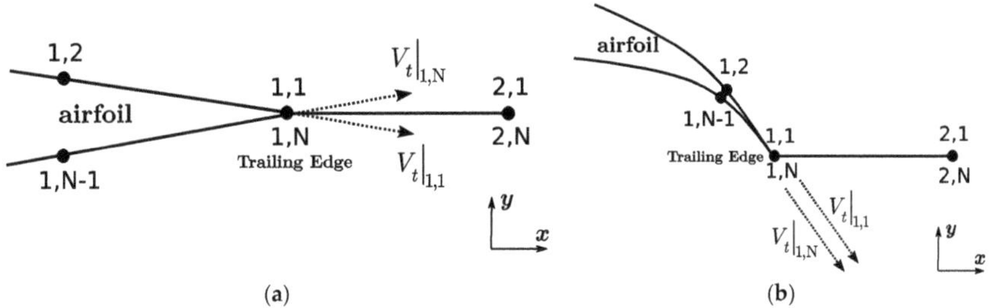



**Figure 5.** Grid notation and tangential velocity components at trailing edge. (**a**) Finite angle trailing edge. (**b**) Cusped trailing edge.

### 2.5. Computation Procedure

According to the mapping scheme adopted in Figure 1, there are four sections where the nodal value of the flow variables $f_{i,j}$ ($f \equiv \psi, \rho, p, u, v, V, c, M, \ldots$) should be calculated.

1. Inside the domain to calculate the variables $f_{i,j}$ ($i = 2, \ldots, M-1, j = 2, \ldots, N-1$).
2. On the airfoil surface to calculate the variables $f_{1,j}$ ($j = 1, \ldots, N$).
3. At the outer boundary (far-field) to calculate the variables $f_{M,j}$ ($j = 1, \ldots, N$).
4. On the branch cut to calculate the variables $f_{i,1}$ ($i = 2, \ldots, M-1$). We know that $f_{i,N} = f_{i,1}$.

The known free-stream variables are $M_\infty, p_\infty, \rho_\infty, T_\infty$, and the angle of attack $\alpha$. Thus, we can write

$$M_{M,j} = M_\infty (j = 1, \ldots, N) \tag{52}$$

$$T_{M,j} = T_\infty (j = 1, \ldots, N) \tag{53}$$

$$\rho_{M,j} = \rho_\infty (j = 1, \ldots, N) \tag{54}$$

$$p_{M,j} = p_\infty (j = 1, \ldots, N) \tag{55}$$

$$c_{M,j} = c_\infty = \sqrt{\gamma_{air} R_{air} T_\infty} = \sqrt{(1.4)(287) T_{M,j}} (j = 1, \ldots, N) \tag{56}$$

$$V_{M,j} = V_\infty = M_\infty c_\infty = M_{M,j} c_{M,j} (j = 1, \ldots, N) \tag{57}$$

from the local isentropic stagnation properties of an ideal gas, we have

$$\frac{\rho_{0_{M,j}}}{\rho_{M,j}} = \left(1 + \frac{\gamma_{air} - 1}{2} M_{M,j}^2\right)^{\frac{1}{\gamma_{air}-1}} (j = 1, \ldots, N) \tag{58}$$

so the local total density $\rho_{0_{M,j}}$ can be computed as

$$\rho_{0_{M,j}} = \rho_{M,j}\left(1 + \frac{\gamma_{air} - 1}{2} M_{M,j}^2\right)^{\frac{1}{\gamma_{air}-1}} (j = 1, \ldots, N) \tag{59}$$

In a similar fashion, we can write

$$p_{0_{M,j}} = p_{M,j}\left(1 + \frac{\gamma_{\text{air}} - 1}{2}M_{M,j}^2\right)^{\frac{\gamma_{\text{air}}}{\gamma_{\text{air}} - 1}} \quad (j = 1, \ldots, N) \tag{60}$$

$$T_{0_{M,j}} = T_{M,j}\left(1 + \frac{\gamma_{\text{air}} - 1}{2}M_{M,j}^2\right)(j = 1, \ldots, N) \tag{61}$$

$$c_{0_{M,j}} = \sqrt{\gamma_{\text{air}}R_{\text{air}}T_{0_{M,j}}} = \sqrt{(1.4)(287)T_{0_{M,j}}}(j = 1, \ldots, N) \tag{62}$$

We know that if the general flow field is isentropic throughout, then the local total density $\rho_0$, the local total pressure $p_0$, and the local total temperature $T_0$ are constant values throughout the flow. Thus

$$c_{0_{i,j}} = c_{0_{M,j}}(i = 1, \ldots, M, j = 1, \ldots, N) \tag{63}$$

$$\rho_{0_{i,j}} = \rho_{0_{M,j}}(i = 1, \ldots, M, j = 1, \ldots, N) \tag{64}$$

$$p_{0_{i,j}} = p_{0_{M,j}}(i = 1, \ldots, M, j = 1, \ldots, N) \tag{65}$$

$$T_{0_{i,j}} = T_{0_{M,j}}(i = 1, \ldots, M, j = 1, \ldots, N) \tag{66}$$

Also, on the outer boundary (far-field) we have

$$u_{M,j} = V_{x_\infty} = V_\infty \cos\alpha(j = 1, \ldots, N) \tag{67}$$

$$v_{M,j} = V_{y_\infty} = V_\infty \sin\alpha(j = 1, \ldots, N) \tag{68}$$

If the number of nodes on the outer boundary (the side $BG$ in Figure 1) is $N$, then, as shown in Figure 6, the node number of two points at top (point $E$) and bottom (point $F$) of the circular outer boundary would be $(M, \frac{N+3}{4})$ and $(M, \frac{3N+1}{4})$, respectively.

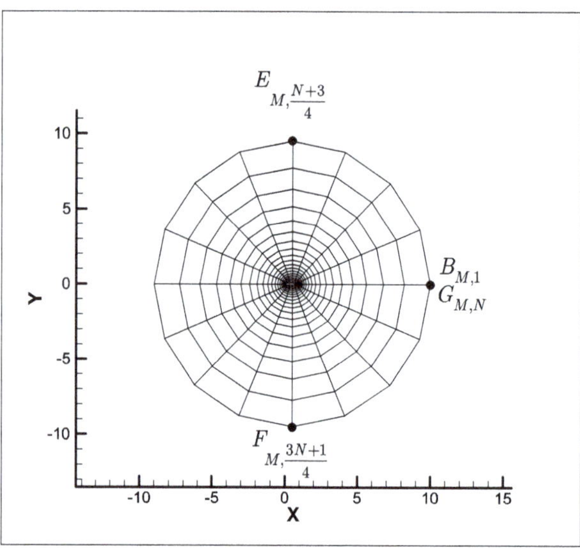


**Figure 6.** Representation of node number of top and bottom points on the outer boundary (far-field).

Then we can express the magnitude of the stream function on the outer boundary $\psi_{M,j}$ in terms of far-field velocity $V_{M,j}$ as (by considering the equal magnitudes but opposite in sign for stream functions at top and bottom points of $E$ and $F$, that is, $\psi_F = -c$ and $\psi_E = c$, $c$ is a constant)

$$
\begin{aligned}
u_{M,\frac{3N+1}{4}} &= \frac{\rho 0_{M,\frac{3N+1}{4}}}{\rho_{M,\frac{3N+1}{4}}} \frac{\partial \psi}{\partial y}\Big|_{M,\frac{3N+1}{4}} = \frac{\rho 0_{M,\frac{3N+1}{4}}}{\rho_{M,\frac{3N+1}{4}}}\left(\frac{\psi_{M,\frac{N+3}{4}} - \psi_{M,\frac{3N+1}{4}}}{y_{M,\frac{N+3}{4}} - y_{M,\frac{3N+1}{4}}}\right) = \\
&\frac{\rho 0_{M,\frac{3N+1}{4}}}{\rho_{M,\frac{3N+1}{4}}}\left(\frac{-\psi_{M,\frac{3N+1}{4}} - \psi_{M,\frac{3N+1}{4}}}{y_{M,\frac{N+3}{4}} - y_{M,\frac{3N+1}{4}}}\right) = -\frac{\rho 0_{M,\frac{3N+1}{4}}}{\rho_{M,\frac{3N+1}{4}}}\left(\frac{2\psi_{M,\frac{3N+1}{4}}}{y_{M,\frac{N+3}{4}} - y_{M,\frac{3N+1}{4}}}\right)
\end{aligned}
\tag{69}
$$

Therefore,

$$
\psi_{M,\frac{3N+1}{4}} = -\frac{\rho_{M,\frac{3N+1}{4}}}{\rho 0_{M,\frac{3N+1}{4}}} u_{M,\frac{3N+1}{4}} \frac{y_{M,\frac{N+3}{4}} - y_{M,\frac{3N+1}{4}}}{2}
\tag{70}
$$

Now the magnitude of the stream function on the outer boundary can be calculated as a (based on the relation $\psi = -\frac{\rho}{\rho_0}V_y x + \frac{\rho}{\rho_0}V_x y$)

$$
\psi_{M,j} = \psi_{M,\frac{3N+1}{4}} + \frac{\rho_{M,j}}{\rho 0_{M,j}} u_{M,j}\left(y_{M,j} - y_{M,\frac{3N+1}{4}}\right) - \frac{\rho_{M,j}}{\rho 0_{M,j}} v_{M,j}\left(x_{M,j} - x_{M,\frac{3N+1}{4}}\right)
\tag{71}
$$

In Equation (69), we should note that the two points $E$ and $F$ have the same $x$-coordinates and hence $\psi = -\frac{\rho}{\rho_0}V_y x + \frac{\rho}{\rho_0}V_x y = 0 + \frac{\rho}{\rho_0}V_x y = \frac{\rho}{\rho_0}V_x y$.

The iterative process to obtain the value of $\psi_{i,j}$ and $\rho_{i,j}$ may be initiated by assuming $\frac{\rho 0_{i,j}}{\rho_{i,j}} = 1$ or $\rho_{i,j} = \rho 0_{i,j}$. This initial assumption implies that the magnitude of the speed of sound is infinite ($c = \infty \Rightarrow \frac{1}{c^2} = 0$) and Equation (5) reduces to Laplace's equation

$$
\psi_{xx} + \psi_{yy} = 0
\tag{72}
$$

The first iteration is concerned with the solution of Laplace's equation which is described thoroughly in [21] and we do not elaborate further on it. After solving the Laplace's equation to find the value of the stream function at each node, $\psi_{i,j}$, in the flow field, the values of the velocity components of $u_{i,j}$ and $v_{i,j}$ can be computed from Equations (2) and (3). Then, the value of the speed of sound at each node, $c_{i,j}$, can be calculated from Equation (4). Finally, the value of the density at each node, $\rho_{i,j}$, can be determined from Equation (7). If the flow is compressible, $\rho_{i,j} \neq \rho 0_{i,j}$ and from the second iteration onwards, instead of Laplace's equation, the stream function given in Equation (6) must be solved by having the density $\rho_{i,j}$ and the stream function $\psi_{i,j}$ from the last iteration (i.e., iteration 1) to get new values of the stream function $\psi_{i,j}$. The iterative process (solving Equations (6) and (7) simultaneously) repeated until successive iterations produce a sufficiently small change in density $\rho_{i,j}$ and the stream function $\psi_{i,j}$. Equation (6) is solved by initially discretizing the partial differential terms in Equations (21)–(26) using relations given in Equations (27)–(31) and then substituting the terms into Equation (6), and finally, solving the obtained algebraic equation using an algebraic solver (such as Maple) to get an explicit algebraic expression in terms of $\psi_{i,j}$. As stated before, the Kutta condition is implemented as

$$
\psi_{1,1} = \psi_{2,1}
\tag{73}
$$

$$
\psi_{1,N} = \psi_{1,1}\,(\text{the same node})
\tag{74}
$$

and the wall boundary condition is implemented as

$$
\psi_{1,j} = \psi_{1,j-1}\,(j = 2,\ldots,N-1)
\tag{75}
$$

and since the branch cut ($AB$ or $HG$ in Figure 1) is inside the flow field, the same procedure employed to obtain an algebraic expression for $\psi_{i,j}$ can also be used to obtain an expression for the stream function on the branch cut, $\psi_{i,1}$. However, some changes are needed in the terms discretized by the finite-difference method (Equations (27)–(31)) as follows (see Figure 7)

$$(f_\xi)_{i,1} = \frac{1}{2}(f_{i+1,1} - f_{i-1,1}) \tag{76}$$

$$(f_\eta)_{i,1} = \frac{1}{2}(f_{i,2} - f_{i,N-1}) \tag{77}$$

$$(f_{\xi\xi})_{i,1} = (f_{i+1,1} - 2f_{i,1} + f_{i-1,1}) \tag{78}$$

$$(f_{\eta\eta})_{i,1} = (f_{i,2} - 2f_{i,1} + f_{i,N-1}) \tag{79}$$

$$(f_{\xi\eta})_{i,1} = \frac{1}{4}(f_{i+1,2} - f_{i-1,2} - f_{i+1,N-1} + f_{i-1,N-1}) \tag{80}$$

where $f \equiv x, y, \psi$.

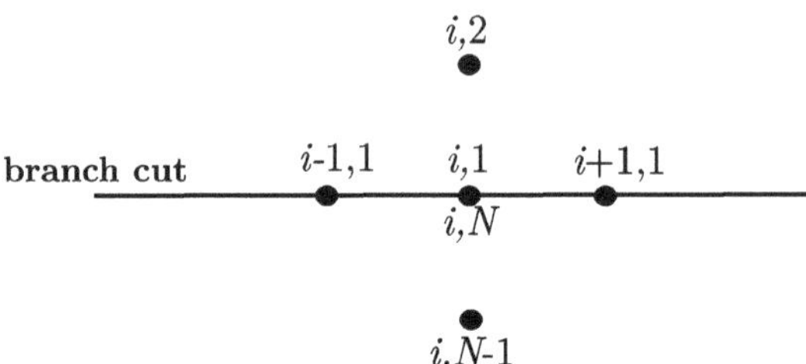


**Figure 7.** Node notation on the branch cut.

and on the branch cut

$$\psi_{i,N} = \psi_{i,1} (i = 2, \ldots, M-1) \tag{81}$$

In order to solve the elliptic grid generation equations (for $x$ and $y$) and the stream function equation, the iterative method *Successive Over Relaxation* (SOR) is used due to its high convergence rate

$$f_{i,j}^{(k)} = \omega f_{i,j}^{(k-1)} + (1-\omega)f_{i,j}^{(k-2)} \tag{82}$$

where $f \equiv x, y, \psi$. In these relation for SOR, $k$ is iteration number, and $\omega$ is *relaxation factor* which has a value of $1 < \omega < 2$. A value of $\omega = 1.8$ is chosen in this study and the code FOILcom.

We should define the *stopping criteria* for convergence of solution of the stream function equation (Equation (6)) and density equation (Equation (7)). These two equations constitute the system of equations to be solved simultaneously. The stopping criteria are defined as follows

$$\lambda_\psi = \sum_{i=2}^{M-1} \sum_{j=2}^{N-1} (\psi_{i,j}^{(k+1)} - \psi_{i,j}^{(k)})^2 \tag{83}$$

$$\lambda_\rho = \sum_{i=2}^{M-1} \sum_{j=2}^{N-1} (\rho_{i,j}^{(k+1)} - \rho_{i,j}^{(k)})^2 \tag{84}$$

where $k$ is iteration number. A value of $\lambda_\psi = \lambda_\rho = 10^{-4}$ can be considered to get sufficiently accurate results. The other parameters of interest can also be computed after obtaining the density $\rho_{i,j}$ and the stream function $\psi_{i,j}$. The components of the velocity at each node, $u_{i,j}$ and $v_{i,j}$, can be computed from

$$u_{i,j} = \frac{\rho_{0_{i,j}}}{\rho_{i,j}} \psi_y \bigg|_{i,j} = \frac{\rho_{0_{i,j}}}{\rho_{i,j}} \frac{1}{J}(-x_\eta \psi_\xi + x_\xi \psi_\eta)\bigg|_{i,j} \tag{85}$$

$$v_{i,j} = -\frac{p_{0_{i,j}}}{\rho_{i,j}}\left.\psi_x\right|_{i,j} = -\frac{p_{0_{i,j}}}{\rho_{i,j}}\frac{1}{J}\left.\left(y_\eta\psi_\xi - y_\xi\psi_\eta\right)\right|_{i,j} \tag{86}$$

Then, the velocity at each node, $V_{i,j}$, is given by

$$V_{i,j} = \sqrt{u_{i,j}^2 + v_{i,j}^2} \tag{87}$$

and the local (nodal) Mach number can be obtained by

$$M_{i,j} = \frac{V_{i,j}}{c_{i,j}} \tag{88}$$

where the local speed of sound, $c_{i,j}$, is calculated from Equation (4). And finally, the local pressure, $p_{i,j}$, is calculated from

$$\frac{p_{0_{i,j}}}{p_{i,j}} = \left(1 + \frac{\gamma_{\text{air}} - 1}{2}M_{i,j}^2\right)^{\frac{\gamma_{\text{air}}}{\gamma_{\text{air}} - 1}} \tag{89}$$

Now the values of drag and lift forces on the airfoil can be computed as follows

$$D = A\cos\alpha + N\sin\alpha \tag{90}$$

$$L = -A\sin\alpha + N\cos\alpha \tag{91}$$

where $A$ and $N$ are axial and normal forces, respectively (Figure 8) [4]. We can write the drag and lift forces as

$$D = \sum_{j=1}^{N-1}\left(p_{1,j}(y_{1,j} - y_{1,j+1})\cos\alpha + p_{1,j}(x_{1,j+1} - x_{1,j})\sin\alpha\right) \tag{92}$$

$$L = \sum_{j=1}^{N-1}\left(-p_{1,j}(y_{1,j} - y_{1,j+1})\sin\alpha + p_{1,j}(x_{1,j+1} - x_{1,j})\cos\alpha\right) \tag{93}$$

and the drag and the lift coefficients may be expressed as

$$c_d = \frac{D}{\frac{1}{2}\rho_\infty V_\infty^2} \tag{94}$$

$$c_l = \frac{L}{\frac{1}{2}\rho_\infty V_\infty^2} \tag{95}$$

Furthermore, the flowchart of the computational procedure is presented in Figure 9.

The theoretical and computational details presented here for compressible flows are implemented in the freely available code FOILcom. Interested readers can refer to the code and use it by changing the airfoil shape, the free stream subcritical Mach number, and the angle of attack. The freely available code FOILincom(DOI: 10.13140/RG.2.2.21727.15524) also takes advantage of the same computational procedure but only for incompressible flow.

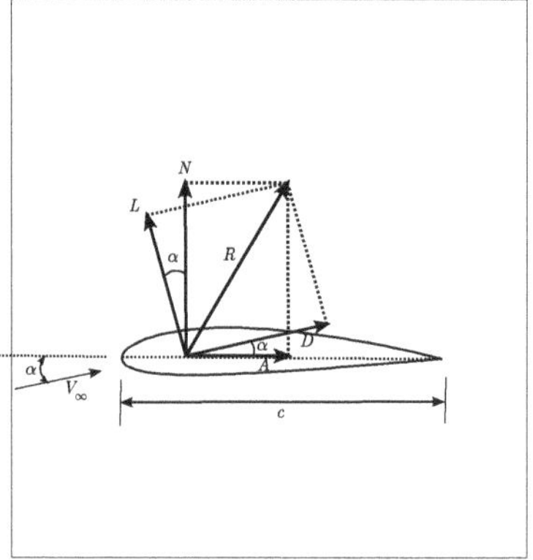

**Figure 8.** Aerodynamic force $R$ and the components into which it splits [4].

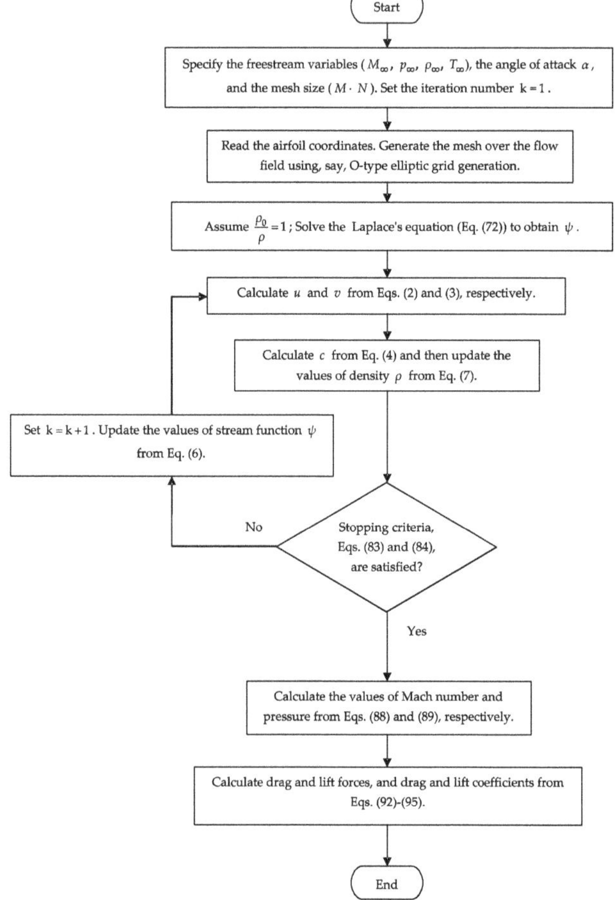


**Figure 9.** Flowchart of computational procedure.

## 3. Results

A few test cases are given to reveal the accuracy and robustness of the numerical scheme. Here, the results from FOILcom are compared with experimental and numerical results (alternative numerical schemes) to show its accuracy and efficiency.

*Test case 1*: Critical Mach number for the NACA 0012 airfoil at zero angle of attack.

The airfoil surface pressure coefficient distribution using FOILcom is shown in Figure 10. The critical Mach number for the NACA 0012 airfoil at zero angle of attack is obtained as 0.722 using FOILcom with a grid of size $110 \times 101$ and $x_{M,1} = 10$ m. The critical Much number in references [4,22] is obtained as 0.725 (the experimental data in [22] is investigated in [4]). A comparison of results is shown in Figure 11 which reveals an excellent agreement.

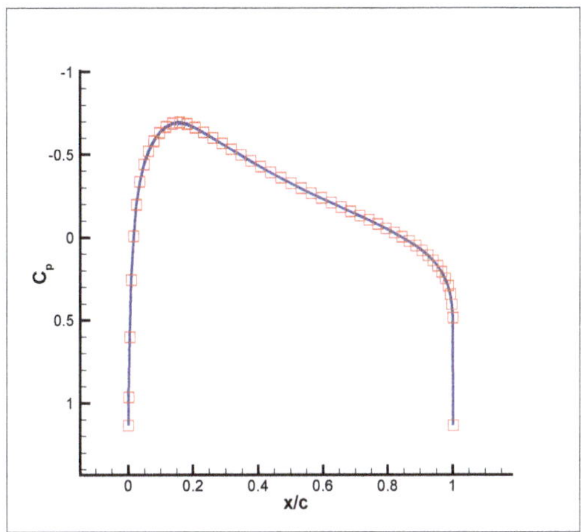

**Figure 10.** The pressure coefficient distribution for NACA0012 using FOILcom ($M_\infty = M_{\text{critical}} = 0.722$, $\alpha = 0$).

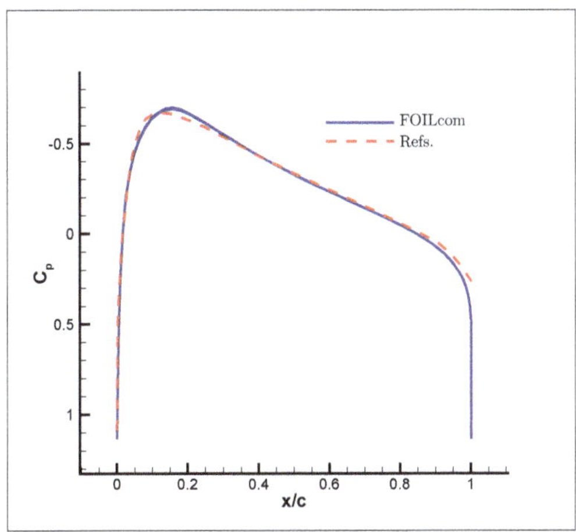


**Figure 11.** The pressure coefficient distribution. Comparison of FOILcom result and the one from References [4,22].

*Test case 2*: The airfoil surface pressure coefficient distribution for the NACA 0012 using a grid of size $110 \times 101$, $M_\infty = 0.5$, and $\alpha = 3°$ (Figures 12 and 13). The results from FOILcom (Figure 12) and both finite volume and meshless methods (the results from finite volume and meshless methods in [23] are presented in one plot due to an excellent agreement between the results) are compared and depicted in Figure 13.

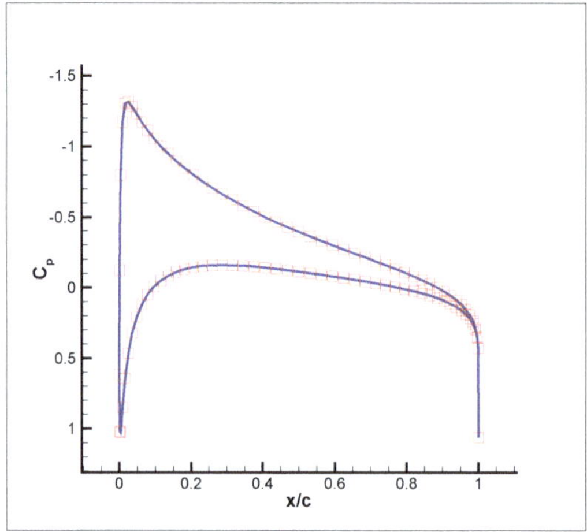

**Figure 12.** The pressure coefficient distribution for NACA0012 using FOILcom ($M_\infty = 0.5$, $\alpha = 3°$).

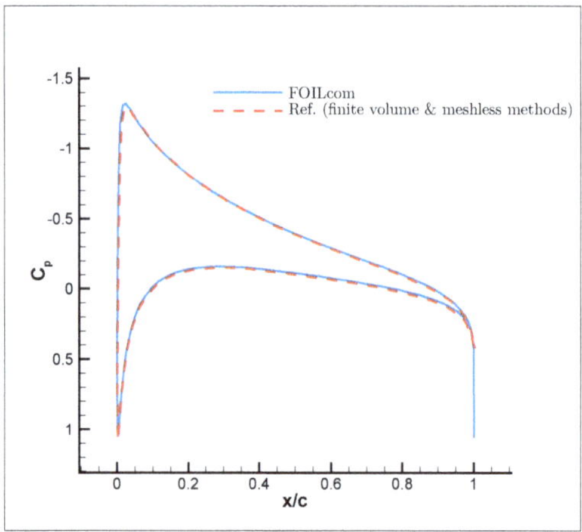



**Figure 13.** The pressure coefficient distribution. Comparison of FOILcom result and the one from Reference [23].

*Test case 3*: The airfoil surface pressure coefficient distribution for the NACA 0012 using a grid of size $110 \times 101$, $M_\infty = 0.63$, and $\alpha = 2°$ (Figures 14 and 15). In this test case, the result from FOILcom is compared to the one from the code FLO42. There is excellent agreement between the results.

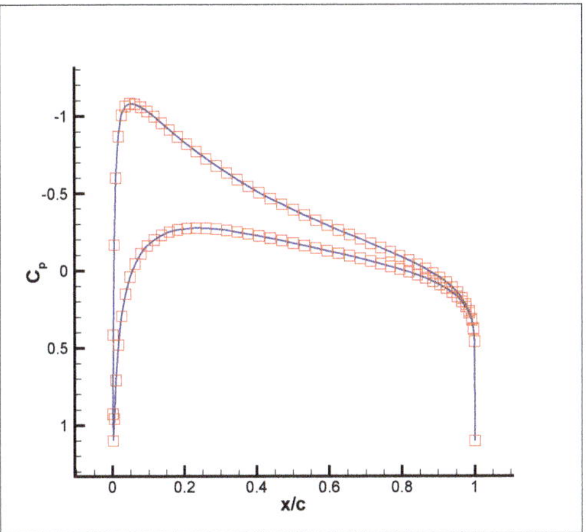

**Figure 14.** The pressure coefficient distribution for NACA0012 using FOILcom ($M_\infty = 0.63$, $\alpha = 2°$).

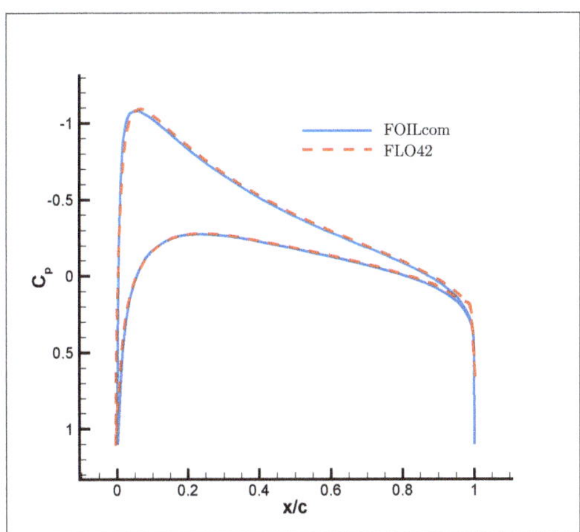

**Figure 15.** The pressure coefficient distribution. Comparison of FOILcom result and the one from FLO42 [24].

*Test case 4*: The airfoil surface pressure coefficient distribution for the NACA 2414. $M_\infty = 0.435$ and $\alpha = 2°$ (Figures 16 and 17). In this test case, 19,740 nodes (mesh size of $140 \times 141$) and 51,792 nodes are used in FOILcom and ANSYS Fluent, respectively. The drag and lift coefficients are calculated as follows:

$$\text{FOILcom} : c_d = -4.362 \times 10^{-3}, c_l = 0.573$$

$$\text{ANSYS Fluent} : c_d = 9.985 \times 10^{-4}, c_l = 0.574$$

which shows perfect agreement. The results are obtained by a FORTRAN compiler and computations are run on a PC with Intel Core i5 and 6G RAM. A tolerance of $10^{-7}$ is used in iterative loops to increase the accuracy of results. The computation time is about 5 min which is high due to the iterative solution of the elliptic grid generation and the stream function equations.

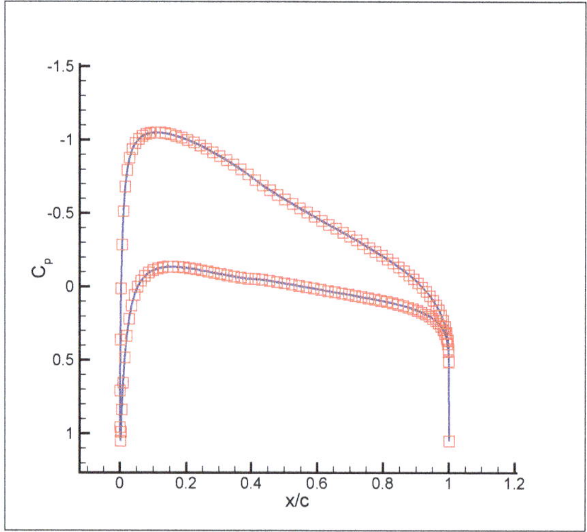

**Figure 16.** The pressure coefficient distribution for NACA2414 using FOILcom ($M_\infty = 0.435$, $\alpha = 2°$).

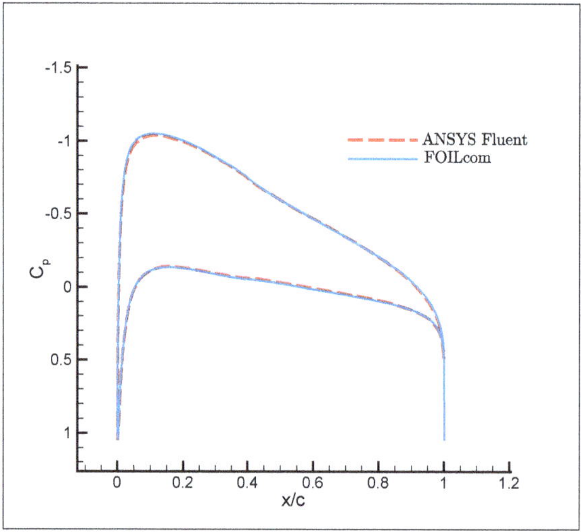


**Figure 17.** The pressure coefficient distribution. Comparison of FOILcom result and the one from ANSYS Fluent (V19.2).

*Test case 5*: The airfoil surface pressure coefficient distribution using different grid sizes for NACA 2214 airfoil and $M_\infty = 0.55$, $\alpha = 2°$.

Grid sizes are:

(a)  $30 \times 41$
(b)  $80 \times 61$
(c)  $120 \times 161$
(d)  $200 \times 201$

The results are shown in Figure 18. In this test case, four different grid sizes are used to obtain the airfoil surface pressure coefficient distribution. As shown in Figure 18, the distribution can be obtained with reasonable accuracy using even a course grid ($30 \times 41$). The effect of grid size on the drag and lift coefficients is depicted in Figure 19. Moreover, the drag and lift coefficients using the four different grid sizes are compared to the ones from XFOIL (Figure 20) and are given in Table 1. As can be seen, the results from FOILcom are in excellent agreement with the result from XFOIL.

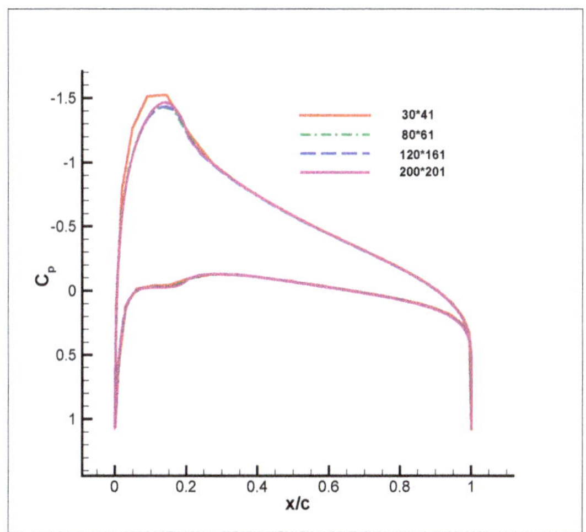


**Figure 18.** The airfoil surface pressure coefficient distribution using different grid sizes for NACA 2214 airfoil ($M_\infty = 0.55$, $\alpha = 2°$). The code FOILcom is used.

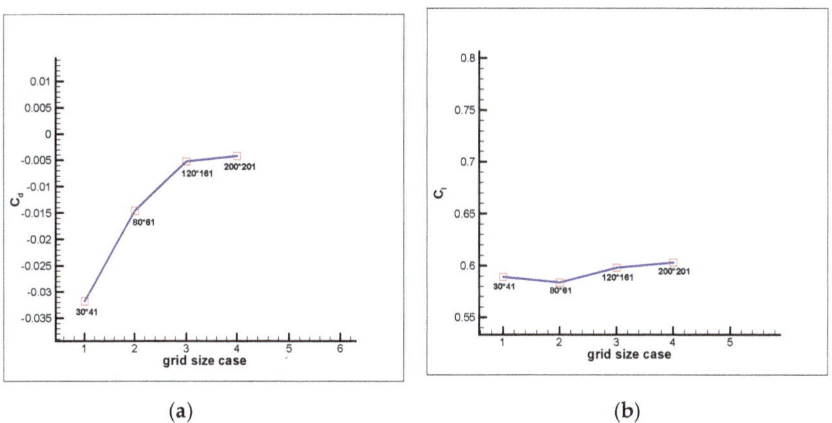


(a)                                        (b)

**Figure 19.** Effect of grid size on drag coefficient (**a**) and lift coefficient (**b**).

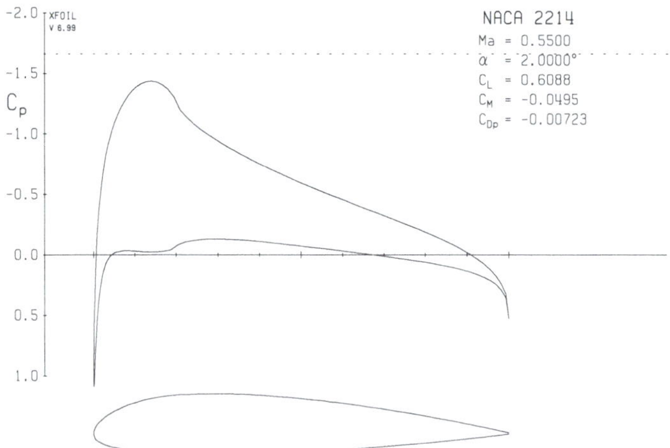


**Figure 20.** The airfoil surface pressure coefficient distribution for NACA 2214 airfoil ($M_\infty = 0.55$, $\alpha = 2°$) using XFOIL.

**Table 1.** Drag and lift coefficients using different grid sizes. Codes FOILcom and XFOIL are used.

| Code and Grid Size | $c_d$ | $c_l$ |
|---|---|---|
| FOILcom: $30 \times 41$ | $-3.1764 \times 10^{-2}$ | 0.5893 |
| FOILcom: $80 \times 61$ | $-1.4563 \times 10^{-2}$ | 0.5839 |
| FOILcom: $120 \times 161$ | $-5.2130 \times 10^{-3}$ | 0.5983 |
| FOILcom: $200 \times 201$ | $-4.2097 \times 10^{-3}$ | 0.6032 |
| XFOIL | $-7.23 \times 10^{-3}$ | 0.6088 |

*Test case 6*: The airfoil surface pressure coefficient distribution for NACA 64-012 airfoil and $M_\infty = 0.3$, $\alpha = 6°$. The mesh size used in FOILcom is $50 \times 141$. The computation time is 46 s. The results from two codes FOILcom and FOILincom along with the Karman-Tsien compressibility correction are compared and depicted in Figure 21. Moreover, the convergence history of $\lambda_\psi$ and $\lambda_\rho$ are shown in Figure 22. As can be seen, in this test case five iterations are needed to satisfy the stopping criteria ($\lambda_\psi = \lambda_\rho = 10^{-4}$).

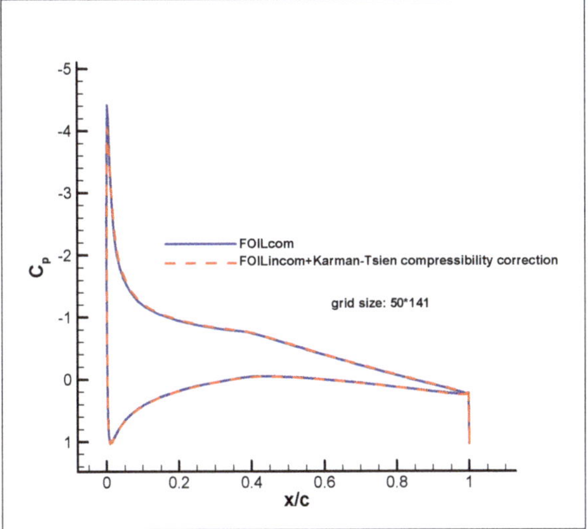

**Figure 21.** Comparison of surface pressure coefficient distributions for NACA 64-012 airfoil ($M_\infty = 0.3$, $\alpha = 6°$) using the codes FOILcom and FOILincom along with the Karman-Tsien compressibility correction.

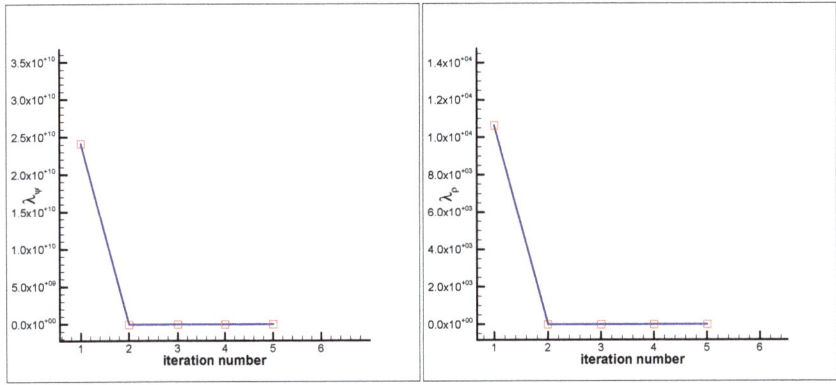

**Figure 22.** Convergence history of $\lambda_\psi$ and $\lambda_\rho$.

*Test case 7*: The airfoil surface pressure coefficient distribution for NACA 2240 airfoil and $M_\infty = 0.3$ at two different angles of attack $\alpha = 3°$ and $\alpha = 6°$. The mesh size used in FOILcom is $120 \times 141$. In this test case, a thick airfoil, NACA 2240, is used to reveal the accuracy of the proposed numerical method in dealing with thick airfoils at high angles of attack. The results from FOILcom, FOILincom along with the Karman-Tsien compressibility correction, and XFOIL are compared and depicted in Figure 23 for the angle of attack $\alpha = 3°$ and Figure 24 for the angle of attack $\alpha = 6°$. Furthermore, the results from XFOIL are given in Figure 25 for the angle of attack $\alpha = 3°$ and Figure 26 for the angle of attack $\alpha = 6°$. The drag and lift coefficients using both codes for two different angles of attack are given in Table 2.

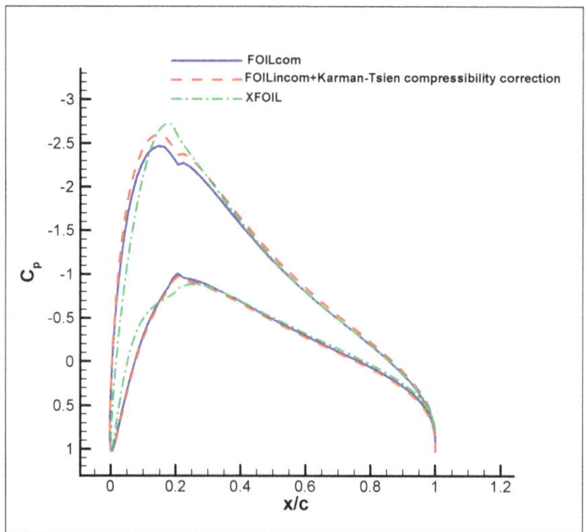

**Figure 23.** Comparison of surface pressure coefficient distributions for NACA 2240 airfoil ($M_\infty = 0.3$, $\alpha = 3°$) using the codes FOILcom, FOILincom along with the Karman-Tsien compressibility correction, and XFOIL.

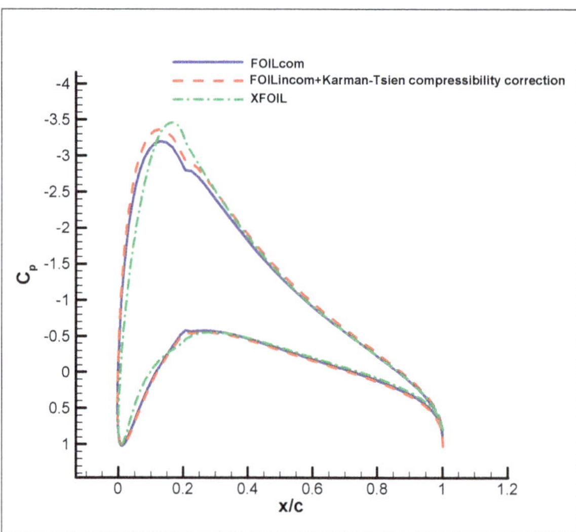

**Figure 24.** Comparison of surface pressure coefficient distributions for NACA 2240 airfoil ($M_\infty = 0.3$, $\alpha = 6°$) using the codes FOILcom, FOILincom along with the Karman-Tsien compressibility correction, and XFOIL.

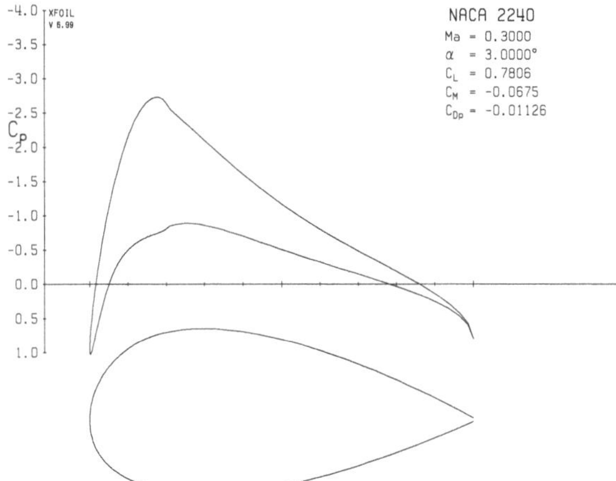

**Figure 25.** The airfoil surface pressure coefficient distribution for NACA 2240 airfoil ($M_\infty = 0.3$, $\alpha = 3°$) using XFOIL.

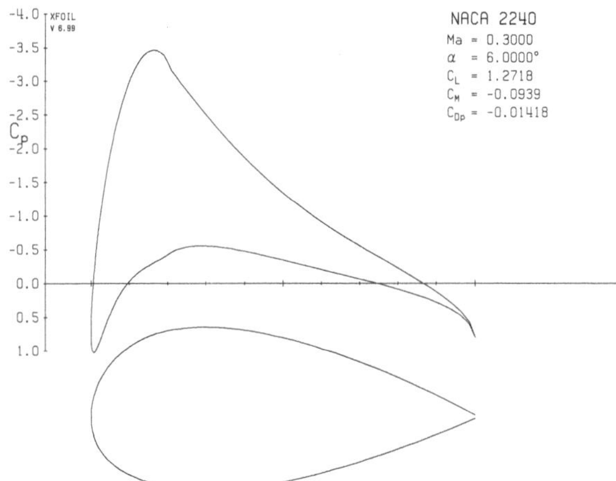

**Figure 26.** The airfoil surface pressure coefficient distribution for NACA 2240 airfoil ($M_\infty = 0.3$, $\alpha = 6°$) using XFOIL.

**Table 2.** Drag and lift coefficients using FOILcom and XFOIL.

| NACA 2240 Airfoil | $c_d$ | $c_l$ |
|---|---|---|
| $M_\infty = 0.3, \alpha = 3°$ | FOILcom: −0.01668<br>XFOIL: −0.01126 | FOILcom: 0.7977<br>XFOIL: 0.7806 |
| $M_\infty = 0.3, \alpha = 6°$ | FOILcom: −0.02684<br>XFOIL: −0.01418 | FOILcom: 1.2714<br>XFOIL: 1.2718 |

## 4. Conclusions

The solution of 2D steady, irrotational, subsonic (subcritical) compressible flow over isolated airfoils using the stream function equation and a novel method to implement the Kutta condition have been presented. The numerical scheme takes advantage of transformation of the flow solver and the boundary conditions from the physical domain to the computational domain. The physical domain was meshed by an O-grid elliptic grid generation method and the transformed flow solver is discretized by finite-difference method, a method chosen for its simplicity and ease of implementation. The numerical scheme is exempt from considering the panels and the quantities such as the vortex panel strength and circulation used in the panel method. An accurate Kutta condition scheme is proposed and implemented into the computational loop by an exact derived expression for the stream function at the airfoil trailing edge. The exact expression is general, and encompasses both the finite-angle and cusped trailing edges. Through several test cases, the proposed numerical scheme was validated by results from the experimental and the other numerical methods. The obtained results revealed that the proposed algorithm is very accurate and robust.

**Author Contributions:** Conceptualization, F.M.; Formal analysis, F.M.; Funding acquisition, F.M.; Investigation, F.M., B.E. and M.S.; Methodology, F.M.; Software, F.M.; Validation, F.M.; Visualization, F.M.; Writing—original draft, F.M.; Writing—review and editing, B.E. and M.S.

**Funding:** This research was supported by funding from the European Union's Horizon 2020 research and innovation programme under the Marie Skłodowska-Curie grant agreement number 663830.

**Conflicts of Interest:** The authors declare no conflict of interest.

## Nomenclature

| | |
|---|---|
| $A$ | axial force (N) |
| $c$ | speed of sound (m/s) |
| $c_d$ | drag coefficient |
| $c_l$ | lift coefficient |
| $D$ | drag force (N) |
| $J$ | Jacobian of transformation |
| $L$ | lift force (N) |
| $M$ | Mach number ($V/c$) |
| $N$ | normal force (N) |
| $p$ | pressure ($N/m^2$) |
| $T$ | temperature (K) |
| $u, v$ | velocity components (m/s) |
| $V$ | velocity (m/s) |
| $x, y$ | Cartesian coordinates in the physical domain (m) |

### Greek symbols

| | |
|---|---|
| $\alpha$ | angle of attack, metric coefficient in 2-D elliptic grid generation |
| $\gamma$ | ratio of specific heats |
| $\lambda$ | stopping criterion |
| $\rho$ | density ($kg/m^3$) |
| $\omega$ | relaxation factor |
| $\xi, \eta$ | Cartesian coordinates in the computational domain |
| $\psi$ | stream function |

### Subscripts

| | |
|---|---|
| $0$ | stagnation condition |
| $\infty$ | free stream condition |
| $i$ | grid index in $\xi$-direction |
| $j$ | grid index in $\eta$-direction |
| $M$ | number of grid points in $\xi$-direction |
| $N$ | number of grid points in $\eta$-direction |

### Superscript

| | |
|---|---|
| k | iteration number |

## Abbreviations

| | |
|---|---|
| PDE | Partial Differential Equation |
| FDM | Finite Difference Method |
| SOR | Successive Over Relaxation |

## References


1. Hess, J.L.; Smith, A.M.O. Calculation of potential flow about arbitrary bodies. *Prog. Aerosp. Sci.* **1967**, *8*, 1–138. [CrossRef]
2. Hess, J. Panel methods in computational fluid dynamics. *Annu. Rev. Fluid Mech.* **1990**, *22*, 255–274. [CrossRef]
3. Erickson, L.L. *Panel Methods: An Introduction*; NASA Ames Research Center: Moffett Field, CA, USA, 1990.
4. Anderson, J.D. *Fundamentals of Aerodynamics*; McGraw-Hill Education: New York, NY, USA, 2016.
5. Drela, M. *Flight Vehicle Aerodynamics*; MIT Press: Cambridge, MA, USA, 2014.
6. Katz, J.; Plotkin, A. *Low-Speed Aerodynamics*; Cambridge University Press: Cambridge, UK, 2001.
7. Kutta, W.M. *Lifting Forces in Flowing Fluids*; G. Braunbeck & Gutenberg: Berlin, Germany, 1902.
8. Lewis, R.I. *Vortex Element Methods for Fluid Dynamic Analysis of Engineering Systems*; Cambridge University Press: Cambridge, UK, 1991.
9. Xu, C. Kutta Condition for sharp edge flows. *Mech. Res. Commun.* **1998**, *25*, 415–420. [CrossRef]
10. Jacob, K.; Riegels, F. The calculation of the pressure distribution over aerofoil sections of finite thickness with and without flaps and slats. *Z. Flugwiss* **1963**, *11*, 357–367.
11. Drela, M. *XFOIL: An analysis and design system for low Reynolds number airfoils. InLow Reynolds Number Aerodynamics*; Springer: Berlin/Heidelberg, Germany, 1989; pp. 1–12.
12. Cummings, R.M.; Mason, W.H.; Morton, S.A.; McDaniel, D.R. *Applied Computational Aerodynamics: A Modern Engineering Approach*; Cambridge University Press: Cambridge, UK, 2015.
13. Hirsch, C. *Numerical Computation of Internal and External Flows, Computational Methods for Inviscid and Viscous Flows*; Wiley: Chichester, UK, 1991.
14. Shapiro, A.H. *The Dynamics and Thermodynamics of Compressible Fluid Flow*; John Wiley & Sons: New York, NY, USA, 1953.
15. Yahya, S.M. *Fundamentals of Compressible Flow: SI Units with Aircraft and Rocket Propulsion*; New Age International: Delhi, India, 2003.
16. Johnson, R.W. *Handbook of Fluid Dynamics*; CRC Press: Boca Raton, FL, USA, 2016.
17. Mohebbi, F. *Optimal Shape Design based on Body-Fitted Grid Generation*; University of Canterbury: Christchurch, New Zealand, 2014.
18. Thompson, J.; Warsi, Z.; Mastin, C. Numerical Grid Generation: Foundations and Applications. 1997. Available online: https://www.hpc.msstate.edu/publications/gridbook/PDFS/NumericalGridGenerationComplete.pdf (accessed on 25 June 2019).
19. Özişik, M.N.; Orlande, H.R.B.; Colaço, M.J.; Cotta, R.M. *Finite Difference Methods in Heat Transfer*; CRC Press: Boca Raton, FL, USA, 2017.
20. Thames, F.C.; Thompson, J.F.; Mastin, C.W. *Numerical Solution of the Navier-Stokes Equations for Arbitrary Two-Dimensional Airfoils*; Mississippi State Univiversity: Mississippi State, MS, USA, 1975.
21. Mohebbi, F.; Sellier, M. On the Kutta condition in potential flow over airfoil. *J. Aerodyn.* **2014**, *2014*, 676912. [CrossRef]
22. Freuler, R.; Gregorek, G. *An Evaluation of Four Single Element Airfoil Analytic Methods*; Ohio State University, General Aviation Airfoil Design and Analysis Center: Columbus, OH, USA, 1979.
23. Katz, A.; Jameson, A. Meshless scheme based on alignment constraints. *AIAA J.* **2010**, *48*, 2501–2511. [CrossRef]
24. Jameson, A.; Caughey, D.; Jou, W.; Steinhoff, J.; Pelz, R. Accelerated finite-volume calculation of transonic potential flows. In *Numerical Methods for the Computation of Inviscid Transonic Flows with Shock Waves*; Springer: Braunschweig, Germany, 1981; pp. 11–27.

 *fluids*

*Article*

# Soliton Solution of Schrödinger Equation Using Cubic B-Spline Galerkin Method

**Azhar Iqbal** [1,2,*], **Nur Nadiah Abd Hamid** [1] and **Ahmad Izani Md. Ismail** [1]

[1]  School of Mathematical Sciences, Universiti Sains Malaysia, Penang 11800, Malaysia; nurnadiah@usm.my (N.N.A.H.); ahmad_izani@usm.my (A.I.M.I.)
[2]  Mathematics and Natural Sciences, Prince Mohammad Bin Fahd University, Al Khobar 31952, Saudi Arabia
*  Correspondence: azhariqbal31@gmail.com

Received: 17 April 2019; Accepted: 7 June 2019; Published: 12 June 2019

**Abstract:** The non-linear Schrödinger (NLS) equation has often been used as a model equation in the study of quantum states of physical systems. Numerical solution of NLS equation is obtained using cubic B-spline Galerkin method. We have applied the Crank–Nicolson scheme for time discretization and the cubic B-spline basis function for space discretization. Three numerical problems, including single soliton, interaction of two solitons and birth of standing soliton, are demonstrated to evaluate to the performance and accuracy of the method. The error norms and conservation laws are determined and found to be in good agreement with the published results. The obtained results show that the approach is feasible and accurate. The proposed method has almost second order convergence. The linear stability of the method is performed using the Von Neumann method.

**Keywords:** non-linear Schrödinger equation; cubic B-spline basis functions; Galerkin method

---

## 1. Introduction

The non-linear Schrödinger (NLS) equation describes how the behavior of quantum states of a physical system changes in time and space. The NLS equation can be used to describe the propagation of optical pulses and waves in water and plasmas, among other things. Due to the presence of nonlinearity and the complex nature of the problem, it is still a challenge for researchers to determine the most suitable method. Many theoretical and numerical studies have been carried out to overcome this difficulty. The B-spline Galerkin method is quite advanced and has been used by researchers to solve other complex problems.

Dag [1] presented the quadratic and cubic B-spline Galerkin finite element method for solving the Burger's equation. This method was found to give satisfactory results for the Burger's equation, particularly when continuity of the solutions was essential. Dag et al. [2] also proposed a cubic B-splines collocation method for solving the one-dimensional Burger's equation. The proposed scheme was easy to implement and did not require any inner iteration to deal with the nonlinear term of the Burger's equation. Gorgulu et al. [3] presented exponential B-splines Galerkin finite element method for the numerical solution of the advection-diffusion equation. In this study, a new algorithm was developed for the numerical solution of differential equations. This algorithm was obtained by utilizing exponential B-spline functions for the Galerkin finite element method. It was reported that the proposed method gave satisfactory results. The exponential B-splines Galerkin method for the numerical solution of the Burger's equation was also applied by Gorgulu et al. [4].

Saka et al. [5] proposed a quartic B-splines Galerkin finite element method for solving the regularized long wave equation. The performance and accuracy of the method were observed. The method has a weakness because of its large number of matrix operations. The quartic B-spline method was found to be very useful for finding the numerical solutions of the differential equation when higher continuity of solutions occurs. Aksan [6] presented the quadratic B-spline finite element method for finding the approximate solution of the nonlinear Burger's equation. The high accuracy of the proposed method was thoroughly examined. Gardner et al. [7] also applied a cubic B-spline finite element method for the numerical solution of the Burger's equation. They noticed that the proposed finite element method gave more accurate results than other approaches.

Sheng et al. [8] solved the NLS equation using a finite difference method based on quartic spline approximation and semidiscretization. The NLS equation was approximated using a finite element method by Zlotnik and Zlotnik [9]. Ersoy et al. [10] studied the numerical solution of the NLS equation using an exponential B-spline with collocation method. In that paper, they used the Crank Nicolson formulation for time integration and exponential cubic B-spline functions for space integration. Mokhtari et al. [11] solved the NLS equation using delta-shaped basis functions. Saka [12] solved the NLS using the collocation method with quintic B-spline functions. The generating function of the Hermite polynomial and the orthogonal Hermite function were introduced with ordinary Hermite polynomials by Cesarano et al. [13]. The nature of Hermite polynomials is described in the Schrödinger Equation. Analytical solutions of the NLS equation for certain initial and boundary conditions were presented by Zakharov and Shabat [14]. The numerical solution of the NLS equation was obtained by Lin [15] using a septic spline function method with the collocation method. The Galerkin finite element method for the NLS equation was applied by Dag, with quadratic B-spline functions as the weight and trial functions over finite elements [16]. The numerical solution of the NLS equation was also obtained by using the collocation method based on cubic B-spline by Gardner et al. [17]. Hu [18] presented a conservative two-grid finite element method for the numerical solution of the NLS equation and applied one Newton iteration for the linearization of the problem by using the coarse-grid solution as the initial guess. The Galerkin B-spline method is not often used for the numerical solution of the NLS equation. A review of some numerical methods for solving the NLS equation indicates that most methods use the collocation finite element method.

In this paper, we study the use of the Galerkin method with cubic B-spline function as the weight and trial functions over finite elements to solve the NLS equation. This paper is organized as follows. In Section 2, we discussed the governing equation and fundamentals of Cubic B-spline Galerkin method. In Section 3, the stability analysis of the numerical scheme is analyzed using the Von Neumann method. The numerical results and test problems are discussed in Section 4. In the last section, the conclusions are given.

## 2. Governing Equation and Cubic B-Spline Galerkin Method

The NLS equation is

$$iu_t + u_{xx} + q|u|^2 u = 0,\tag{1}$$

The equation in (1) is called a self-focusing NLS equation when $q > 0$ and a defocusing NLS equation when $q < 0$. The subscripts $x$ and $t$ denote the spatial variable and time variable respectively; $q$ is a positive real parameter, and $i = \sqrt{-1}$. $u(x,t)$ is a complex-valued function and defined over the whole real line. The initial condition is

$$u(x,0) = f(x), \quad x \in [a,b],\tag{2}$$

and the boundary conditions are as follows:

$$u(a,t) = u(b,t) = 0.\tag{3}$$

The solution $u(x,t)$ is decomposed into its real and imaginary parts as

$$u(x,t) = r(x,t) + is(x,t),\tag{4}$$

Substituting Equation (4) into Equation (1), the following coupled partial differential equations are obtained:

$$\begin{cases} s_t - r_{xx} - q\left(r^2 + s^2\right)r = 0, \\ r_t + s_{xx} + q\left(r^2 + s^2\right)s = 0. \end{cases} \tag{5}$$

The solution domain $[a, b]$ is divided equally into N finite elements by the nodes $x_k$ such that $a = x_0 < x_1 < \cdots x_N = b$ and $h = \frac{b-a}{N} = x_{k+1} - x_k$. In this paper, we choose cubic B-spline function as the weight and trial function. The cubic B-splines $\varnothing_k(x)$, $(k = -1, \ldots, N+1)$ are defined over the interval $[a, b]$ as follows [1]:

$$\varnothing_k(x) = \frac{1}{h^3} \begin{cases} (x - x_{k-2})^3 & x \in [x_{k-2}, x_{k-1}], \\ h^3 + 3h^2(x - x_{k-1}) + 3h(x - x_{k-1})^2 - 3(x - x_{k-1})^3 & x \in [x_{k-1}, x_k], \\ h^3 + 3h^2(x_{k+1} - x) + 3h(x_{k+1} - x)^2 - 3(x_{k+1} - x)^3 & x \in [x_k, x_{k+1}], \\ (x_{k+2} - x)^3 & x \in [x_{k+1}, x_{k+2}], \\ 0 & otherwise. \end{cases} \tag{6}$$

The approximate solution $u_N(x, t)$ to the analytical solution $u(x, t)$ for Equation (1) is written in terms of the cubic B-spline function as

$$u_N(x, t) = s_N(x, t) + r_N(x, t) = \sum_{j=-1}^{N+1} \delta_j(t)\varnothing_j(x), \tag{7}$$

where $\delta_j(t) = \rho_j(t) + \psi_j(t)$,

$$s_N(x, t) = \sum_{j=-1}^{N+1} \rho_j(t)\varnothing_j(x),$$

$$r_N(x, t) = \sum_{j=-1}^{N+1} \psi_j(t)\varnothing_j(x). \tag{8}$$

The coefficients $\rho_j(t)$ and $\psi_j(t)$ are unknown time-dependent parameters that will be determined from the boundary and weighted residual conditions. Since each cubic B-spline covers 4 elements, each element $[x_k, x_{k+1}]$ is covered by 4 splines. Using the local coordinate's transformation $\eta = x - x_k$ for $0 \leq \eta \leq h$, the cubic B-spline function over the element $[x_k, x_{k+1}]$ can be written as

$$\begin{matrix} \varnothing_{k-1} \\ \varnothing_k \\ \varnothing_{k+1} \\ \varnothing_{k+2} \end{matrix} = \frac{1}{h^3} \begin{cases} (h - \eta)^3 \\ 4h^3 - 3h^2\eta + 3h(h - \eta)^2 - 3(h - \eta)^3 \\ h^3 + 3h^2\eta + 3h\eta^2 - 3\eta^3 \\ \eta^3 \end{cases}, \quad 0 \leq \eta \leq h. \tag{9}$$

Here, all other cubic B-spline functions are zero over the finite element $[x_k, x_{k+1}]$. The approximation function (Equation (8)) over the finite element can be defined in terms of the basis functions (Equation (9)) as

$$s_N(x, t) = \sum_{j=k-1}^{k+2} \rho_j(t)\varnothing_j(x).$$

$$r_N(x, t) = \sum_{j=k-1}^{k+2} \psi_j(t)\varnothing_j(x). \tag{10}$$

Using the B-splines basis defined in Equation (6) and the trial function stated in Equation (8), the nodal values of $s_k$, $r_k$, $s'_k$ and $r'_k$ at the knots $x_k$ are determined in terms of time-dependent element parameters $\rho_j$ and $\psi_j$ as:

$$
\begin{aligned}
s_k &= s(x_k) = \rho_{k-1} + 4\rho_k + \rho_{k+1}, \\
r_k &= r(x_k) = \psi_{k-1} + 4\psi_k + \psi_{k+1}, \\
hs'_k &= hs'(x_k) == 3(\rho_{k+1} - \rho_{k-1}), \\
hr'_k &= hr'(x_k) == 3(\psi_{k+1} - \psi_{k-1}), \\
h^2 s''_k &= h^2 s''(x_k) == 6(\rho_{k-1} - 2\rho_k + \rho_{k+1}), \\
h^2 r''_k &= h^2 r''(x_k) = 6(\psi_{k-1} - 2\psi_k + \psi_{k+1}),
\end{aligned}
\tag{11}
$$

where the prime denotes differentiation with respect to $x$.

Applying the Galerkin approach to Equation (5) with the weight function $W(x)$, the weak form of Equation (5) over the finite element $[x_k, x_{k+1}]$ is

$$
\begin{cases}
\displaystyle\int_{x_k}^{x_{k+1}} [Ws_t + W_x r_x - z_k Wr]dx = 0, \\[2em]
\displaystyle\int_{x_k}^{x_{k+1}} [Wr_t - W_x s_x + z_k Ws]dx = 0,
\end{cases}
\tag{12}
$$

where

$$
z_k = q(r^2 + s^2).
\tag{13}
$$

Using the weight function $W$ as the cubic B-spline and substituting Equations (10) into (12) with some manipulations, the following differential equations are obtained:

$$
\begin{cases}
\displaystyle\sum_{j=k-1}^{k+2}\left[\left(\int_0^h \varnothing_m \varnothing_j d\eta\right)\dot{\rho}_j + \left(\int_0^h \varnothing'_m \varnothing'_j d\eta\right)\psi_j - \left(z_k \int_0^h \varnothing_m \varnothing_j d\eta\right)\psi_j\right] = 0, \\[2em]
\displaystyle\sum_{j=k-1}^{k+2}\left[\left(\int_0^h \varnothing_m \varnothing_j d\eta\right)\dot{\psi}_j + \left(\int_0^h \varnothing'_m \varnothing'_j d\eta\right)\rho_j - \left(z_k \int_0^h \varnothing_m \varnothing_j d\eta\right)\rho_j\right] = 0,
\end{cases}
\tag{14}
$$

Equation (14) can be written in matrix form as

$$
\begin{cases}
A^e \dot{\rho}^e + B^e \psi^e - z_k C^e \psi^e = 0, \\
A^e \dot{\psi}^e - B^e \rho^e + z_k C^e \rho^e = 0,
\end{cases}
\tag{15}
$$

where $\rho^e = (\rho_{k-1}, \rho_k, \rho_{k+1}, \rho_{k+2})^T$ and $\psi^e = (\psi_{k-1}, \psi_k, \psi_{k+1}, \psi_{k+2})^T$ are the element parameters and dot ($\cdot$) represents differentiation with respect to $t$. The element matrix $A^e_{mj}$, $B^e_{mj}$ and $C^e_{mj}$ are given by

$$
A^e_{mj} = C^e_{mj} = \int_0^h \varnothing_m \varnothing_j d\eta, \quad B^e_{mj} = \int_0^h \varnothing'_m \varnothing'_j d\eta.
\tag{16}
$$

where $m, j = k-1, k, k+1, k+2$. The element matrix (16) is calculated as

$$A^e_{mj} = C^e_{mj} = \int_0^h \varnothing_m \varnothing_j d\eta = \tfrac{h}{140} \begin{bmatrix} 20 & 129 & 60 & 1.0 \\ 129 & 1188 & 933 & 60 \\ 60 & 933 & 1188 & 129 \\ 1.0 & 60 & 129 & 20 \end{bmatrix},$$

$$B^e_{mj} = \int_0^h \varnothing'_m \varnothing'_j d\eta = \tfrac{1}{10h} \begin{bmatrix} 18 & 21 & -36 & -3.0 \\ 21 & 102 & -87 & -36 \\ -36 & -87 & 102 & 21 \\ -3.0 & -36 & 21 & 18 \end{bmatrix}.$$

$z_k$ values are calculated by writing $s = \frac{s_m + s_{m+1}}{2}$ and $r = \frac{r_m + r_{m+1}}{2}$ in Equation (13). Using the values of $s_N$ and $r_N$ at the points $x_k$, we obtain

$$z_k = q \left[ \frac{(\rho_{k-1} + 5\rho_k + 5\rho_{k+1} + \rho_{k+2})^2}{4} + \frac{(\psi_{k-1} + 5\psi_k + 5\psi_{k+1} + \psi_{k+2})^2}{4} \right]. \tag{17}$$

Assembling the contributions from all of the elements, Equation (15) becomes

$$\begin{cases} A\dot{\rho} + B\psi - C(z_k)\psi = 0, \\ A\dot{\psi} - B\rho + C(z_k)\rho = 0, \end{cases} \tag{18}$$

where $\rho = (\rho_{-1}, \rho_0, \dots, \rho_k, \rho_{k+1})^T$ and $\psi = (\psi_{-1}, \psi_0, \dots, \psi_k, \psi_{k+1})^T$ are global element parameters and the matrices $A$, $B$ and $C(z_k)$, which are global matrices with a generalized kth row, have the following form:

$$A = \tfrac{h}{140}(1, 120, 1191, 2416, 1191, 120, 1),$$
$$B = \tfrac{1}{10h}(-3, -72, -45, 240, -45, -72, -3),$$
$$C(z_k) = \frac{h}{140}(z_{1k}, 60z_{1k} + 60z_{2k}, 129z_{1k} + 933z_{2k} + 129z_{3k}, 20z_{1k} + 1188z_{2k}$$
$$+ 1188z_{3k} + 20z_{4k}, 129z_{2k} + 933z_{3k} + 129z_{4k}, 60z_{3k} + 60z_{4k}, z_{4k}). \tag{19}$$

where

$$z_{1k} = \tfrac{q}{4}\left[(\rho_{k-2} + 5\rho_{k-1} + 5\rho_k + \rho_{k+1})^2 + (\psi_{k-2} + 5\psi_{k-1} + 5\psi_k + \psi_{k+1})^2\right],$$
$$z_{2k} = \tfrac{q}{4}\left[(\rho_{k-1} + 5\rho_k + 5\rho_{k+1} + \rho_{k+2})^2 + (\psi_{k-1} + 5\psi_k + 5\psi_{k+1} + \psi_{k+2})^2\right],$$
$$z_{3k} = \tfrac{q}{4}\left[(\rho_k + 5\rho_{k+1} + 5\rho_{k+2} + \rho_{k+3})^2 + (\psi_k + 5\psi_{k+1} + 5\psi_{k+2} + \psi_{k+3})^2\right],$$
$$z_{4k} = \tfrac{q}{4}\left[(\rho_{k+1} + 5\rho_{k+2} + 5\rho_{k+3} + \rho_{k+4})^2 + (\psi_{k+1} + 5\psi_{k+2} + 5\psi_{k+3} + \psi_{k+4})^2\right]. \tag{20}$$

Discretization by the Crank-Nicolson method gives us $\rho = \left(\rho^n + \rho^{(n+1)}\right)/2$ and $\psi = \left(\psi^n + \psi^{n+1}\right)/2$. Similarly, the time discretization by finite difference scheme gives us $\dot{\rho} = \left(\rho^{n+1} - \rho^n\right)/\Delta t$ and $\dot{\psi} = (\psi^{n+1} - \psi^n)/\Delta t$. Substituting these discretizations into Equation (18), we obtain the following system:

$$\begin{cases} A\rho^{n+1} + 0.5\Delta t(B - C(z_k))\psi^{n+1} = A\rho^n - 0.5\Delta t(B - C(z_k))\psi^n, \\ A\psi^{n+1} - 0.5\Delta t(B - C(z_k))\rho^{n+1} = A\psi^n + 0.5\Delta t(B - C(z_k))\rho^n. \end{cases} \tag{21}$$

Using the boundary conditions in Equation (21) yields a septadiagonal matrix. This system is solved by an appropriate method in MATLAB.

First, the initial vectors of parameters $\rho^0 = \left(\rho^0{}_0, \ldots, \rho^0{}_N\right)$ and $\psi^0 = \left(\psi^0{}_0, \ldots, \psi^0{}_N\right)$ are calculated to solve System (21) with the use of the initial condition using the relations

$$s_N(x,0) = \sum_{j=-1}^{N+1} \varnothing_j(x)\rho^o_j \text{ and } r_N(x,0) = \sum_{j=-1}^{N+1} \varnothing_j(x)\psi^o_j, \tag{22}$$

where all parameters $\rho^0$ and $\psi^0$ are determined. $s_N$ and $r_N$ require the following relations to be satisfied at points $x_k$:

$$s_N(x_k,0) = s(x_k,0),$$
$$r_N(x_k,0) = r(x_k,0), k = 0, 1, \ldots, N. \tag{23}$$

The initial vectors $\rho^0$ and $\psi^0$ can be calculated using the initial and boundary conditions from the following matrix equations:

$$\begin{bmatrix} 4 & 1 & 0 & 0 & & & & 0 \\ 1 & 4 & 1 & 0 & & & & \\ 0 & 1 & 4 & 1 & & & & \\ & & \ddots & \ddots & \ddots & & & \\ & & & 1 & 4 & 1 & & \\ & & & 0 & 1 & 4 & 1 & \\ 0 & & & & & 0 & 1 & 4 \end{bmatrix} \begin{bmatrix} \rho^0{}_0 \\ \rho^0{}_1 \\ \rho^0{}_2 \\ \vdots \\ \rho^0{}_{N-2} \\ \rho^0{}_{N-1} \end{bmatrix} = \begin{bmatrix} s^0{}_0 \\ s^0{}_1 \\ s^0{}_2 \\ \vdots \\ s^0{}_{N-3} \\ s^0{}_{N-2} \\ s^0{}_{N-1} \end{bmatrix},$$

and

$$\begin{bmatrix} 4 & 1 & 0 & 0 & & & & 0 \\ 1 & 4 & 1 & 0 & & & & \\ 0 & 1 & 4 & 1 & & & & \\ & & \ddots & \ddots & \ddots & & & \\ & & & 1 & 4 & 1 & & \\ & & & 0 & 1 & 4 & 1 & \\ 0 & & & & & 0 & 1 & 4 \end{bmatrix} \begin{bmatrix} \psi^0{}_0 \\ \psi^0{}_1 \\ \psi^0{}_2 \\ \vdots \\ \psi^0{}_{N-2} \\ \psi^0{}_{N-1} \end{bmatrix} = \begin{bmatrix} r^0{}_0 \\ r^0{}_1 \\ r^0{}_2 \\ \vdots \\ r^0{}_{N-3} \\ r^0{}_{N-2} \\ r^0{}_{N-1} \end{bmatrix}.$$

This system is solved using a suitable method in MATLAB. The approximate numerical solution of $s_N(x,t)$ and $r_N(x,t)$ is obtained from the $\rho^n$ and $\psi^n$ using Equation (21).

### 3. Stability Analysis

Stability analysis of the numerical scheme is carried out using the von Neumann method. The coupled system of Equation (5) is written as

$$\frac{\partial U}{\partial t} + M \frac{\partial^2 U}{\partial x^2} + G(U)U = 0, \tag{24}$$

where $U = \begin{bmatrix} s \\ r \end{bmatrix}, M = \begin{bmatrix} 0 & -1 \\ 1 & 0 \end{bmatrix}$ and $G(U) = \begin{bmatrix} 0 & -\mu_1 \\ \mu_2 & 0 \end{bmatrix}$.

To perform linear stability analysis, we further linearize the nonlinear term in Equation (24) by taking $G(U) = \mu M$. Here $\mu = \max[\mu_1, \mu_2]$. Therefore, the linearized form of Equation (24) is as follows:

$$\frac{\partial U}{\partial t} + M \frac{\partial^2 U}{\partial x^2} + \mu M U = 0. \tag{25}$$

The discretization of linear Equation (24) by the proposed scheme is given by

$$A U_m^{n+1} - MD U_m^{n+1} = A U_m^n + MD U_m^{n+1}, \tag{26}$$

where $D = B - \mu I$.

We use the following Fourier modes analysis for Scheme (25)

$$U_m^n = PG^n e^{imv}. \tag{27}$$

Here $i = \sqrt{-1}$, $P \in R^2$ and $G \in R^{2\times2}$ is the amplification matrix. Equation (26) is substituted into Equation (25) to obtain the value of the amplification $G$. After tedious algebra, we obtain the value of $G$ as follows

$$G = \left[\frac{A + MD}{A - MD}\right] \tag{28}$$

Matrix $M$ is a skew symmetric matrix, and therefore both the matrices $A + MD$ and $A - MD$ have the same eigen values. Thus, the maximum value of $|G| = 1$. Hence, the linearized Scheme (25) is unconditionally stable.

## 4. Numerical Results and Test Problems

Three test problems, including the single soliton, interaction of two solitons and birth of standing soliton, are presented to evaluate the effectiveness and performance of the proposed method. The accuracy of the proposed method is examined using the error norms $L_2$ and $L_\infty$ and conservation laws defined as

$$L_2 = \|u^{exact} - u_N\|_2 = \sqrt{h \sum_{k=0}^{N} |u_k^{exact} - (u_N)_k|^2},$$

$$L_\infty = \|u^{exact} - u_N\|_\infty = \max_{0 \le k \le N} |u_k^{exact} - (u_N)_k|. \tag{29}$$

Moreover, Equation (1) must satisfy the two conservation laws

$$I_1 = \int_a^b |u|^2 dx,$$

$$I_2 = \int_a^b \left(|u_x|^2 - \tfrac{1}{2} q|u|^4\right) dx. \tag{30}$$

### 4.1. Single Soliton Solution to the NLS Equation

The analytical single soliton solution to the NLS equation is given as [2]:

$$u(x,t) = \beta\left(\sqrt{\frac{2}{q}}\right) expi\left[\frac{1}{2}Sx - \frac{1}{4}\left(S^2 - \beta^2\right)t\right] sech(\beta x - \beta St), \tag{31}$$

where $S$ is the speed of the soliton solution whose magnitude depends on the parameter $\beta$.

The numerical solution is computed using the following parameters: $q = 2, S = 4, \beta = 1$, $x_L = -20$, and $x_R = 20$. The conservative quantities $I_1$ and $I_2$ and error norms $L_\infty$ and $L_2$ are calculated. Numerical simulations were carried out at different space steps and time steps for comparison with the published results of the previous methods. The results are documented in Tables 1 and 2. The numerical simulations and the absolute numerical error are shown at different times in Figure 1.

Table 2 displays a comparison between the results obtained by the proposed method and published results. The results of the proposed method are in agreement with analytical solutions within very satisfactory limits, and the proposed method exhibits the same results as the quintic B-spline collocation method and quadratic B-spline method, as can be seen from the comparison in Table 2. We found good results even for large step sizes. It is noted that the error norms of the Galerkin cubic B-spline are lower than Explicit method [3], Implicit/explicit [3], and Split-step Fourier method [3] when we compare the results of the proposed method with those referenced in [1,4,19].

**Table 1.** Error norms and conservation laws for single soliton: with $h = 0.04, q = 2, S = 4, \beta = 1$.

| | $\Delta t = 0.0025$ | | | | $\Delta t = 0.005$ | | | |
| --- | --- | --- | --- | --- | --- | --- | --- | --- |
| $t$ | $L_\infty$ | $L_2$ | $I_1$ | $I_2$ | $L_\infty$ | $L_2$ | $I_1$ | $I_2$ |
| 0.5 | 0.0000896 | 0.0000914 | 2.0 | 7.33333332 | 0.00018 | 0.00018 | 2.0 | 7.33333327 |
| 1 | 0.0000896 | 0.0000914 | 2.0 | 7.33333332 | 0.00018 | 0.00018 | 2.0 | 7.33333327 |
| 1.5 | 0.0000896 | 0.0000914 | 2.0 | 7.33333332 | 0.00018 | 0.00018 | 2.0 | 7.33333327 |
| 2 | 0.0000896 | 0.0000914 | 2.0 | 7.33333331 | 0.00018 | 0.00018 | 2.0 | 7.33333327 |
| 2.5 | 0.0000896 | 0.0000914 | 2.0 | 7.33333330 | 0.00018 | 0.00018 | 2.0 | 7.33333325 |
| 3 | 0.0000896 | 0.0000914 | 2.0 | 7.33333223 | 0.00018 | 0.00018 | 2.0 | 7.33333219 |
| 3.5 | 0.0000896 | 0.0000914 | 1.99999 | 7.33327430 | 0.00018 | 0.00018 | 1.99999 | 7.33327425 |

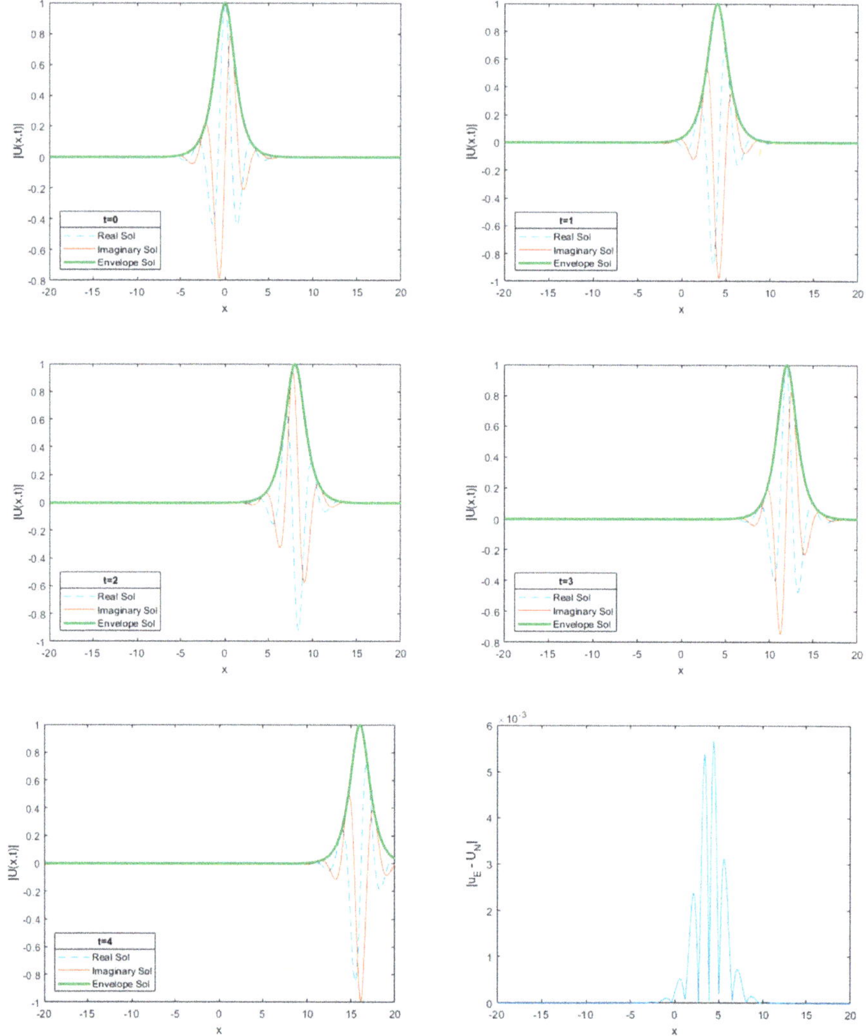

**Figure 1.** Single soliton solution and errors |Numerical − Analytical|.

**Table 2.** Comparisons of the present results with those of Taha et al. [20]., amplitude = 1 at time = 1.

| $h$ | $\Delta t$ | $L_\infty$ | $L_2$ | $I_1$ | $I_2$ | $L_\infty$ [20] |
|---|---|---|---|---|---|---|
| 0.05 | 0.0025 | 0.0001086 | 0.0001106 | 2.0 | 7.33333330 | 0.008 |
| 0.05 | 0.000625 | 0.0000272 | 0.0000277 | 2.0 | 7.33333333 | 0.006 |
| 0.05 | 0.001 | 0.0000435 | 0.0000443 | 2.0 | 7.33333333 | 0.006 |
| 0.08 | 0.002 | 0.0000628 | 0.0000639 | 2.0 | 7.33333330 | 0.005 |
| 0.3125 | 0.00026 | 0.0003174 | 0.0004973 | 2.00001 | 7.33332035 | 0.005 |
| 0.3125 | 0.020 | 0.0024417 | 0.0038252 | 2.00010 | 7.33320599 | 0.005 |
| 0.05 | 0.0005 | 0.0000217 | 0.0000221 | 2.0 | 7.33333333 | 0.008 |
| 0.3125 | 0.0026 | 0.0003174 | 0.0004973 | 2.00001 | 7.33332035 | 0.006 |

The rate of convergence for spatial and temporal directions is calculated using the formula [21]:

$$\text{rate of convergence} \approx log_2 \frac{error(2h, 2\Delta t)}{error(h, \Delta t)},$$

where $error(2h, 2\Delta t)$ is the error norms $L_\infty$ and $L_2$ in spatial and temporal directions. The error norms $L_\infty, L_2$ and order of convergence rate at time $t = 1$ are shown in Table 3.

**Table 3.** Rate of convergence in spatial and temporal directions at $t = 1$ with $h = \Delta t$.

| $h$ | $L_\infty$ | Order | $L_2$ | Order |
|---|---|---|---|---|
| 0.050000 | 0.0020492 | – | 0.0019146 | - |
| 0.025000 | 0.0005730 | 1.83 | 0.0005302 | 1.85 |
| 0.012500 | 0.0001507 | 1.93 | 0.000139 | 1.93 |
| 0.001625 | 0.0000386 | 1.97 | 0.0000356 | 1.96 |
| 0.003125 | 0.00000965 | 2.00 | 0.0000089 | 2.00 |

In Table 4, the results are documented for when the soliton amplitude is equal to 2. It was found that the results of the error norms at different times increase very slightly, as shown in Table 4. One soliton solution with amplitude = 2 is calculated, and simulations are shown in Figure 2.

**Table 4.** Error norms and conservation laws for single soliton with $h = 0.05, \Delta t = 0.0025$ $h = 0.05$, $q = 2, S = 4, \beta = 2$.

| $t$ | $L_2$ | $L_\infty$ |
|---|---|---|
| 0.5 | 0.0005580 | 0.0007595 |
| 1 | 0.0005580 | 0.0007595 |
| 1.5 | 0.0005580 | 0.0007595 |
| 2 | 0.0005580 | 0.0007595 |
| 2.5 | 0.0005580 | 0.0007595 |
| 3 | 0.0005580 | 0.0007595 |
| 3.5 | 0.0005580 | 0.0007595 |
| 4 | 0.0005580 | 0.0007595 |

Table 5 displays the approximate results obtained using the proposed method compared with the published methods listed in the reference for single soliton when the amplitude is 2. These results are in good agreement with the analytical solution.

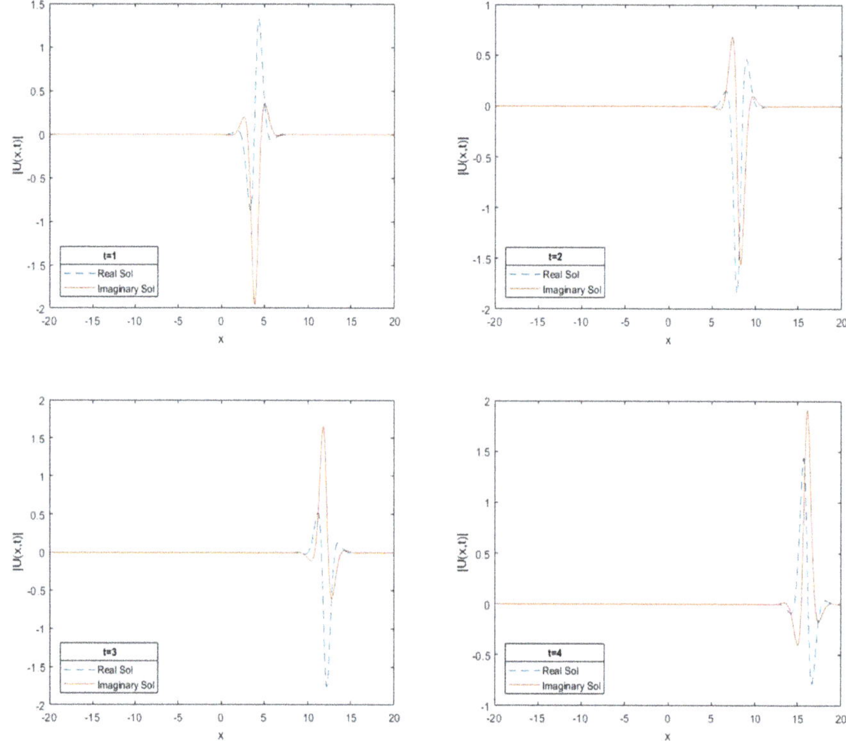

**Figure 2.** Single soliton solution with amplitude 2.

**Table 5.** Comparisons of single soliton with those of Taha et al. [20], amplitude = 2 at time = 1.

| $h$ | $\Delta t$ | $L_\infty$ | $L_2$ | $L_\infty$ [20] |
|---|---|---|---|---|
| 0.02 | 0.0025 | 0.0003909 | 0.0002859 | 0.0011 |
| 0.066 | 0.002 | 0.0009658 | 0.0007168 | 0.006 |
| 0.02 | 0.0001 | 0.0000156 | 0.0000114 | 0.009 |
| 0.03 | 0.00022 | 0.0000143 | 0.0000105 | 0.008 |
| 0.1 | 0.0025 | 0.0016852 | 0.0012732 | 0.0004 |
| 0.1563 | 0.0011 | 0.0010143 | 0.0007852 | 0.008 |

### 4.2. The Interaction of Two Solitons for the NLS Equation

In the second problem, we discuss the interaction of two solitons moving in opposite directions, have the same amplitude of magnitude 1 with initial conditions as follows [16,17,20]

$$u(x,0) = \sum_{k=1}^{2} u_k(x,0),$$

$$u_k(x,0) = \beta_k \left( \sqrt{\tfrac{2}{q}} \right) expi\left[ \tfrac{1}{2} S(x - x_k) \right] sech(\beta_k(x - x_k)).$$

(32)

where $\beta_k, q$ and $x_k$ are constants.

The numerical solution is computed by using the following parameters $x_1 = 10, x_2 = -10,$ $q = 2, \beta_1 = \beta_2 = 1, S_1 = -4$ and $S_2 = 4$. The first soliton is placed at $x = 10$ and is moving to the left with speed 4, and the second soliton is placed on the other side at $x = -10$, traveling to the right with speed 4. Both solitons are moving in opposite directions, and they collide and separate. The interactions of these two solitons are visualized in Figure 3. The error norms $L_\infty$ and $L_2$ are computed at different times and at different space steps and time steps. The calculated results are tabulated in Table 6.

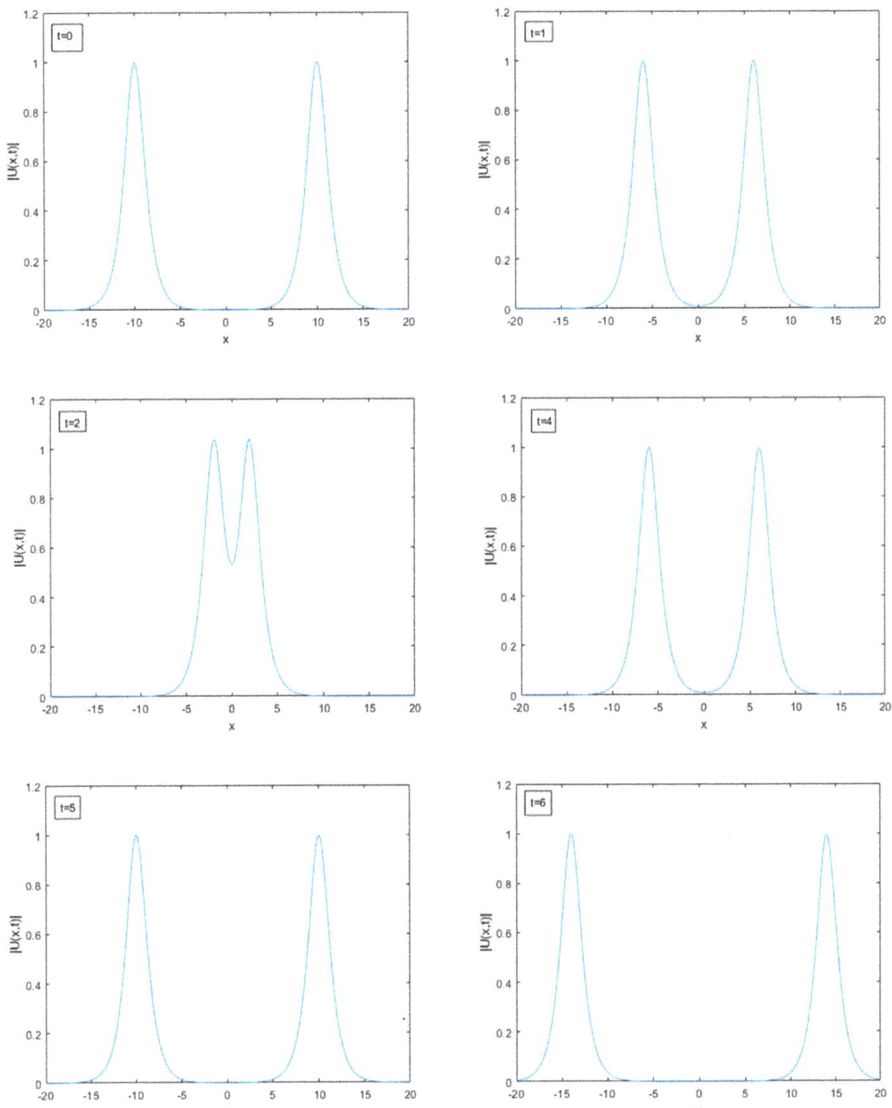

**Figure 3.** Interaction of two solitons, amplitude = 1 at different space and time steps.

The error norms $L_\infty$ and $L_2$ are very small compared with those of Taha et al. [20], as shown in Table 7. It can be clearly seen that the error norms $L_\infty$ and $L_2$ obtained by the present method are smaller than those of previous methods [20], as shown in Table 7.

**Table 6.** Two-solitons, amplitude = 1 at different space and time steps.

| $t$ | $h$ | $\Delta t$ | $L_\infty$ | $L_2$ |
|---|---|---|---|---|
| 1 | 0.0667 | 0.0025 | 0.0001601 | 0.0002367 |
| 1.5 | 0.133 | 0.01 | 0.0012135 | 0.0018299 |
| 2 | 0.133 | 0.01 | 0.0011711 | 0.0017397 |
| 3.5 | 0.133 | 0.01 | 0.0012135 | 0.0018299 |
| 4.5 | 0.133 | 0.01 | 0.0012144 | 0.0018317 |
| 0.5 | 0.625 | 0.005 | 0.0018529 | 0.0030593 |
| 1 | 0.625 | 0.005 | 0.0017864 | 0.0030560 |
| 1 | 0.05 | 0.001 | 0.0000486 | 0.0000716 |
| 1.5 | 0.05 | 0.001 | 0.0000485 | 0.0000715 |
| 3 | 0.5 | 0.0036 | 0.0011024 | 0.0017556 |
| 3 | 0.5 | 0.0025 | 0.0007655 | 0.0012192 |
| 5 | 0.1 | 0.0025 | 0.0002342 | 0.0003495 |

**Table 7.** Comparisons of two solitons with Taha et al. [20], amplitude = 1.

| Galerkin Cubic B-spline (Present Method) | | | | Taha et al. [20] |
|---|---|---|---|---|
| $t$ | $h$ | $\Delta t$ | $L_\infty$ | $L_\infty$ |
| 1.6 | 0.05 | 0.001 | 0.0000485 | 0.00173 |
| 1.8 | 0.07 | 0.07 | 0.0047374 | 0.00158 |
| 1 | 0.05 | 0.0025 | 0.0001214 | 0.00096 |
| 1 | 0.05 | 0.001 | 0.0000486 | 0.00141 |
| 1 | 0.625 | 0.0071 | 0.0025367 | 0.00122 |
| 1 | 0.130 | 0.0036 | 0.0004304 | 0.00141 |

*4.3. Birth of Standing Soliton with the Maxwellian Initial Condition*

If

$$\int_{-\infty}^{\infty} u(x,0) \geq \pi, \tag{33}$$

a soliton should appear over time with initial values greater than $\pi$, otherwise the soliton will decay away [22].

Consider the birth of soliton with the Maxwellian initial condition given by [16]:

$$u(x,0) = A\, exp\left(-x^2\right), \tag{34}$$

The values of all parameters are chosen to be $h = 0.08$, $\Delta t = 0.004$ and $q = 2$ for the domain $[-45, 45]$ to exhibit the birth of a soliton. The numerical simulations are shown at different times for the values of $A = 1.78$ over the domain $[-45,45]$ in Figure 4. With $A = 1$, the soliton decays away as expected.

The conserved quantities $I_1$ and $I_2$ are computed using the Maxwellian initial condition (34). Analytical conserved quantities can be computed as:

$$I_1 = A^2 \sqrt{\tfrac{\pi}{2}} = 3.97100051267043,$$
$$I_2 = \tfrac{1}{4}A^2\left(2\sqrt{2} - qA^2\right)\sqrt{\pi} = -4.92561762132093,$$

The conserved quantities $I_1$ and $I_2$ are tabulated in Table 8. The numerical results obtained are compared with the published results of Mokhtari et al. [11] and the exact solution, as shown in Table 8. The proposed method conserves $I_1$ to 7 decimal places and $I_2$ to 3 decimal places.

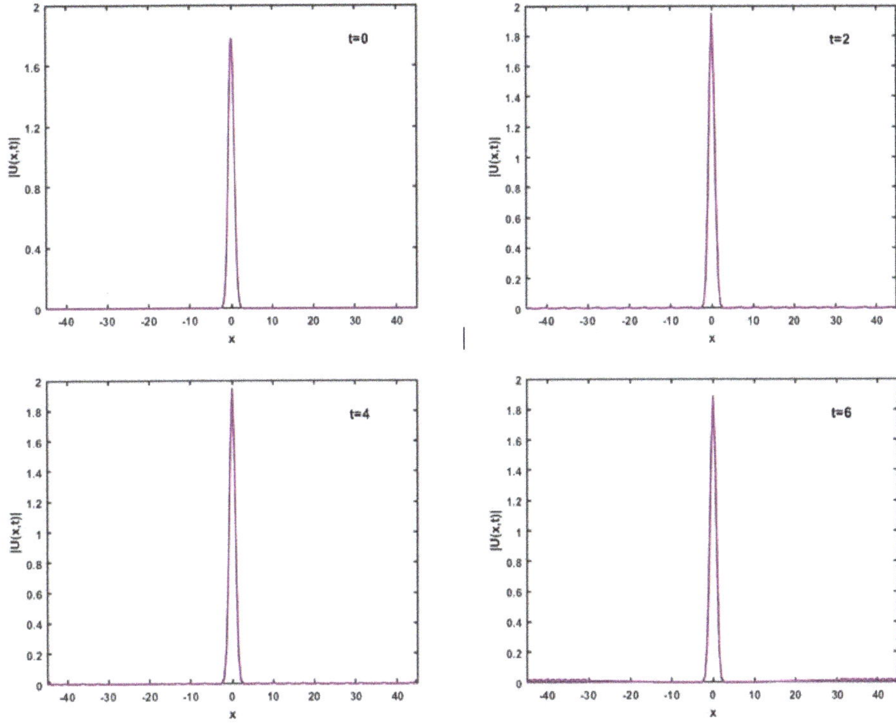

**Figure 4.** Birth of a standing soliton for $A = 1.78$.

**Table 8.** Comparison of conserved quantities of the birth of a standing soliton with Mokhtari et al. [11], $h = 0.08, \Delta t = 0.004$ and $q = 2$.

| | Galerkin Cubic B-spline (Present Method) | | | | Mokhtari et al. [11] | |
| --- | --- | --- | --- | --- | --- | --- |
| $t$ | $I_1$ | $I_2$ | $EI_1$ | $EI_2$ | $EI_1$ | $EI_2$ |
| 1 | 3.97100052134110 | $-4.925146$ | $8.67067 \times 10^{-9}$ | $4.7 \times 10^{-4}$ | $3.5 \times 10^{-11}$ | $4.8 \times 10^{-10}$ |
| 2 | 3.97100050351461 | $-4.925347$ | $2.08442 \times 10^{-8}$ | $2.7 \times 10^{-4}$ | $3.8 \times 10^{-11}$ | $6.9 \times 10^{-10}$ |
| 3 | 3.97100050291892 | $-4.925138$ | $2.02485 \times 10^{-8}$ | $4.8 \times 10^{-4}$ | $1.1 \times 10^{-10}$ | $-6.4 \times 10^{-9}$ |
| 4 | 3.9710005812876 | $-4.925325$ | $6.86172 \times 10^{-8}$ | $2.9 \times 10^{-4}$ | $-2.0 \times 10^{-8}$ | $-4.5 \times 10^{-6}$ |
| 5 | 3.97100055661924 | $-4.925367$ | $4.39488 \times 10^{-8}$ | $2.5 \times 10^{-4}$ | $-9.2 \times 10^{-7}$ | $-4.7 \times 10^{-5}$ |
| 6 | 3.97100053521529 | $-4.925432$ | $2.25449 \times 10^{-8}$ | $1.9 \times 10^{-4}$ | $-6.0 \times 10^{-6}$ | $-9.8 \times 10^{-5}$ |

## 5. Conclusions

The approximate solution of the NLS equation was investigated using the Galerkin finite element method with a cubic B-spline shape function. Three numerical problems, including single soliton, interaction of two solitons, and birth of a standing soliton with the Maxwellian initial condition, were demonstrated to evaluate to the performance and accuracy of the method. Furthermore, we simulated the numerical solution by choosing different parameters for motion of the single soliton, the interaction of two solitons and the birth of the standing soliton. The error norms $L_\infty$ and $L_2$ and conservation laws $I_1$ and $I_2$ were determined and compared with published results [11,16,17,20]. It was seen that all of our results for the problems were computed to be reasonably low, and were found to be in good agreement with the analytical solution. The present method was shown to be unconditionally stable. The method has almost a second-order convergence. In conclusion, the proposed scheme with cubic B-spline presents an acceptable result for the NLS equation.

**Author Contributions:** Formal analysis, A.I. and N.N.A.H.; Investigation, A.I., N.N.A.H. and A.I.M.I.; Methodology, A.I.; Supervision, N.N.A.H. and Ahmad A.I.M.I.; Writing – original draft, A.I.; Writing – review & editing, N.N.A.H. and A.I.M.I.

**Funding:** This work was supported by USM Short Term Grant of account number 304.PMATHS.6315077.

**Conflicts of Interest:** The authors declare no conflict of interest.

## References

1. Dag, I.; Saka, B.; Boz, A. B-spline Galerkin methods for numerical solutions of the Burgers equation. *Appl. Math. Comput.* **2005**, *166*, 506–522.
2. Dag, I.; Irk, D.; Saka, B. A numerical solution of the Burger's equation using cubic B-splines. *Appl. Math. Comput.* **2005**, *163*, 199–211.
3. Gorgulu, M.Z.; Dag, I. Galerkin method for the numerical solution of the advection-diffusion equation by using exponential B-splines. *arXiv* **2016**, arXiv:1604.04267v1.
4. Gorgulu, M.Z.; Dag, I.; Irk, D. Wave propogation by way of exponential B-spline Galerkin method. *J. Phys.* **2016**, *766*, 012031.
5. Saka, B.; Dag, I. A numerical solution of the RLW equation by Galerkin method using quartic B-splines. *Commun. Numer. Meth. Engng.* **2008**, *24*, 1339–1361. [CrossRef]
6. Aksan, E. Quadratic B-spline finite element method for numerical. *Appl. Math. Comput.* **2006**, *174*, 884–896.
7. Ali, A.; Gardner, G.; Gardner, L. A collocation solution for Burgers using cubic B-spline finite elements. *Comput. Meth. Appl.* **1992**, *100*, 325–337. [CrossRef]
8. Sheng, Q.; Khaliq, A.Q.M.; Al-Said, E.A. Solving the generalized nonlinear Schrödinger equation via quartic spline approximation. *J. Comput. Phys.* **2001**, *166*, 400–417. [CrossRef]
9. Zlotnik, A.A.; Zlotnik, I.A. Finite element method with discrete transparent boundary conditions for the time-dependent 1D schrödinger equation. *Kinet. Relat. Models* **2012**, *5*, 639–663. [CrossRef]
10. Ersoy, O.; Dag, I.; Sahin, A. Numerical investigation of the solution of Schrödinger equation with exponential cubic B-spline finite element method. *arXiv* **2016**, arXiv:1607.00166v1.
11. Mokhtari, R.; Isvand, D.; Chegini, N.G.; Salaripanah, A. Numerical solution of the Schrödinger equations. *Nonlin. Dyn.* **2013**, *74*, 77–93. [CrossRef]
12. Saka, B. A quintic B-spline finite-element method for solving the nonlinear Schrödinger equation. *Phys. Wave Phenom.* **2012**, *20*, 107–117. [CrossRef]
13. Cesarano, C.; Fornaro, C.; Fornaro, C.; Vazquez, L. Operational results in bi-orthogonal Hermite functions. *Acta Mathemat. Univ. Comenian.* **2016**, *85*, 43–68.
14. Zakharov, V.E.; Shabat, A.B. Exact theory of two-dimensional self-focusing and one-dimensional self-modulation of waves in nonlinear media. *J. Exp. Theor. Phys.* **1972**, *34*, 62–68.
15. Lin, B. Septic spline function method for nonlinear Schrödinger equations. *Appl. Analys.* **2015**, *94*, 279–293. [CrossRef]
16. Dag, I. A quadratic B-spline finite element method for solving nonlinear Schrödinger equation. *Comput. Meth. Appl. Mech. Eng.* **1999**, *174*, 247–258.
17. Gardner, L.R.T.; Gardner, G.A.; Zaki, S.I.; El Sahrawi, Z. B-spline finite element studies of the non-linear Schrödinger equation. *Comput. Methods Appl. Mech. Eng.* **1993**, *108*, 303–318. [CrossRef]
18. Hu, H. Two-grid method for two-dimensional nonlinear Schrödinger equation by finite element method. *Numeric. Methods Part. Differ. Eqs.* **2018**, *75*, 900–917.
19. Prenter, P. *Splines and Variational Methods*; Wiley: New York, NY, USA, 1975.
20. Taha, T.R.; Ablowitz, M.J. Analytical and numerical aspects of certain nonlinear evolution equations II: Numerical nonlinear Schrödinger equations. *J. Comput. Phys.* **1984**, *55*, 203–230. [CrossRef]
21. Qu, W.; Liang, Y. Stability and convergence of the Crank-Nicolson scheme for a class of variable-coefficient tempered fractional diffusion equations. *Advc. Differ. Eqs.* **2017**, *2017*, 108. [CrossRef]

22. Delfour, M.; Fortin, M.; Payne, G. Finite-difference solutions of non-linear Schrödinger equation. *J. Comput. Phys.* **1981**, *44*, 277–288. [CrossRef]

*Article*

# Effect of Overburden Height on Hydraulic Fracturing of Concrete-Lined Pressure Tunnels Excavated in Intact Rock: A Numerical Study

**Moses Karakouzian [1], Mohammad Nazari-Sharabian [1,*] and Mehrdad Karami [2]**

[1]  Department of Civil and Environmental Engineering and Construction, University of Nevada Las Vegas, Las Vegas, NV 89154, USA; mkar@unlv.nevada.edu

[2]  Department of Civil Engineering, Isfahan University of Technology, Isfahan 8415683111, Iran; karami.mehrdad.ce@gmail.com

*  Correspondence: nazarish@unlv.nevada.edu; Tel.: +1-702-205-9336

Received: 2 June 2019; Accepted: 17 June 2019; Published: 19 June 2019

**Abstract:** This study investigated the impact of overburden height on the hydraulic fracturing of a concrete-lined pressure tunnel, excavated in intact rock, under steady-state and transient-state conditions. Moreover, the Norwegian design criterion that only suggests increasing the overburden height as a countermeasure against hydraulic fracturing was evaluated. The Mohr–Coulomb failure criterion was implemented to investigate failure in the rock elements adjacent to the lining. A pressure tunnel with an inner diameter of 3.6 m was modeled in Abaqus Finite Element Analysis (FEA), using the finite element method (FEM). It was assumed that transient pressures occur inside the tunnel due to control gate closure in a hydroelectric power plant, downstream of the tunnel, in three different closure modes: fast (14 s), normal (18 s), and slow (26 s). For steady-state conditions, the results indicated that resistance to the fracturing of the rock increased with increasing the rock friction angle, as well as the overburden height. However, the influence of the friction angle on the resistance to rock fracture was much larger than that of the overburden height. For transient-state conditions, the results showed that, in fast, normal, and slow control gate closure modes, the required overburden heights to failure were respectively 1.07, 0.8, and 0.67 times the static head of water in the tunnel under a steady-state condition. It was concluded that increasing the height of overburden should not be the absolute solution to prevent hydraulic fracturing in pressure tunnels.

**Keywords:** pressure tunnel; hydraulic fracturing; transient flow; finite element method (FEM); Abaqus Finite Element Analysis (FEA)

---

## 1. Introduction

High-pressure water tunnels, usually excavated in rock, convey water from the upstream reservoir toward the turbines in hydropower plants, and turn the energy of the flowing water into electricity. These tunnels can be either steel-lined or concrete-lined; steel-lined pressure tunnels are safer, but more costly to construct. Concrete-lined pressure tunnels are faced with serious challenges in design and construction. In these tunnels, due to the lining permeability in areas where the rock is jointed, water seeps into the joints and expands the cracks. In cases where the rock is intact and not jointed, high internal pressure can fracture the rock. This phenomenon is called hydraulic fracturing, which makes the rock mass on the tunnel unstable.

Researchers studied various aspects of hydraulic fracturing in pressure tunnels, some of which are mentioned below. In 1986, Schleiss investigated leakage from pressure tunnels in the presence of groundwater, and presented the leakage relationships under two conditions: cracked and non-cracked concrete linings. The author concluded that dynamic pressure caused by various factors, such as a water hammer or earthquake, had a significant impact on the rate of leakage and expansion of rock joints [1]. Fernández and Alvarez (1994) used the finite element method (FEM) to study water leakage from a tunnel and effective stresses in the surrounding rock. The authors considered the ratio of total stress to effective stress in the rock as a safety factor against hydraulic fracturing [2]. Furthermore, using the FEM and a closed-form analytical solution, Bobet and Nam (2007) studied effective stresses around pressure tunnels under steady-state operation. The authors assumed there was no groundwater presence around the tunnel, and presented a new analytical solution for stresses and displacements in the surrounding rock and the lining. Their solution considered the interaction between the lining, surrounding rock, and pore pressure in the rock environment, as a result of leakage through the lining [3]. Moreover, Hachem and Schleiss (2010) evaluated pressure wave velocities in steel-lined pressure tunnels, with and without considering the fluid–structure interactions. The authors showed changes in pressure wave velocity, due to the characteristics of the surrounding rock, the thickness of the lining, and wave frequency [4].

Using the FEM, and in order to optimize the thickness of the concrete lining, Olumide and Marence (2012) simulated a pressure tunnel in a two-dimensional elasto-plastic space in order to study the effects of seepage on the tunnel's concrete lining. The authors investigated cracks in the lining by coupling in situ stresses in the rock with the stresses caused by leakage, and validated their numerical model by the analytical method proposed by Schleiss (1986) [5]. Furthermore, Simanjuntak et al. (2014) studied displacements and stresses in the concrete lining and rock mass using the FEM, considering non-uniform in situ stresses around a pressure tunnel [6]. The authors showed that, in non-uniform stress conditions around the tunnel, when the coefficient of lateral rock pressure ($k_0$) is $\leq 1$, cracks form in the crown of the tunnel due to high internal pressure. In 2015, Zhou et al. considered the interactions between fluid and the concrete lining in pressure tunnels in order to study stresses in a tunnel lining. The authors considered the hydro-mechanical behavior of the surrounding rock and the tunnel lining by simulating the fluid inside rock joints and cracks in the concrete lining [7]. More recently, Pachoud and Schleiss (2016) investigated the impacts of the transient pressure wave in pressure tunnels due to water hammer. The authors presented analytical formulas for stresses and displacements in the environment around the tunnel [8]. In 2018, Zareifard presented an analytical solution for the design of pressure tunnels, considering seepage from a tunnel [9]. It should be noted that, in addition to the negative impacts of hydraulic fracturing in water resources and related structures, it is widely used in the oil and gas industry as a novel method for the exploitation of underground reservoirs [10].

Based on the literature review, many researchers investigated cracking in the rock or tunnel concrete lining in steady-state conditions with constant overburden, yet they did not consider the interactive effects of overburden height and transient pressures on hydraulic fracturing in the rock environment surrounding a tunnel. In the present study, in order to fill this gap in knowledge, the impact of overburden height on the bearing capacity of a tunnel in a steady-state condition was studied. For the next step, the effect of transient pressure caused by rapid gate operations on hydraulic fracturing in a tunnel in critical conditions (e.g., load rejection of the powerhouse) was investigated. The Mohr–Coulomb failure criterion was used to investigate the rock failure both in steady-state and transient-state conditions.

## 2. Materials and Methods

Using the FEM and considering the environment around the tunnel as an intact rock with elasto-plastic behavior, stresses around a tunnel were studied using the Abaqus Finite Element Analysis (FEA) software [11]. Because cracks were not explicitly modeled in the numerical model, principal stresses were used as indicators of potential fracturing. Since this solution involves the hydraulics of water flow in the tunnel and the impacts of internal pressures on the surrounding rock, the Hammer software and Abaqus FEA were linked for modeling and analyzing the system. The Hammer software [12], which works based on the method of characteristics (MOC), was employed to analyze the changes in the internal pressure inside the tunnel, as a function of time. The hydraulic analysis results, from the Hammer software in transient-state conditions, were transferred to the Abaqus FEA in order to analyze stresses in the surrounding rock. It was assumed that, at first, the internal pressure was in the steady-state condition. Afterward, because of sudden gate closure, transient pressures built up in the tunnel. The flowchart below summarizes the main steps in this study (Figure 1).

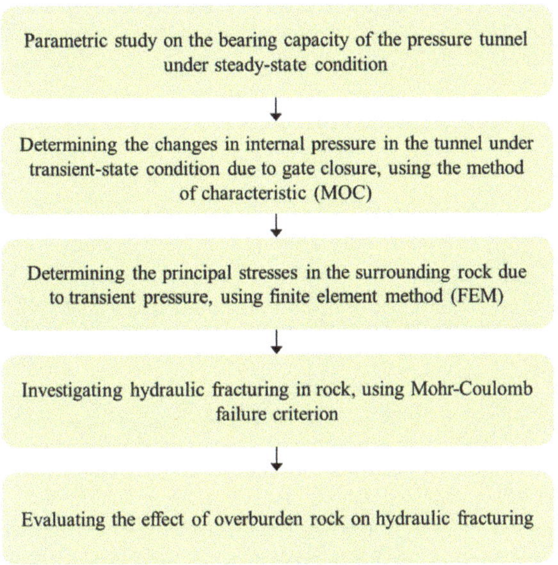


**Figure 1.** The flowchart of this study.

### 2.1. Governing Equations

Using the general Navier–Stokes equation (Equation (1)) and simplifying assumptions—(a) fluid inside the tunnel is non-viscous, (b) fluid velocity fluctuations are ignored, and (c) water has linear compressibility—the hydrodynamic pressure inside the tunnel can be obtained using Equations (1) and (2) [13].

$$\rho\left(\frac{\partial v}{\partial t} + v.\nabla v\right) = -\nabla p + \mu\nabla^2 v + B, \tag{1}$$

$$\frac{\rho}{K}\frac{\partial^2 p}{\partial t^2} = \nabla^2 p, \tag{2}$$

where $v$ is the flow velocity (m/s), $\mu$ is the fluid viscosity (Pa·s), $B$ is the vector of body forces (N), $p$ is the fluid hydrodynamic pressure in the tunnel (Pa), $K$ is the bulk modulus (Pa), $t$ is time (s), and $\rho$ is the water density (kg/m$^3$). The only boundary condition governing the fluid inside the tunnel is the boundary between the water and the lining. This boundary condition is defined as in Equation (3) [13].

$$\frac{\partial p}{\partial n} = -\rho\ddot{u}_{s_n}, \tag{3}$$

where $n$ is a perpendicular vector to the shared surface between the water and the lining, and $\ddot{u}_s$ is the second derivative of the lining elements' displacements in contact with the fluid (m/s$^2$). The above equation is based on the assumption that the radial displacements of the fluid and lining are similar at the contact surfaces. The connection between the hydrodynamic pressures inside the tunnel $\{p\}$

and forces applied to the model caused by the hydrodynamic pressure $\{f\}$ are linked by matrix $[Q]$ (Equation (4)) [14].

$$[Q]\{p\} = \{f\}. \tag{4}$$

The coupling matrix $[Q]$ links the hydrodynamic pressures due to gate closure, with forces generated in the concrete lining. This equation is solved by the Abaqus FEA, by applying the boundary and initial conditions. Finally, the displacements and stresses in the rock due to hydrodynamic pressures are calculated. The forces in the concrete lining due to water pressure are transferred to the surrounding rock. As a result, the equilibrium equation in the surrounding rock is as follows [14]:

$$[M]\{\ddot{u}_r\} + [C]\{\dot{u}_r\} + [K]\{u_r\} = \{R_r\}, \tag{5}$$

$$\{R_r\} = \{f\} - \{f_s\}, \tag{6}$$

where $M$, $C$, and $K$ are mass, damping, and hardness matrices, respectively; $R_r$ is the vector of external forces on the rock; $\ddot{u}_r$ and $\dot{u}_r$ are the second and first derivatives of the rock elements, respectively; and $u_r$ is the displacement of the rock elements. Equation (6) shows that part of the hydrodynamic force is resisted by the concrete lining ($f_s$). Therefore, the relationship between hydrodynamic pressure and displacements in the surrounding rock is obtained by linking Equations (4) and (5).

*2.2. FEM and Effective Parameters*

In the FEM modeling, the following factors were considered in the simulations:

- A circular tunnel was excavated in intact rock by a tunnel boring machine (TBM);
- Water pore pressure was considered in the concrete lining and in the rock;
- The Mohr–Coulomb failure criterion was implemented in order to study stresses in the rock;
- A damage plasticity behavior was considered in the concrete lining.

Using a trial-and-error procedure, the model boundaries were selected in a way so as to not affect stress distribution around the tunnel. Therefore, the model boundaries were set at 6d × 25d (d = tunnel diameter) (Figure 2). More details on the grid and boundary sensitivity analyses are presented in the Appendix A. Table 1 shows the mechanical properties of the lining. It was assumed that the tunnel was located above the groundwater level.

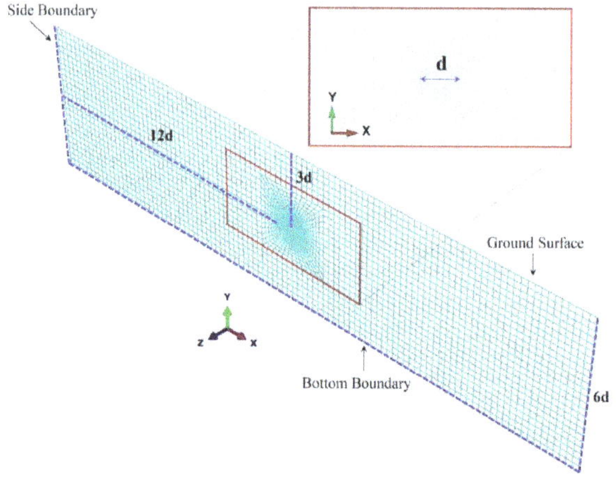


**Figure 2.** The geometry of the surrounding rock and the excavation section.

**Table 1.** Mechanical properties of the lining [15,16].

| Parameter | Value |
|---|---|
| Elasticity (GPa) | 25 |
| Tensile strength (MPa) | 3 |
| Cohesion (MPa) | 4.5 |
| Inner friction angle (°) | 45 |
| Poisson's ratio (-) | 0.25 |
| Density (kg/m$^3$) | 2400 |
| Axial compressive strength (MPa) | 28.3 |
| Hydraulic conductivity (m/s) | $1 \times 10^{-8}$ |

The only degree of freedom inside a water pressure tunnel is pressure in the fluid nodes. In Abaqus FEA, acoustic elements possess this feature and are appropriate for simulating the fluid movement inside a tunnel. Since acoustic environments are elastic, shear stresses do not exist in these environments and pressure is proportional to volume strain. In addition, the effects of inertia and compressibility are considered in acoustic elements [17].

In this study, the interactions between the fluid, lining, and surrounding rock were considered, and the fluid density was set at 1000 kg/m$^3$, with a bulk modulus of 2.07 GPa. Taking into account the interaction between the fluid and lining, the structure surfaces and fluid that are in contact were tied together in Abaqus FEA. Applying the same reasoning, the external nodes of the lining were tied with the nodes of the surrounding rock due to structure–rock interactions.

The environment around the tunnel has in situ stresses. By defining the height, specific weight, and the lateral stress coefficients of the rock, equilibrium conditions between the horizontal and vertical stresses were achieved, which resulted in zero ground settlement before tunnel excavation.

Since the transient pressure inside the tunnel depends on factors such as gate operations, this pressure was applied to the model as a quasi-static load (by gradually increasing the loading, so that the impacts of inertia became negligible). This means that, at different times, different pressures were applied to the lining. In order to apply the time-variable pressure to all of the nodes in the acoustic environment, the amplitude function in Abaqus FEA was used.

The plane strain elements with linear interpolation function were employed to mesh the surrounding rock and concrete lining. In Abaqus FEA, these types of elements are called CPE4R (continuum plain strain element 4-node reduced integration). Meshing was such that, by approaching the tunnel section, the mesh size decreased to achieve more precise results. Moreover, plane stress family elements were used for meshing the lining. In Abaqus FEA, these elements are called CPS4R (continuum plain stress element 4-node reduced integration). Furthermore, in the lining and rock elements, the reduced integration method was implemented, and AC2D4 elements (acoustic continuum 2-dimension 4-node) were used to simulate water transient pressure inside the tunnel [11,18].

The first step in this research study was simulating in situ stresses. In order to have zero ground-surface settlement before tunnel excavation, these stresses should satisfy Equations (7) and (8) [19].

$$\sigma_v = \gamma_r h, \tag{7}$$

$$\sigma_h = \gamma_r h k_0, \tag{8}$$

where $\gamma_r$ is the specific weight of the rock, $h$ is the height of the overburden rock, and $k_0$ is the coefficient of lateral rock pressure. In the next step, the tunnel excavation was simulated. It was assumed that the tunnel was excavated by a tunnel boring machine. At this stage, the release of stress and displacements occurred in the rock. After the tunnel cross-section, the concrete lining was simulated. In this step, a concrete lining with the specifications presented in Table 1 was modeled. Interaction between the rock and the lining was considered using tangential and normal stiffnesses. The final step was applying steady-state conditions and hydrodynamic water pressure to the lining. It was assumed that the transient pressure inside the tunnel would occur due to the turbine gate closure in a typical hydropower plant (Figure 3).

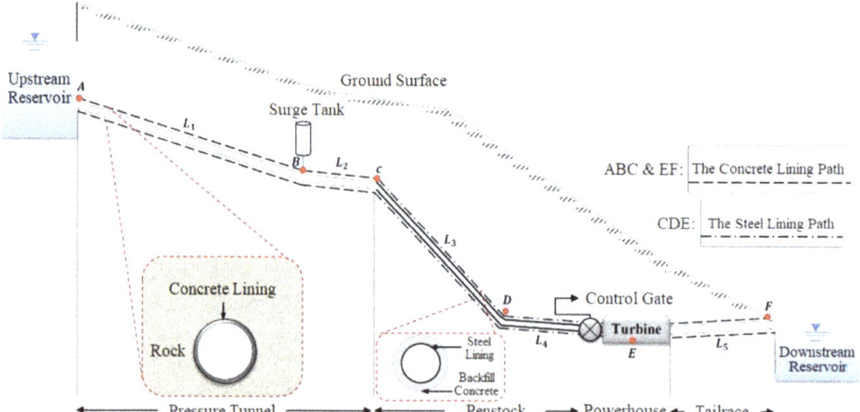


**Figure 3.** Simplified hydropower plan.

Using the Hammer software, hydraulic analyses were performed in steady-state and transient-state conditions, by defining the characteristics of the duct (water tunnel), reservoir, and gate closure schedule. The parameters used for these analyses and their values are presented in Tables 2 and 3. Water velocity was considered to be 6 m/s, according to the recommended value by the United States Bureau of Reclamation [20]. Moreover, in order to create critical conditions in the tunnel to determine its bearing capacity, it was assumed that the surge tank was disabled. The maximum flow rate was considered to be 60 m$^3$/s, and, according to the gate closure time in typical water tunnels, the gate closure times were considered to be 14 s (fast), 18 s (normal), and 26 s (slow). Furthermore, based on the duct characteristics, the pressure wave velocity according to the characteristics of the duct was determined by Equation (9), where $D$ is the inner diameter of the tunnel, $E$ is the elastic concrete modulus, and $t_c$ is the lining thickness [21].

$$ a = \sqrt{\frac{1}{\rho_w\left(\left(\frac{1}{K_w}\right) + \left(\frac{D\lambda}{Et_c}\right)\right)}}. \tag{9}$$

**Table 2.** Geometric and hydraulic characteristics of the duct.

| Inner Diameter (m) | Thickness (mm) | Hazen Williams Coefficient (-) | Initial Discharge (m$^3$/s) | Pressure Wave Velocity (m/s) | Upstream Head (m) | Initial Flow Velocity (m/s) |
|---|---|---|---|---|---|---|
| 3.6 | 200 | 100 | 60 | 911 | 80 | 6 |

It was assumed that the lining could expand in the longitudinal direction, which results in a $\lambda$ (confinement coefficient) value of 1 [21].

The location for investigating stresses in the rock was considered to be near the penstock (Node C) in the tunnel path. Finally, the pressure values at this point, obtained by the Hammer software, were input to the Abaqus FEA, under hydrodynamic pressure loadings.

**Table 3.** Node elevations and properties of node E (turbine).

| Elevations (m) | Node | Properties of Node E (Turbine) | |
|---|---|---|---|
| | | The diameter of the spherical valve (m) | 1.5 |
| A (inlet) | 2305 | Efficiency (%) | 90 |
| B | 2263.1 | Moment of inertia (N·m$^2$) | $10^7$ |
| C | 2262.62 | Speed (rpm) | 580 |
| D | 1753.5 | Specific speed | 115 |
| E | 1714.9 | Gate closure schedule curve | Variable |
| F (outlet) | 1770 | | |

## 2.3. Parametric Study in Steady-State Conditions

In this study, the bearing capacity of the pressure tunnel under a steady-state condition was investigated. The bearing capacity of a concrete-lined pressure tunnel is dependent either on the tensile strength of the surrounding rock only, or both the concrete and the rock. Generally, in pressure tunnels that are excavated in intact rock, the concrete lining does not contribute to the stability of the tunnel during excavation and only provides an appropriate bed for water to be conveyed toward the turbine.

Crack formation in the surrounding rock (hydraulic fracturing), due to high internal pressure, shows that the pressure tunnel reached its ultimate capacity; therefore, a parametric study was performed to investigate the impact of overburden height on the ultimate bearing capacity of the pressure tunnel in a steady-state condition. Figure 4 shows the effective parameters in the bearing capacity of a tunnel. $P_e(p_i)$ is the water pressure on the outer surface of the concrete lining due to seepage. According to Equation (10), this parameter is dependent on internal pressure, and it was calculated using the FEM in the present study [22].

$$P_e(p_i) = \frac{p_i}{1 + \left( \dfrac{k_r ln\left(\frac{r_e}{r_i}\right)}{k_c ln\left(\frac{R}{r_e}\right)} \right)}, \tag{10}$$

where $t$ (thickness of the lining), $k_0$, $k_r$ (hydraulic conductivity of the rock), and $\gamma_r$ (specific weight of the rock) were considered constant for all analyses, and their values are presented in Table 4. Moreover, the values for $E_r$, $C$, $\varphi$, and $h$ are presented in Table 5.

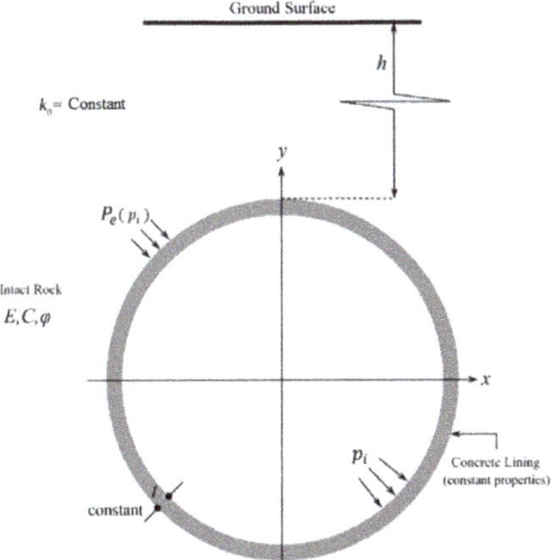


**Figure 4.** Effective parameters in the parametric study.

**Table 4.** Constant values in present study.

| Parameter | Value |
| --- | --- |
| $t$ (cm) | 20 |
| $k_0$ (-) | 0.5 |
| $k_r$ (m/s) | $1 \times 10^{-7}$ |
| $\gamma_r$ (kN) | 28 |

**Table 5.** Parameters and their values in parametric analyses [23–25].

| Parameter | Range |
| --- | --- |
| $E_r$ (GPa) | 2–10 |
| $C$ (MPa) | 0.635–1 |
| $\varphi$ (°) | 27.57–35.47 |
| $h$ (m) | 10–40 |

In each analysis, the bearing capacity of the tunnel was determined by increasing the internal pressure ($p_i$) to the point at which the first crack was formed in the surrounding rock. In this situation, the corresponding $p_i$ was considered as the failure threshold, which indicates the ultimate bearing capacity of the tunnel.

## 2.4. Verification of the FEM

Tunsakul et al. (2014) investigated fracture propagation in a rock mass surrounding a tunnel, under high internal pressure, in an experimental model [26]. In order to verify the numerical model developed in the present study, an experiment with $k_0 = 0.5$ was considered. Other assumptions based on the experimental model were as follows: no concrete lining exists ($t = 0$); the surrounding rock is impermeable ($P_e(p_i) = 0$). According to Tunsakul et al. (2014), the cracking location depends on the coefficient of lateral rock pressure ($k_0$). When $k_0 < 1$, cracking occurs in the crown of the tunnel, and, for $k_0 \geq 1$, cracks form at the sides of the tunnel. The FEM results were in agreement with the findings of Tunsakul et al. (2014), since cracks formed in the crown of the tunnel (Figure 5).

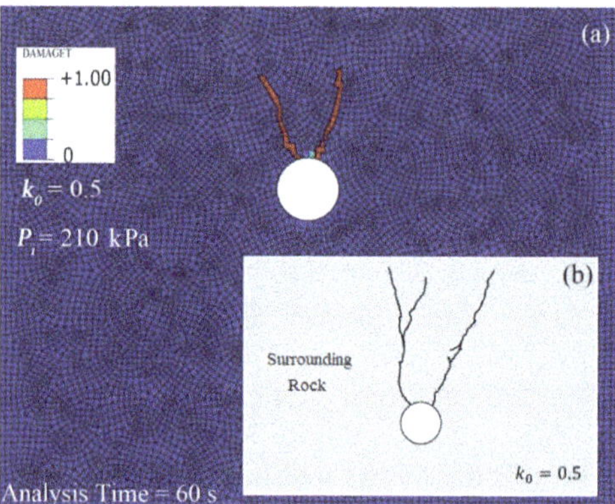


**Figure 5.** Comparing the formation and location of cracks in the numerical model (**a**) vs. the experimental model by Tunsakul et al. (2014) (**b**).

## 3. Results and Discussion

### 3.1. Changes in Pore Pressure

Since concrete is not completely impermeable, water passing through the pores penetrates into the surrounding rock. The permeability of rock varies between 1 (completely permeable) and $10^{-12}$ m/s (almost impermeable). Figure 6 shows a comparison between the pore pressures obtained from the numerical model and the value obtained from Bouvard's and Pinto's (1969) analytical solutions (Equation (10)) [22]. According to Figure 6, the pore pressure obtained using the analytical solution is greater than the value obtained from the numerical solution.

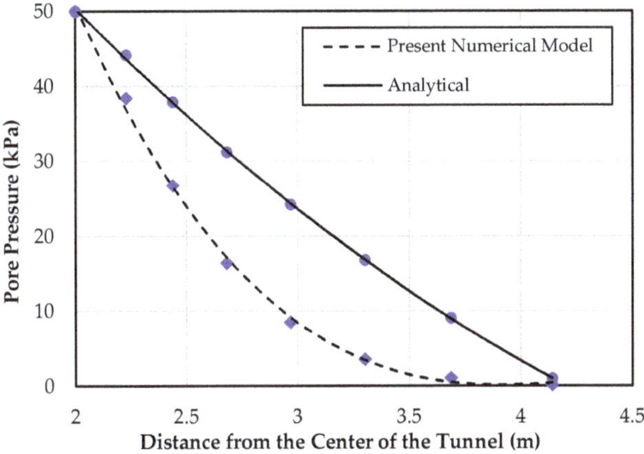


**Figure 6.** Numerical vs. analytical results for pore pressures in the rock.

The hydro-mechanical interaction is caused by jointing in rocks due to water pressure. As a result, water penetration toward the rock increases, which consequently leads to more leakage from the tunnel, and increases the pore pressure inside the rock joints. In analyses performed by Bouvard and Pinto, it was assumed that the rock was a porous medium, without effective stresses due to overburden rock. However, in the present numerical modeling, effective stresses due to overburden rock, which reduce the pore pressure in the rock, were considered as important factors in the simulation.

### 3.2. Bearing Capacity of the Tunnel in Normal Operating Conditions (Steady-State Conditions)

Based on the mechanical characteristics of the rock, and using the Mohr–Coulomb failure criterion, the failure of elements in the rock environment surrounding the tunnel was investigated. In this regard, maximum and minimum principal stresses were calculated and are presented in Figure 7. Since the problem was symmetric, half of the rock elements surrounding the tunnel were investigated.

In the Mohr–Coulomb failure criterion, the critical line divides the two-dimensional (2D) principal stresses into the safe zone (below the critical line), and the area where stresses cause failure in the element (above the critical line). Due to water pressure on the lining, and the transmission of part of this pressure to the rock, tensile stresses appear around the tunnel. Figure 7 shows that the combined stresses in elements surrounding the tunnel are located below the critical line, indicating that these elements do not fail. In Figure 7, the elements located near the critical line are considered as critical elements, which means higher water pressures will cause these elements to fail. Therefore, this internal pressure is the ultimate value that the rock can endure.

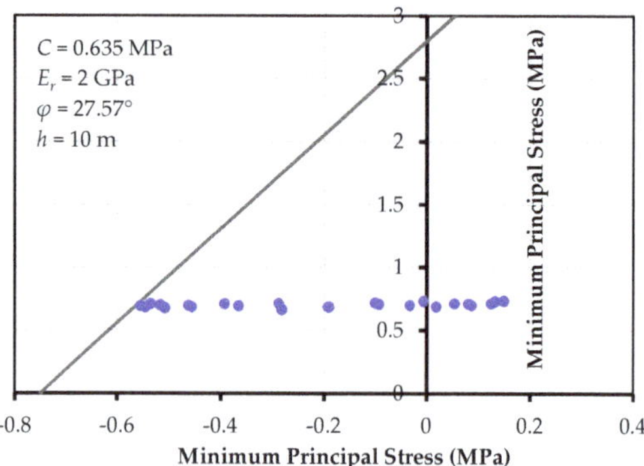

**Figure 7.** The principal stresses in elements surrounding the tunnel, and their locations relative to the critical line.

Considering the ultimate bearing capacity ($P_i{}^U$) obtained from the FEM, the Norwegian design criterion (NC) was evaluated to determine the overburden required to prevent hydraulic fracturing, due to the internal static head in the pressure tunnel (Equation (11)).

$$h > \frac{\gamma_w \cdot H}{\gamma_r \cdot \cos \beta},$$ (11)

where $h$ is the minimum overburden height to prevent hydraulic fracturing, $H$ is the static water height in the tunnel, $\beta$ is the slope of the valley that is equal to zero in this study, and $\gamma_w$ and $\gamma_r$ are the specific gravities of water and rock, respectively. The NC estimated the ultimate bearing capacity of the tunnel as 28 m water (for $\gamma_w$ = 10 kN, $\gamma_r$ = 28 kN, and $h$ = 10 m).

Figure 8 shows (a) the ultimate bearing capacity of the tunnel under a steady-state condition based on the FEM results and the NC, (b) stress distribution in the rock, and (c) vertical displacements in the rock. According to the FEM results, the internal pressure equal to 8.7 bar (87 m water) is the maximum bearing capacity of the tunnel. According to Figure 8d, at first, the concrete lining cracks, and then cracking occurs in the surrounding rock. Afterward, a parametric study was performed and the effect of overburden height on the bearing capacity of the tunnel was investigated (Figure 9 and Table 6).

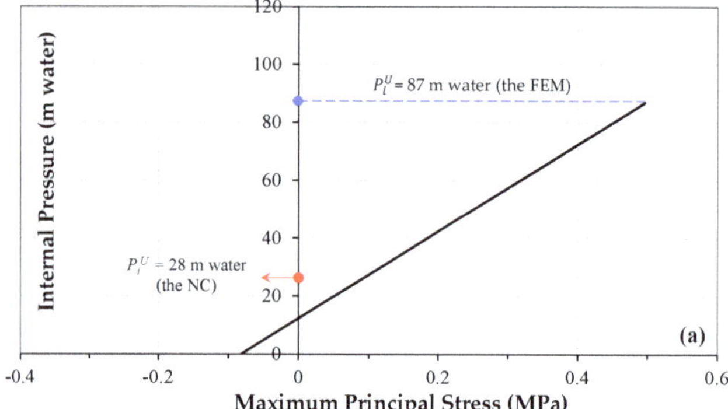

**Figure 8.** *Cont.*

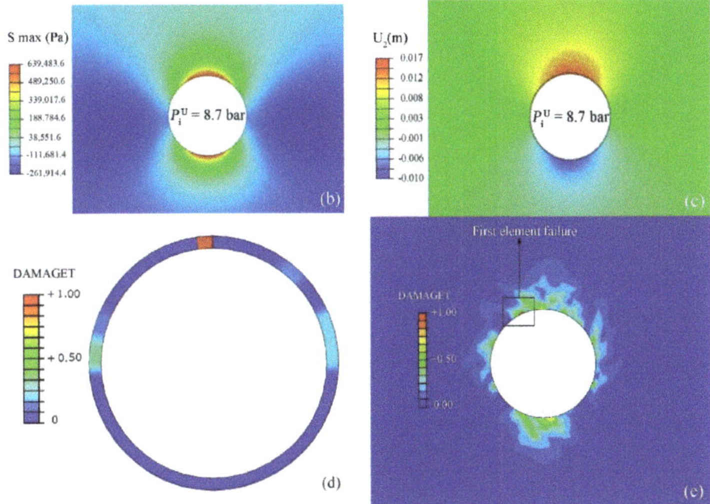



**Figure 8.** (**a**) The bearing capacity of the tunnel; (**b,c**) maximum stress and vertical displacement contours in the surrounding rock in a steady-state condition; (**d,e**) crack formation in the lining and rock.

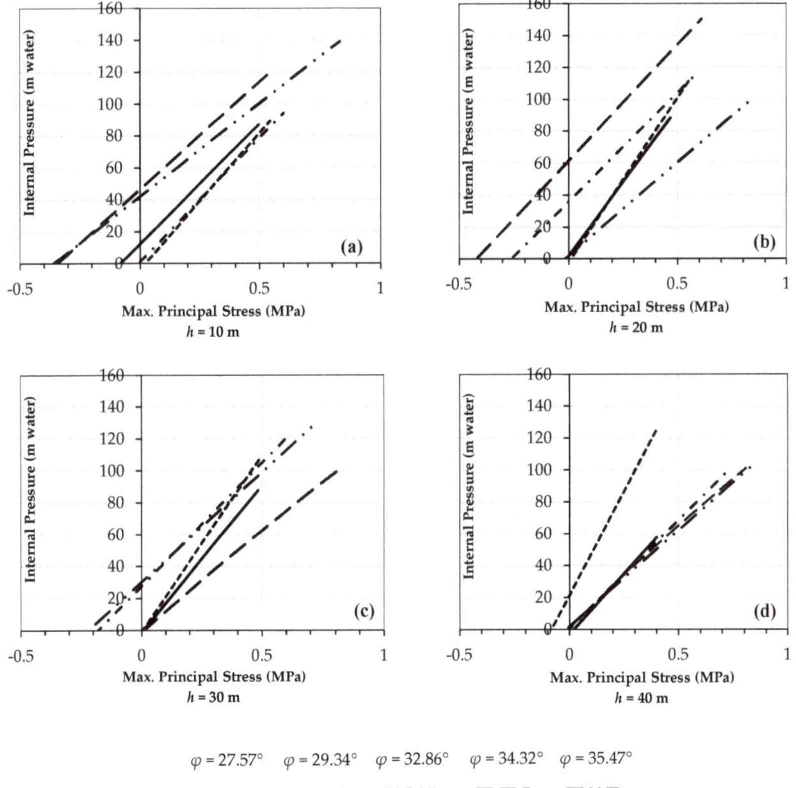



**Figure 9.** Effect of overburden height on the bearing capacity of the tunnel. (**a**) $h = 10$ m; (**b**) $h = 20$ m; (**c**) $h = 30$ m; (**d**) $h = 40$ m.

**Table 6.** The ultimate bearing capacity of the tunnel ($P_i^U$), under different scenarios. FEM—finite element model.

| Scenario No. | $E_r$ (MPa) | $h$ (m) | $C$ (MPa) | $\varphi$ (°) | $P_i^U$ (m) Based on the FEM |
|---|---|---|---|---|---|
| M-1 | 2177.94 | | 0.635 | 27.57 | 87.3373 |
| M-2 | 3076.42 | | 0.701 | 29.34 | 87.2097 |
| M-3 | 6138.26 | 10 | 0.851 | 32.86 | 93.5221 |
| M-4 | 8185.51 | | 0.929 | 34.32 | 120.7 |
| M-5 | 10,304.94 | | 1.003 | 35.47 | 138.62 |
| M-6 | 2177.94 | | 0.635 | 27.57 | 83.0878 |
| M-7 | 3076.42 | | 0.701 | 29.34 | 116.1696 |
| M-8 | 6138.26 | 20 | 0.851 | 32.86 | 118.795 |
| M-9 | 8185.51 | | 0.929 | 34.32 | 114.442 |
| M-10 | 10,304.94 | | 1.003 | 35.47 | 100.166 |
| M-11 | 3076.42 | | 0.701 | 29.34 | 89.491 |
| M-12 | 2177.94 | | 0.635 | 27.57 | 110.301 |
| M-13 | 6138.26 | 30 | 0.851 | 32.86 | 120.613 |
| M-14 | 8185.51 | | 0.929 | 34.32 | 100.115 |
| M-15 | 10,304.94 | | 1.003 | 35.47 | 119.295 |
| M-16 | 2177.94 | | 0.635 | 27.57 | 58.334 |
| M-17 | 3076.42 | | 0.701 | 29.34 | 126.234 |
| M-18 | 6138.26 | 40 | 0.851 | 32.86 | 98.698 |
| M-19 | 8185.51 | | 0.929 | 34.32 | 100.362 |
| M-20 | 10,304.94 | | 1.003 | 35.47 | 101.633 |

For validation of the results, an actual pressure tunnel project (Herlansfoss, Norway) was considered. In this project, the rock surrounding the tunnel is made of non-cracked schist with a friction angle equal to 30°. At chainages 150–200 m of the tunnel path, the tunnel is horizontal, with overburden height and an internal pressure of 45 m and 136 m water, respectively. In this range, no failure of the rock due to the internal pressure is observed [23]. The $P_i^U$ values based on the FEM results (Figure 9), the NC, and in the actual project, are equal to 138, 126, and 136 m water, respectively, indicating that the FEM results are reliable. According to Table 6 and Figure 9, changes in $P_i^U$ with increasing the overburden height are more dependent on $\varphi$ than $E$ and $C$.

In this study, the NC estimated the ultimate bearing capacity of the concrete-lined tunnel with 20 cm of the lining thickness and 3.6 m of diameter in intact rock, less than the value obtained from the numerical model in most cases (Figure 10). According to Figure 10, for all friction angles, up to 30 m of overburden height, the NC estimated $P_i^U$ less than the FEM. By increasing the overburden height to values greater than 30 m, for all quantities of friction angles, $P_i^U$ obtained from the FEM results approached the value obtained from the NC. Therefore, the NC overestimates the overburden required to prevent hydraulic fracturing in pressure tunnels located near the ground surface ($h \le 30$ m).

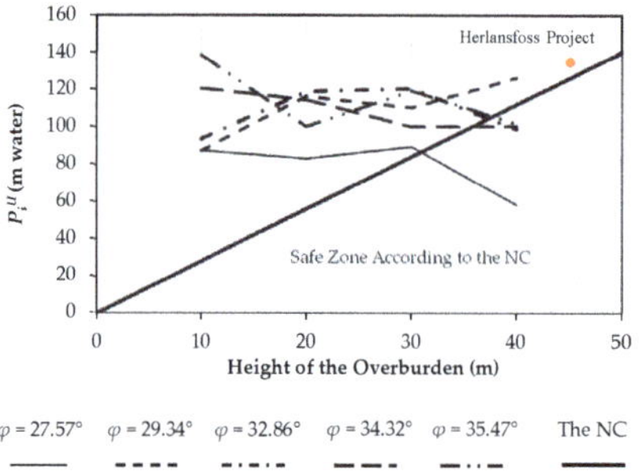


**Figure 10.** Comparing $P_i^U$ based on the finite element model (FEM) results, the Norwegian design criterion (NC), and an actual pressure tunnel project, under steady-state conditions.

### 3.3. Applying Pressure Fluctuations in Transient-State Conditions

While the gate is being closed, the velocity upstream of the gate decreases, which results in a pressure build-up in the tunnel. The maximum pressure occurs when the gate is fully closed. Figure 11 shows the pressure fluctuations upstream of the gate during the total time of hydraulic analysis. In this study, it was assumed that the gate closure was done at a constant rate. Under sudden load rejection, which happens when the gate is closed quickly at a constant rate, critical conditions occur in the hydropower tunnel. This often occurs in special situations, such as a turbine generator shut down due to an earthquake. If the gate is closed in several stages, with different closure rates, the maximum transient pressure and the consequent damage to the lining will decrease.

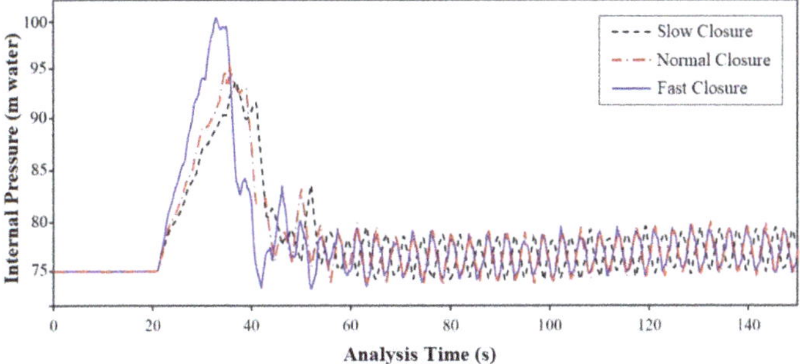


**Figure 11.** Pressure fluctuations upstream of the gate during the hydraulic analysis.

In this study, the maximum pressure that was reached after complete gate closure was considered as the hydrodynamic load in the simulations; therefore, the rock resisted maximum water pressure, and the probability of hydraulic fracturing increased. The pressure–time curves at three different gate closure times are presented in Figure 12.

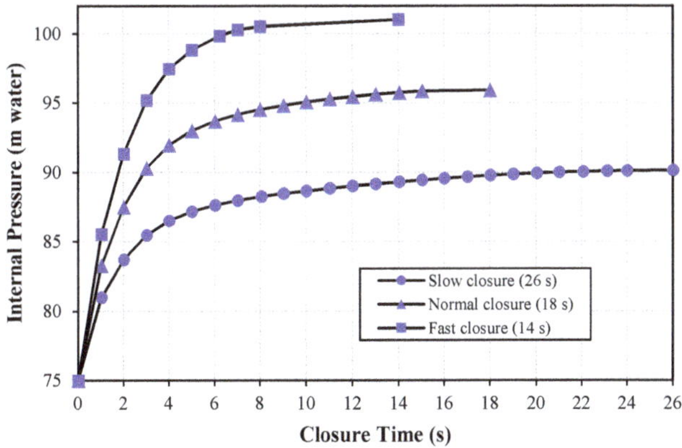


**Figure 12.** Changes in pressure upstream of the gate, in different scenarios: 14 s, 18 s, and 26 s.

Figure 12 shows that the curves have a rising trend. According to Figure 12, by increasing the gate closure time, the maximum hydrodynamic pressure was reduced; during the first few seconds, kinetic energy converts into pressure energy at a faster rate; during the last seconds of gate closure, as water upstream of the gate reaches a steady-state condition, the pressure energy decreases. Transferring these loads to the Abaqus FEA and employing the Mohr–Coulomb failure criterion, the stresses in the rock elements surrounding the tunnel were calculated for different gate closure times, and are presented in Figure 13. According to Figure 13, the stress in almost all of the elements is above the critical line, which indicates that these elements failed. When the gate closes in 14 seconds, hydrodynamic pressures impose greater forces on the surrounding rock. Increasing the gate closure time decreases the pressures resulting from the transient-state condition. Figure 13c shows that some elements are in the safe zone, which demonstrates that increasing the gate closure time decreases the possibility of hydraulic fracturing in the rock.

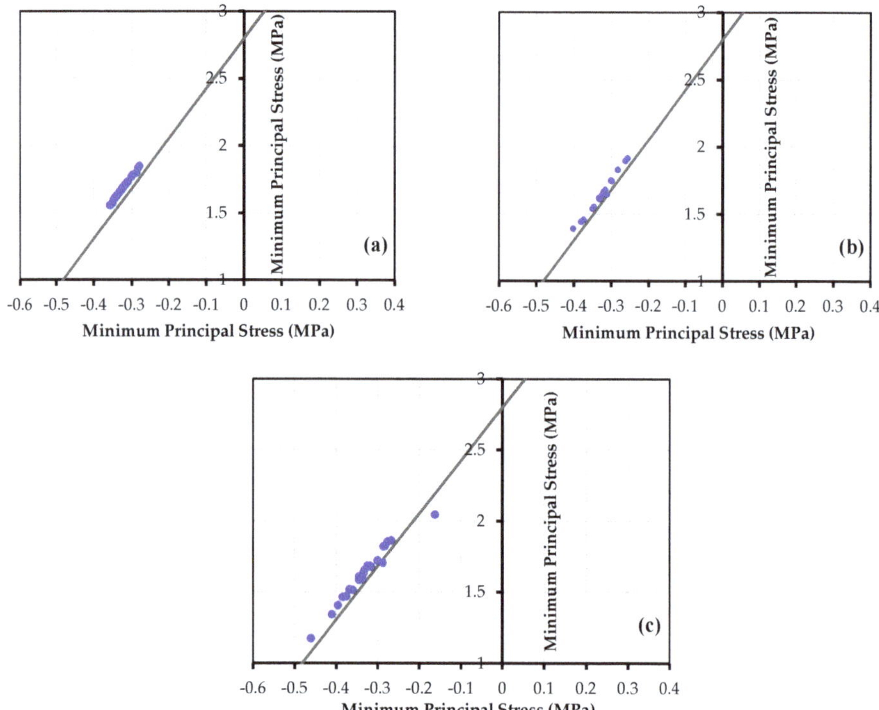

**Figure 13.** Stress in the elements by closing the gate in (**a**) 14 s, (**b**) 18 s, and (**c**) 26 s.

The Effect of Increasing the Overburden Height on Preventing Hydraulic Fracturing

In order to investigate the impact of the overburden rock on hydraulic fracturing, the critical gate closure time of 14 s was considered. In this regard, the failure of the elements in the rock around the lining was investigated by considering the overburden height as 10, 20, 40, and 70 m. Figure 14 shows the stresses in these elements, with respect to the critical line, under different overburdens.

According to Figure 14, by increasing the overburden from 0.13H to 0.26H, stresses in some elements are higher than the critical value; therefore, these elements are located above the critical line. On the other hand, in some elements, stresses are less than the critical value, and they are located below the critical line. By increasing the overburden to 0.26H, as expected, the risk of hydraulic fracturing decreased as more elements appeared below the critical line. By increasing the overburden to 0.53H, more elements appeared below the critical line, but a number of elements were still above the critical line. For the case of an overburden equal to 0.93H, all elements appeared in the safe zone below the critical line, since the elements on the sides of the tunnel section are under the influence of stress concentration. As a result, they appear in the safe zone earlier than the other elements, due to an increase in the overburden.

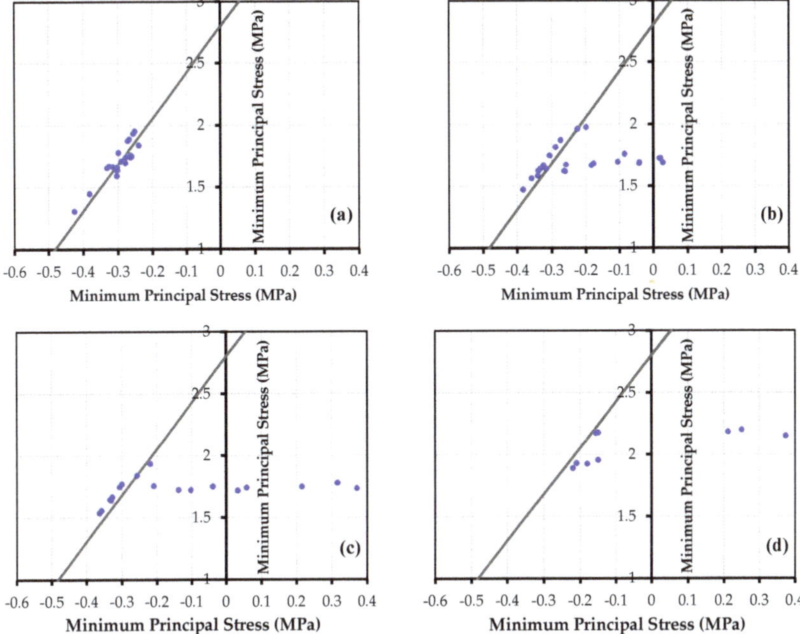

**Figure 14.** The main stresses in surrounding rock by increasing overburden: (**a**) 0.13H, (**b**) 0.26H, (**c**) 0.53H, and (**d**) 0.93H (gate closure time in all cases was 14 s).

Furthermore, Figure 15 shows the relationship between the overburden and the percentage of fractured elements around the tunnel. According to Figure 15, an increase in the overburden is not a safe solution to prevent the hydraulic fracturing phenomenon. In other words, due to rock overburden, the environment around the tunnel has in situ stresses. If the combination of stresses caused by the transient pressure inside the tunnel and the rock in situ stresses reaches a critical state, the elements will fail. The occurrence of hydraulic fracturing in one element will lead to a critical state in other elements. Therefore, it should be noted that increasing the overburden does not necessarily put all elements in the safe zone.

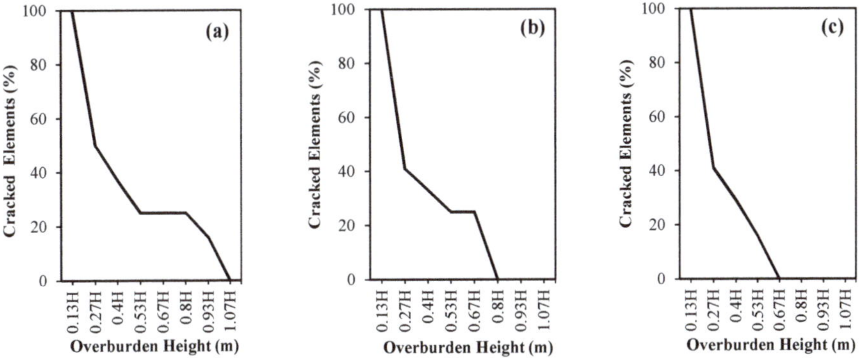

**Figure 15.** Percentage of cracked elements due to increased overburden height for gate closure times of (**a**) 14 s, (**b**) 18 s, and (**c**) 26 s.

## 4. Conclusions

Different aspects of hydraulic fracturing in pressure tunnels were investigated in previous studies, and various strategies were proposed to prevent this serious issue. One strategy is to increase the height of overburden rock. In the present study, using the FEM, and considering the fluid–structure interactions, a pressure tunnel with an inner diameter of 3.6 m was modeled in intact elasto-plastic rock, and the Mohr–Coulomb failure criterion was employed to investigate failure in the rock environment surrounding the tunnel. A parametric study was performed to evaluate the bearing capacity of the tunnel, and the effect of transient pressure on the failure of the rock elements around the tunnel was investigated. Moreover, the applicability of the Norwegian design criterion was evaluated. Hydraulic analyses in transient hydraulic states were performed using Hammer software. The output of this software, including pressure–time values, were used in Abaqus FEA in order to analyze stresses in the rock.

The results showed the following:

- Firstly, the concrete lining cracked, and then the elements in the surrounding rock failed;
- Initial cracks were formed in the crown of the tunnel;
- Increasing the overburden height had a less significant impact than a higher friction angle of the rock, on preventing the hydraulic fracturing of the rock elements;
- The Norwegian design criterion is not an appropriate measure to prevent hydraulic fracturing in pressure tunnels with a typical diameter of about 3 m in intact rock with an approximate specific weight of 28 kN, and low overburden height ($h \leq 30$ m);
- The rate of gate closure is a significant factor causing damage to the tunnel's structure;
- Increasing the gate closure time caused the maximum hydrodynamic pressure to decrease upstream of the gate, which resulted in a fewer number of failed elements in the rock around the tunnel;
- Maximum transient pressures occurred in the early stages of gate closure and, consequently, hydraulic fracturing occurred during that time;
- Analyses of the effects of different overburden heights indicated that increasing the overburden height would not always decrease the fracturing of rock elements;
- Based on the Mohr–Coulomb failure criterion results, by increasing the overburden height, fewer elements in the rock environment surrounding the tunnel failed, which can be attributed to the combination of principal stresses.

In order to prevent hydraulic fracturing in the rock, construction of a grouted zone around the pressure tunnel, along with increasing the overburden height, is recommended. The grout integrates the rock, in case any cracks exist, and improves its mechanical properties. In addition, the grouted zone results in decreased pore pressure in the rock, and reduces the possibility of hydraulic fracturing.

**Author Contributions:** Conceptualization, M.K. (Karakouzian), M.K. (Karami) and M.N.-S.; methodology, M.K. (Karakouzian), M.K. (Karami) and M.N.-S.; software, M.K. (Karami) and M.N.-S.; validation, M.K. (Karami) and M.N.-S.; writing—original draft preparation, M.N.-S. and M.K. (Karami); writing—review and editing, M.N.-S.; supervision, M.K. (Karakouzian).

**Funding:** This research received no external funding.

**Conflicts of Interest:** The authors declare no conflicts of interest.

## Appendix A

*Grid and Boundary Sensitivity Analyses*

Based on a trial-and-error procedure, a grid sensitivity analysis was performed in the model, in order to ensure convergence of the finite element equations, and the optimum pattern of meshing was determined. The mesh sizes used near the tunnel cross-section were smaller than those near the model boundaries (Figure 2).

Moreover, a sensitivity analysis was performed on the side boundaries of the model. According to Figure A1, for $x \geq 6D$ (D is the tunnel diameter) the side boundaries do not affect stresses in the rock environment surrounding the tunnel. In order to ensure preventing errors in the model calculations, this distance was considered twice the value obtained from grid sensitivity analysis (Figure 2).

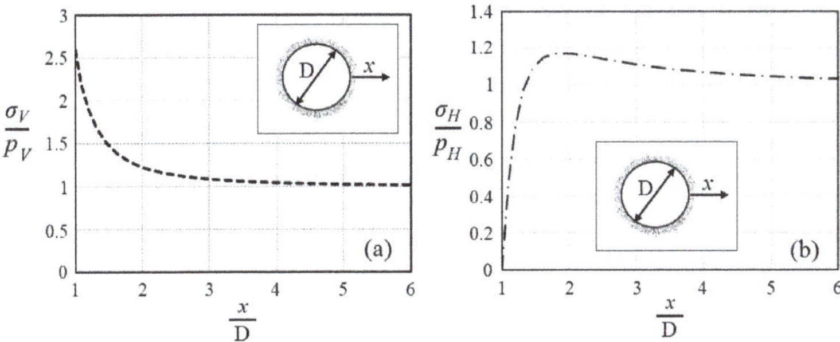

**Figure A1.** Sensitivity analysis on the model boundaries: (**a**) changes in stress in the rock environment, vertically; (**b**) changes in stress in the rock environment, horizontally; $x$ is the distance from the tunnel lining, $\sigma_H$ is the horizontal stress in the rock, $\sigma_V$ is the vertical stress in the rock, $P_H$ is the initial horizontal stress in the rock, and $P_V$ is the initial vertical stress in the rock.

## References


1. Schleiss, A.J. Design of pervious pressure tunnels. *Int. Water Power Dam Constr.* **1986**, *38*, 21–26.
2. Fernández, G.; Alvarez, T.A., Jr. Seepage-induced effective stresses and water pressures around pressure tunnels. *J. Geotech. Eng.* **1994**, *120*, 108–128. [CrossRef]
3. Bobet, A.; Nam, S.W. Stresses around pressure tunnels with semi-permeable liners. *Rock Mech. Rock Eng.* **2007**, *40*, 287–315. [CrossRef]
4. Hachem, F.E.; Schleiss, A.J. A review of wave celerity in frictionless and axisymmetrical steel-lined pressure tunnels. *J. Fluids Struct.* **2011**, *27*, 311–328. [CrossRef]
5. Olumide, B.A.; Marence, M. Finite element model for optimum design of plain concrete pressure tunnels under high internal pressure. *Int. J. Sci. Technol.* **2012**, *1*, 216–223.
6. Simanjuntak, T.D.Y.F.; Marence, M.; Mynett, A.E.; Schleiss, A.J. Pressure tunnels in non-uniform in situ stress conditions. *Tunnell. Undergr. Space Technol.* **2014**, *42*, 227–236. [CrossRef]
7. Zhou, Y.; Su, K.; Wu, H. Hydro-mechanical interaction analysis of high pressure hydraulic tunnel. *Tunnell. Undergr. Space Technol.* **2015**, *47*, 28–34. [CrossRef]
8. Pachoud, A.J.; Schleiss, A.J. Stresses and displacements in steel-lined pressure tunnels and shafts in anisotropic rock under quasi-static internal water pressure. *Rock Mech. Rock Eng.* **2016**, *49*, 1263–1287. [CrossRef]
9. Zareifard, M.R. An analytical solution for design of pressure tunnels considering seepage loads. *Appl. Math. Model.* **2018**, *62*, 62–85. [CrossRef]
10. Jia, B.; Tsau, J.S.; Barati, R. A review of the current progress of $CO_2$ injection EOR and carbon storage in shale oil reservoirs. *Fuel* **2019**, *236*, 404–427. [CrossRef]
11. Simulia. *ABAQUS Theory Manual*; Dassault Systèmes: Providence, RI, USA, 2012.
12. Bentley Systems. HAMMER V8i User's Guide. Available online: https://www.bentley.com/en/products/product-line/hydraulics-and-hydrology-software/hammer (accessed on 15 January 2019).
13. Gasch, T.; Facciolo, L.; Eriksson, D.; Rydell, C.; Malm, R. *Seismic Analyses of Nuclear Facilities With Interaction Between Structure and Water: Comparison between Methods to Account for Fluid-Structure-Interaction (FSI)*; Elforsk: Frederiksberg, Denmark, 2013.
14. Chopra, A. *Dynamic of Structures*, 4th ed.; Pearson Prentice Hall: Upper Saddle River, NJ, USA, 2012.
15. Uchida, Y.; Kurihara, N.; Rokugo, K.; Koyanagi, W. Determination of tension softening diagrams of various kinds of concrete by means of numerical analysis. *Fract. Mech. Concr. Struct.* **1995**, *1*, 17–30.
16. Lee, J.; Fenves, G.L. Plastic-damage model for cyclic loading of concrete structures. *J. Eng. Mech.* **1998**, *124*, 892–900. [CrossRef]

17. Karami, M.; Kabiri-Samani, A.; Nazari-Sharabian, M.; Karakouzian, M. Investigating the effects of transient flow in concrete-lined pressure tunnels, and developing a new analytical formula for pressure wave velocity. *Tunnell. Undergr. Space Technol.* **2019**, *91*, 102992. [CrossRef]

18. Karakouzian, M.; Karami, M.; Nazari-Sharabian, M.; Ahmad, S. Flow-induced stresses and displacements in jointed concrete pipes installed by pipe jacking method. *Fluids* **2019**, *4*, 34. [CrossRef]

19. Jaeger, J.C.; Cook, N.G.W.; Zimmerman, R.W. *Fundamentals of Rock Mechanics*, 4th ed.; Blackwell Publishing: Hoboken, NJ, USA, 2007.

20. Williams, O. *Engineering and Design—Tunnels and Shafts in Rock*; US Army Corps of Engineers: Washington, DC, USA, 1997.

21. Halliwell, A.R. Velocity of a water-hammer wave in an elastic pipe. *J. Hydraul. Div.* **1963**, *89*, 121.

22. Bouvard, M.; Pinto, N. Aménagement Capivari-Cachoeira étude du puits en charge. *La Houille Blanche* **1969**, *7*, 747–760. [CrossRef]

23. Brekke, T.L.; Ripley, B.D. Design of pressu.re tunnels and shafts. *Anal. Design Methods* **1995**, 349–369. [CrossRef]

24. Hoek, E.; Carranza-Torres, C.; Corkum, B. Hoek-Brown failure criterion—2002 edition. *Proc. NARMS-Tac* **2002**, *1*, 267–273.

25. RocLab, Rocscience Inc. Available online: https://www.rocscience.com/ (accessed on 15 January 2019).

26. Tunsakul, J.; Jongpradist, P.; Soparat, P.; Kongkitkul, W.; Nanakorn, P. Analysis of fracture propagation in a rock mass surrounding a tunnel under high internal pressure by the element-free Galerkin method. *Comp. Geotechnol.* **2014**, *55*, 78–90. [CrossRef]

 *fluids*

*Article*

# Reynolds Stress Perturbation for Epistemic Uncertainty Quantification of RANS Models Implemented in OpenFOAM

**Luis F. Cremades Rey [1], Denis F. Hinz [2,*,†] and Mahdi Abkar [1,*]**

[1]  Department of Engineering, Aarhus University, 8000 Aarhus C, Denmark; luiscremadesrey@hotmail.com
[2]  Kamstrup A/S, Industrivej 28, Stilling, 8660 Skanderborg, Denmark
*   Correspondence: dfhinz@gmail.com (D.F.H.); abkar@eng.au.dk (M.A.)
†   Current address: SGL Carbon GmbH, Werner-von-Siemens-Strasse 18, 86405 Meitingen, Germany.

Received: 3 May 2019; Accepted: 18 June 2019; Published: 22 June 2019

**Abstract:** Reynolds-averaged Navier-Stokes (RANS) models are widely used for the simulation of engineering problems. The turbulent-viscosity hypothesis is a central assumption to achieve closures in this class of models. This assumption introduces structural or so-called epistemic uncertainty. Estimating that epistemic uncertainty is a promising approach towards improving the reliability of RANS simulations. In this study, we adopt a methodology to estimate the epistemic uncertainty by perturbing the Reynolds stress tensor. We focus on the perturbation of the turbulent kinetic energy and the eigenvalues separately. We first implement this methodology in the open source package OpenFOAM. Then, we apply this framework to the backward-facing step benchmark case and compare the results with the unperturbed RANS model, available direct numerical simulation data and available experimental data. It is shown that the perturbation of both parameters successfully estimate the region bounding the most accurate results.

**Keywords:** computational fluid dynamics; RANS closures; uncertainty quantification; Reynolds stress tensor; backward-facing step; OpenFOAM

## 1. Introduction

The motion of a fluid in the turbulent regime is fully described by the Navier-Stokes equations. A numerical solution encompassing all spatial and temporal scales is referred to as direct numerical simulation (DNS). Due to the significant computational cost of DNS, approximations like large-eddy simulation (LES) and Reynolds-averaged Navier-Stokes (RANS) have been developed and are more widely used, in particular for practical engineering problems. RANS models use ensemble averages of the physical quantities, thereby decreasing the computational cost of the simulations. Due to the assumptions that are required to achieve closure, all RANS closures naturally introduce structural, or epistemic, uncertainty.

A systematic approach of the epistemic uncertainty quantification (EUQ) in RANS models, focusing on the Reynolds stress tensor, was first proposed by Emory et al. [1]. They introduced perturbation on the Reynolds stress tensor using the barycentric map proposed by Banerjee et al. [2], as a mean to visualize the degree of anisotropy. The same method was used later in a simulation by Gorle et al. [3] comparing RANS with LES results for an under-expanded jet in a supersonic cross flow. Another contribution, proposed by Gorle et al. [3] and further developed by Gorle and Iaccarino [4], was to introduce the perturbation not only in the momentum equations but also in the turbulent scalar fluxes in the scalar transport equation.

Gorle et al. [5] further developed the methodology with the goal of defining the uncertainty in the separation region. They introduced the idea of a marker, which identifies the regions of the flow where the introduction of perturbations would be useful. In this case the marker is designed to recognize regions with parallel shear flow. Emory et al. [6] recognized the importance of testing the method for various cases and successfully applied EUQ to plane channel flow, square duct flow and a shock/boundary layer interaction with flow separation. In their work, they showed that the method is model-independent.

The latest contribution to EUQ taken into consideration for this research was provided by Iaccarino et al. [7]. In their studies, they proposed a combined perturbation of different quantities in the normalized Reynolds stress tensor. They focused on the simultaneous perturbation of eigenvalues and eigenvectors, which is refered as eigenspace perturbation. With this approach, the ellipsoid describing the eigenspace not only varies in its shape but also in its orientation.

This article is organized as follows. Section 2 presents the mathematical details of the EUQ approach. This methodology is applied to a turbulent flow over a backward-facing step. Section 3 presents the behaviour of the pressure coefficient, friction coefficient, mean velocity field and Reynolds stress tensor after perturbing the eigenvalues and the turbulent kinetic energy separately. These results are compared to the DNS data [8] and the experimental results [9]. A summary and conclusions are provided in Section 4.

## 2. Methodology

### 2.1. Epistemic Uncertainty

RANS models decompose the quantities into an averaged part $\bar{u}$ and a fluctuating part $u'$ such that

$$u = \bar{u} + u'. \tag{1}$$

Substituting (1) into the Navier–Stokes equations and applying Reynolds averaging leads to the RANS equations (see e.g., [10–12])

$$\partial_t \bar{u}_i + \bar{u}_j \partial_j \bar{u}_i = \partial_i \bar{p} + \nu \partial_j \partial_j \bar{u}_i - \partial_j R_{ij} \tag{2}$$

and the Reynolds-averaged incompressibility constraint

$$\partial_i \bar{u}_i = 0, \tag{3}$$

where $t$ is time, $\bar{u}_i$ is the mean velocity in the $i-$direction, where $i = 1, 2, 3$ respectively represents the streamwise ($x$), wall-normal ($y$) and lateral ($z$) directions, $\bar{p}$ is the mean kinematic pressure, $\nu$ is the kinematic viscosity, and

$$R_{ij} = \overline{u'_i u'_j} \tag{4}$$

is the Reynolds stress tensor. The Reynolds stress tensor is an extra term that arises as a result of the Reynolds decomposition and contains the fluctuating parts of the velocity.

The closure problem in RANS amounts to finding approximations for $R_{ij}$ that do not include the fluctuating part, since that information is not available in an actual RANS simulation. Yet, any model for $R_{ij}$ will be an approximation that introduces an additional epistemic uncertainty. Most popular RANS closure models are based on the Boussinesq turbulent-viscosity hypothesis (or eddy-viscosity hypothesis) that approximates the Reynolds stress tensor as a function of the stretching tensor (mean strain-rate tensor $S_{ij} = (\partial_i \bar{u}_j + \partial_j \bar{u}_i)/2$), the turbulent kinetic energy ($k = \overline{u'_i u'_i}/2$) and the turbulent viscosity (effect of turbulent eddies on the flow $\nu_t$) such that

$$R_{ij} = \frac{2}{3} k \delta_{ij} - 2\nu_t S_{ij}. \tag{5}$$

### 2.2. Decomposition of the Reynolds Stress Tensor

The Reynolds stress tensor can be decomposed into components that establish its amplitude, shape and orientation. This is done by defining the turbulence anisotropy tensor $a_{ij}$ as

$$a_{ij} = R_{ij} - \frac{2}{3}k\delta_{ij}. \tag{6}$$

Equation (6) can be normalized by dividing by $2k$, yielding

$$b_{ij} = \frac{a_{ij}}{2k} = \frac{R_{ij}}{2k} - \frac{1}{3}\delta_{ij}. \tag{7}$$

The eigenvalues of $R_{ij}$ are non-negative due to the condition of realizability established by Speziale et al. [13] and the Cauchy–Schwartz inequality holds for the off-diagonal components. Hence, the values are constrained within the intervals

$$b_{ii} \in [-1/3, 2/3] \quad \forall i = 1, ..., 3 \tag{8}$$

and

$$b_{ij} \in [-1/2, 1/2] \quad \forall i \neq j. \tag{9}$$

In view of (7), the Reynolds stress tensor can be written as

$$R_{ij} = \frac{2}{3}k\delta_{ij} + b_{ij}2k = 2k(\frac{1}{3}\delta_{ij} + b_{ij}). \tag{10}$$

The eigendecomposition of the normalized anisotropy tensor yields (see e.g., [14])

$$b_{in}v_{nl} = v_{in}\Lambda_{nl}, \tag{11}$$

where

- $\Lambda_{nl}$ is a diagonal tensor containing the eigenvalues of the anisotropy tensor in an order such that $\lambda_1 > \lambda_2 > \lambda_3$,
- $v_{ij}$ is a tensor containing the eigenvectors of the anisotropy tensor in the same order as the eigenvalues.

Isolating $b_{ij}$ in (11) yields

$$b_{ij} = v_{in}\Lambda_{nl}v_{lj}. \tag{12}$$

Substituting (12) into (10), the Reynolds stress tensor becomes

$$R_{ij} = 2k(\frac{1}{3}\delta_{ij} + v_{in}\Lambda_{nl}v_{lj}). \tag{13}$$

### 2.3. Perturbation of the Reynolds Stress Tensor

In Section 2.2, the Reynolds stress tensor is rearranged into components that define the amplitude ($k$), shape ($\Lambda$) and orientation ($v$). Perturbation of these components yields an estimate on the epistemic uncertainty. The perturbation yields

$$R_{ij}^* = 2k^*(\frac{1}{3}\delta_{ij} + v_{in}^*\Lambda_{nl}^*v_{lj}^*), \tag{14}$$

where

- $v^* = v + \Delta v$ is the perturbation on the orientation of the Reynolds stresses,
- $\Lambda^* = \Lambda + \Delta\Lambda$ is the perturbation on the anisotropy of the Reynolds stresses,

- and $k^* = [k/n_k, n_k k]$, is the amplitude of the perturbation of the turbulent kinetic energy. It is established as a range with a minimum and a maximum $k^*$.

In this article, we focus on the perturbation of the turbulent kinetic energy $k$ and the eigenvalues $\Lambda$. The perturbation of $k$ is realized through the parameter $n_k \geq 1$ that determines the limits of perturbation. The maximum perturbation corresponds to $k^* = n_k k$ and the minimum to $k^* = k/n_k$. The perturbation of the eigenvalues is realized using the barycentric map proposed by Banerjee et al. [2] and illustrated in Figure 1.

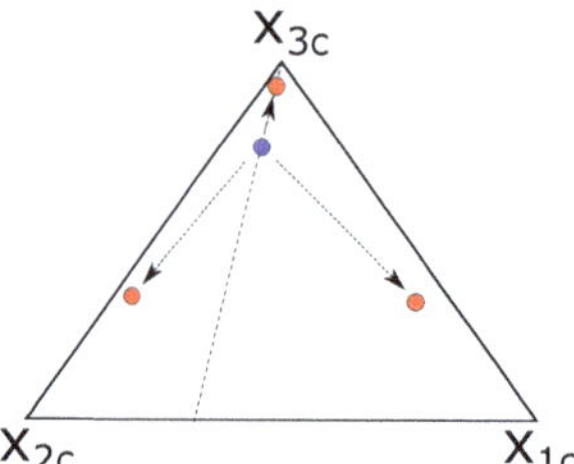


**Figure 1.** Barycentric map. The limiting states are represented in the vertices as one-component $X_{1c}$, two-component $X_{2c}$ and three-component (isotropic) $X_{3c}$ turbulence. The anisotropy is zero at $X_{3c}$ and maximum at $X_{1c}$. The dashed line denotes plane shear flow, where at least one of the eigenvalues $\lambda_l$ is zero. The arrows show perturbations toward limiting states of turbulence [6].

The barycentric map can represent any stress tensor as a function of the limiting states: one-component, two-component and three-component turbulence. The limiting states

$$
\begin{aligned}
C_{1c} &= \lambda_1 - \lambda_2, \\
C_{2c} &= 2(\lambda_2 - \lambda_3), \\
C_{3c} &= 3\lambda_3 + 1,
\end{aligned}
\tag{15}
$$

are functions of the eigenvalues associated with the Reynolds stress tensor. The limiting states are normalized such that

$$
C_{1c} + C_{2c} + C_{3c} = 1.
\tag{16}
$$

The limiting state of the tensor can be represented in a two-dimensional coordinate system with coordinates

$$
\begin{aligned}
x &= C_{1c}x_{1c} + C_{2c}x_{2c} + C_{3c}x_{3c}, \\
y &= C_{1c}y_{1c} + C_{2c}y_{2c} + C_{3c}y_{3c},
\end{aligned}
\tag{17}
$$

or

$$
\begin{aligned}
\mathbf{x} &= B\lambda_l \\
&= \mathbf{x}_{1c}(\lambda_1 - \lambda_2) + \mathbf{x}_{2c}(2\lambda_2 - 2\lambda_3) + \mathbf{x}_{3c}(3\lambda_3 + 1),
\end{aligned}
\tag{18}
$$

where $B^{-1}\mathbf{x}_{1c} = (2/3, -1/3, -1/3)^T$, $B^{-1}\mathbf{x}_{2c} = (1/6, 1/6, -1/3)^T$ and $B^{-1}\mathbf{x}_{3c} = (0,0,0)^T$.

Once the coordinates of the anisotropy tensor are located in the barycentric map, the perturbation is based on two parameters, the direction of the perturbation $x^{(t)}$ and the magnitude of perturbation $\delta_B$. The direction of the perturbation defines the chosen corner in the barycentric map and magnitude describes the distance of displacement to the chosen corner in the barycentric map in a range of $[0,1]$ as shown in Figure 1,

$$
\mathbf{x}^* = \mathbf{x} + \delta_B(\mathbf{x}^{(t)} - \mathbf{x}).
\tag{19}
$$

The perturbed eigenvalues are calculated as [6]

$$\lambda_l^* = B^{-1}\mathbf{x}^*$$

$$= (1 - \delta_B)B^{-1}\mathbf{x} + \delta_B B^{-1}\mathbf{x}^{(t)} \tag{20}$$

$$= (1 - \delta_B)\lambda_l + \delta_B B^{-1}\mathbf{x}^{(t)}.$$

Here, it is noteworthy to mention that, in the RANS simulation of two-dimensional flows using eddy-viscosity models, at least one of the eigenvalues ($\lambda_l$) of the normalized anisotropy tensor ($b_{ij}$) is zero. Hence, all the points in the flow domain will be located along the line representing the plane shear flow (see Figure 1) [6].

## 3. Results and Analysis

We applied the previously explained methodology to the backward-facing step at $Re_h = 5100$ (Re is dependent on the step height $h$, the inlet velocity $u_0$ and the kinematic viscosity $\nu$, $Re_h = u_0 h/\nu$), for which DNS [15] and experimental data [9] are available. The backward facing step offers a simple two-dimensional case, similar to the canonical channel flow but adds a certain component of complexity in the physics right after the expansion appears.

The calculation was performed using the open source software OpenFOAM. The implementation of the Reynolds stress perturbation in OpenFOAM is explained in detail in the Appendix A. The boundary and initial conditions of the case were the same as the ones used by Le and Moin [15] and Jovic and Driver [9], including the mean velocity profile imposed at the inlet by Spalart [16]. They are summarized in Table 1. The $k$-$\omega$ SST model is chosen as RANS closure. The transport equations and their initial conditions were calculated following Ferziger and Peric [17] and Versteeg and Malalasekera [18]. Figure 2 shows the dimensions of the flow domain and its corresponding areas.

**Table 1.** Description of the boundary conditions imposed in each part of the flow domain.

| Boundary | Velocity ($\bar{u}$) | Pressure ($\bar{p}$) |
|---|---|---|
| Upper Wall | No-stress wall | Zero gradient |
| Lower Wall | No-slip condition | Zero gradient |
| Inlet | Non-uniform inlet | Zero gradient |
| Outlet | Zero gradient | Uniform, $\bar{p} = 0$ |

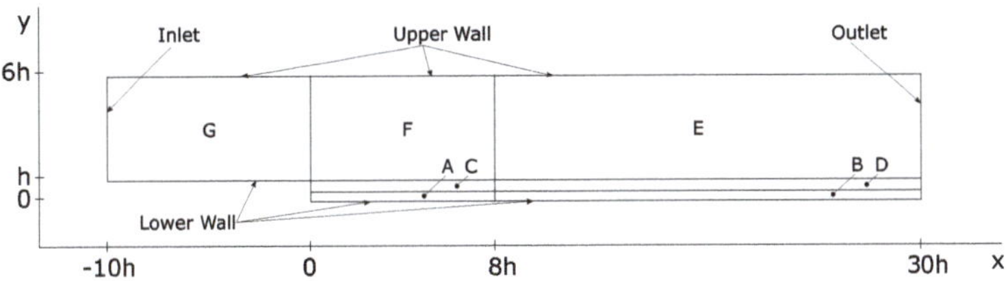


**Figure 2.** Dimensions of the flow domain as a function of the step height h. The domain is divided in seven blocks in order to design and optimize the mesh. It also shows the location of the boundaries.

The topology and mesh resolution is similar to the one described in [19]. Three different mesh resolutions were considered to assess the sensitivity of the results to the grid. A relevant parameter to be taken into account, in this case, is the dimensionless wall distance $y^+$ which is defined as

$$y^+ \equiv \frac{y}{\delta_\nu} = \frac{u_\tau y}{\nu}. \tag{21}$$

Keeping this value below one ($y^+ < 1$) allowed the simulation to capture the physics of the viscous sub-layer. Figure 3 shows the mesh topology. It can be seen that the grid gets finer when it approaches the bottom of the domain and also when it gets closer to the step, which is the most relevant part of the domain. Table 2 summarizes the topology and refinement used in each of the blocks of the domain. The mesh is designed such that the size of the first and last cells can be computed as a function of the grading and the number of cells. The grading is the ratio of the size of the last cell divided by the size of the first cell.

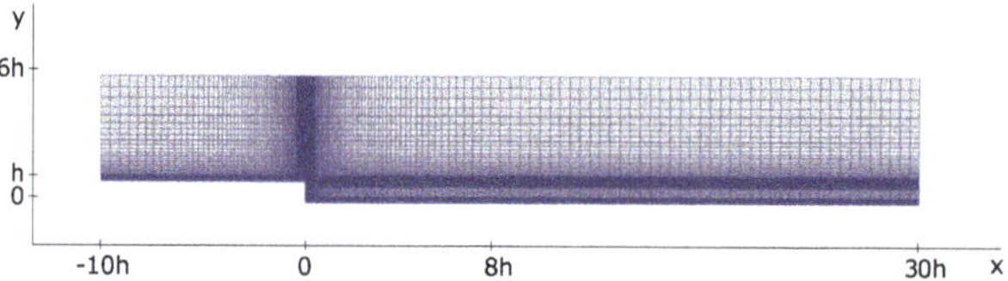

**Figure 3.** Mesh topology in the flow domain.

**Table 2.** Mesh refinement and topology for each of the three types of refinements applied. The *Cells* columns represent the amount of cells in each block in the x- and y-direction, respectively. The *Grading* columns represent the refinement in each block in the x- and y-direction, respectively.

| | Coarse Mesh | | | | Intermediate Mesh | | | | Fine Mesh | | | |
|---|---|---|---|---|---|---|---|---|---|---|---|---|
| | Cells | | Grading | | Cells | | Grading | | Cells | | Grading | |
| Block | x | y | x | y | x | y | x | y | x | y | x | y |
| A | 40 | 11 | 50 | 11.29 | 80 | 21 | 50 | 10.61 | 160 | 42 | 50 | 10.91 |
| B | 21 | 11 | 1 | 11.29 | 41 | 21 | 1 | 10.61 | 81 | 42 | 1 | 10.91 |
| C | 40 | 11 | 50 | 0.09 | 80 | 21 | 50 | 0.09 | 160 | 42 | 50 | 0.09 |
| D | 21 | 11 | 1 | 0.09 | 41 | 21 | 1 | 0.09 | 81 | 42 | 1 | 0.09 |
| E | 21 | 20 | 1 | 100 | 41 | 40 | 1 | 100 | 81 | 80 | 1 | 100 |
| F | 40 | 20 | 50 | 100 | 80 | 40 | 50 | 100 | 160 | 80 | 50 | 100 |
| G | 40 | 20 | 0.02 | 100 | 80 | 40 | 0.02 | 100 | 160 | 80 | 0.02 | 100 |

Figure 4 illustrates the zoom into the corner where the step starts and the expansion begins. It can be seen that the size of the cells does not change sharply. It changes smoothly in both directions and it gets finer towards the lower wall.

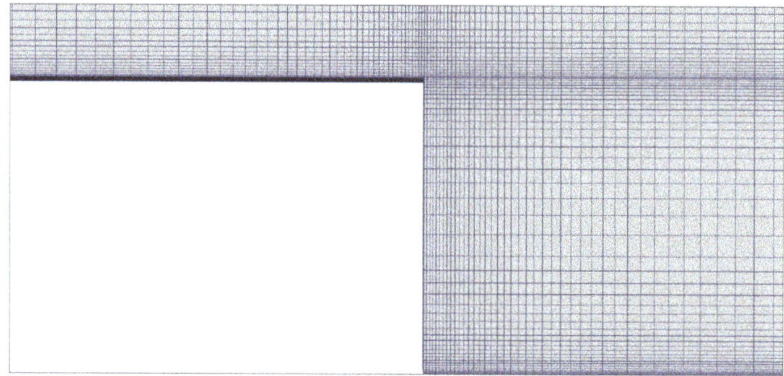

**Figure 4.** Mesh topology in the flow domain zoomed in where the step is located and expansion starts, $x/h = 0$.

In Figure 5, all three refinements were compared for the pressure coefficient, friction coefficient, mean velocity and Reynolds stresses at $x/h = 4$, $x/h = 6$ and $x/h = 10$. The comparisons between different mesh refinements show the grid convergence for the intermediate and the fine grids. The rest of the work is based on the fine mesh described in Table 2. While the focus on the present article are the structural uncertainties in the RANS closures, it is worth mentioning that numerical uncertainties can make up a significant contribution to the overall model error. For a summary of different uncertainty contributions in computational fluid dynamics refer to the ASME standard [20] or more specifically to Roache [21] and Roache [22] in which the estimation of the numerical uncertainty associated with finite grid sizes is addressed.

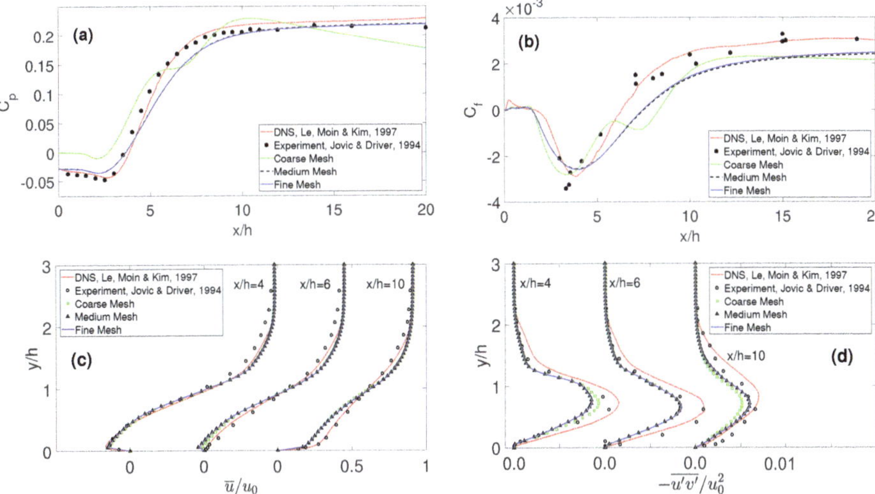

**Figure 5.** (**a**) Convergence of the pressure coefficients calculated at the bottom of the flow domain after and before the expansion, computed as $(\bar{p} - p_0)/(0.5\rho u_0^2)$. (**b**) Convergence of the friction coefficients calculated at the bottom of the flow domain after and before the expansion, computed as $\tau_w/(0.5\rho u_0^2)$. (**c**) Convergence of the mean streamwise velocity profiles calculated at $x/h = 4$, $x/h = 6$, $x/h = 10$ and normalized with respect to the inlet mean velocity ($\bar{u}/u_0$). (**d**) Convergence of the Reynolds shear stress component calculated at $x/h = 4$, $x/h = 6$, $x/h = 10$ and normalized with respect to the inlet mean velocity ($-\overline{u'v'}/u_0^2$).

The following subsections show the behaviour of: pressure coefficient, friction coefficient, mean velocity profile and Reynolds stress computed at several positions in the domain (perturbed and unperturbed). The corresponding DNS and experimental data are also included for comparison.

*3.1. Pressure Coefficient*

The pressure coefficient was computed at the bottom of the domain throughout the stream-wise direction as

$$C_p = \frac{\bar{p} - p_0}{0.5\rho u_0^2},$$

(22)

where $p_0$ is the reference wall static pressure, $p$ is the mean pressure computed at any location in the domain and $u_0$ is the upstream freestream reference velocity. We can see in Figure 6 that the results are bounded in a well defined area, thus decreasing the level of uncertainty. We can also see that a small amount of perturbation does not capture properly the physics of the problem. When the amount of perturbation increases, the bounds widen and they are able to cover better the discrepancy between the RANS model and the reference DNS and experimental data.

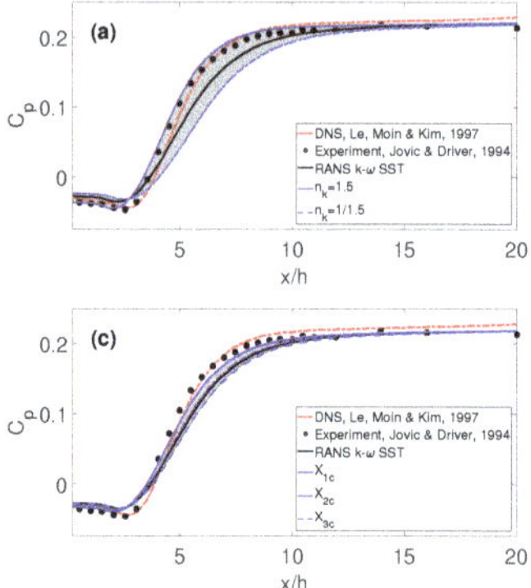
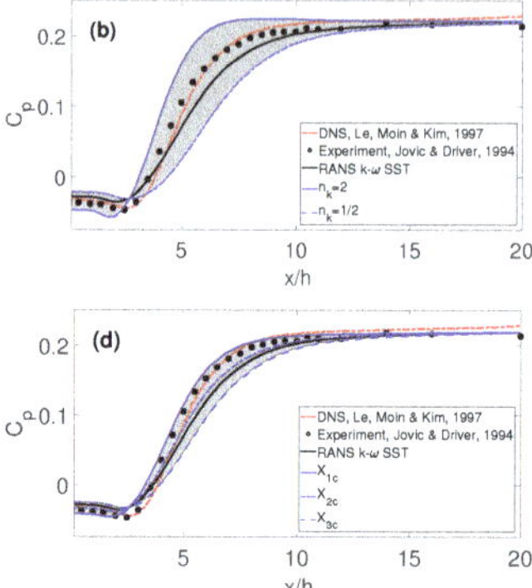

**Figure 6.** Pressure coefficients at $y/h = 0$, computed as: $(\bar{p} - p_0)/(0.5\rho u_0^2)$. (**a**) Turbulent kinetic energy perturbation $n_k = 1.5$, (**b**) turbulent kinetic energy perturbation $n_k = 2$, (**c**) eigenvalue perturbation with $\delta_B = 0.1$ and (**d**) eigenvalue perturbation with $\delta_B = 0.25$. $X_{1c}$, $X_{2c}$ and $X_{3c}$ represent perturbations toward the three corners of the barycentric map. direct numerical simulation (DNS) data [15] ($-\cdot$); experiment data [9] ($\bullet$); Reynolds-averaged Navier-Stokes (RANS) $k$-$\omega$ SST model ($-$). The uncertainty bounds are shown with gray areas.

### 3.2. Friction Coefficient

The friction coefficient

$$C_f = \frac{\tau_w}{0.5\rho u_0^2},\tag{23}$$

with $\tau_w$ the wall shear stress was computed at the bottom of the domain, throughout the stream-wise direction. The results seen in Figure 7 show similar behavior as the one seen in Figure 6 with the exception that the turbulent kinetic energy perturbation for $n_k = 2$ almost entirely defines the bounds of discrepancy between RANS and DNS and experimental results.

### 3.3. Mean Velocity in the x-Direction

The mean streamwise velocity $\bar{u}/u_0$ throughout the $y$-axis was computed at $x/h = 4$, $x/h = 6$ and $x/h = 10$, respectively. The comparison between the models with perturbed kinetic energy, perturbed eigenvalue and the non perturbed model in Figure 8 show the pattern seen in previous parameters. The higher the amount of perturbation, the wider the bounds, thus the proportion of data covered in the bounds is higher. These results show the capability of the framework to capture the uncertainty in the simulation results.

### 3.4. Reynolds Shear Stress

Reynolds shear stress was computed vertically at $x/h = 4$, $x/h = 6$ and $x/h = 10$, respectively, and normalized with respect to the upstream reference velocity. As can be seen from Figure 9, there is some discrepancies between the unperturbed RANS model and the reference data from experiments and DNS. By introducing the structural uncertainties to the RANS, the model can bound the experimental data. It is also observed that the model uncertainty is not uniformly distributed in the domain. In some regions of the flow, the model uncertainty is larger due to the shortcoming of the model assumptions in capturing the physics of the flow. Also, it can be seen that increasing the amount of perturbation in both the magnitude and shape widen the gray regions that envelop the baseline results. For instance, increasing the amount of perturbation for turbulent kinetic energy from $n_k = 1.5$ to $n_k = 2$ widens the uncertainty bound (gray region) by a factor of about 2. A relatively similar trend is observed when the injected uncertainty into the shape of the stress tensor increases from $\delta_B = 0.1$ to $\delta_B = 0.25$.

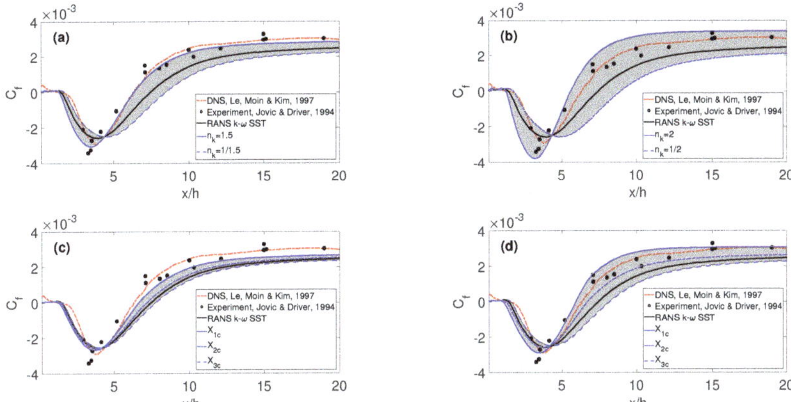

**Figure 7.** Friction coefficients at $y/h = 0$. Computed as: $\tau_w / (0.5\rho u_0^2)$. (**a**) Turbulent kinetic energy perturbation $n_k = 1.5$, (**b**) turbulent kinetic energy perturbation $n_k = 2$, (**c**) eigenvalue perturbation with $\delta_B = 0.1$ and (**d**) eigenvalue perturbation with $\delta_B = 0.25$. $X_{1c}$, $X_{2c}$ and $X_{3c}$ represent perturbations toward the three corners of the barycentric map. DNS data [15] (— ·); experiment data [9] (•); RANS $k$-$\omega$ SST model (—). The uncertainty bounds are shown with gray areas.

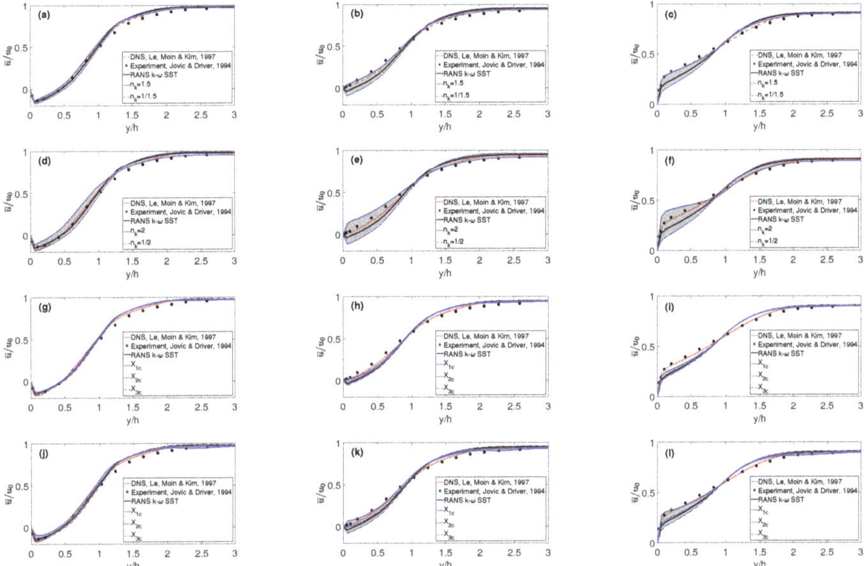

**Figure 8.** Mean streamwise velocity profiles calculated at $x/h = 4$, $x/h = 6$ and $x/h = 10$ and normalized by the inlet mean velocity ($\bar{u}/u_0$). (**a**) $k$ perturbation with $n_k = 1.5$ at $x/h = 4$, (**b**) $k$ perturbation with $n_k = 1.5$ at $x/h = 6$, (**c**) $k$ perturbation with $n_k = 1.5$ at $x/h = 10$, (**d**) $k$ perturbation with $n_k = 2$ at $x/h = 4$, (**e**) $k$ perturbation with $n_k = 2$ at $x/h = 6$, (**f**) $k$ perturbation with $n_k = 2$ at $x/h = 10$, (**g**) eigenvalue perturbation with $\delta_B = 0.1$ at $x/h = 4$, (**h**) eigenvalue perturbation with $\delta_B = 0.1$ at $x/h = 6$, (**i**) eigenvalue perturbation with $\delta_B = 0.1$ at $x/h = 10$, (**j**) eigenvalue perturbation with $\delta_B = 0.25$ at $x/h = 4$, (**k**) eigenvalue perturbation with $\delta_B = 0.25$ at $x/h = 6$ and (**l**) eigenvalue perturbation with $\delta_B = 0.25$ at $x/h = 10$. $X_{1c}$, $X_{2c}$ and $X_{3c}$ represent perturbations toward the three corners of the barycentric map. DNS data [15] (— ·); experimental data [9] (•); RANS $k$-$\omega$ SST model (—). The uncertainty bounds are shown with gray areas.

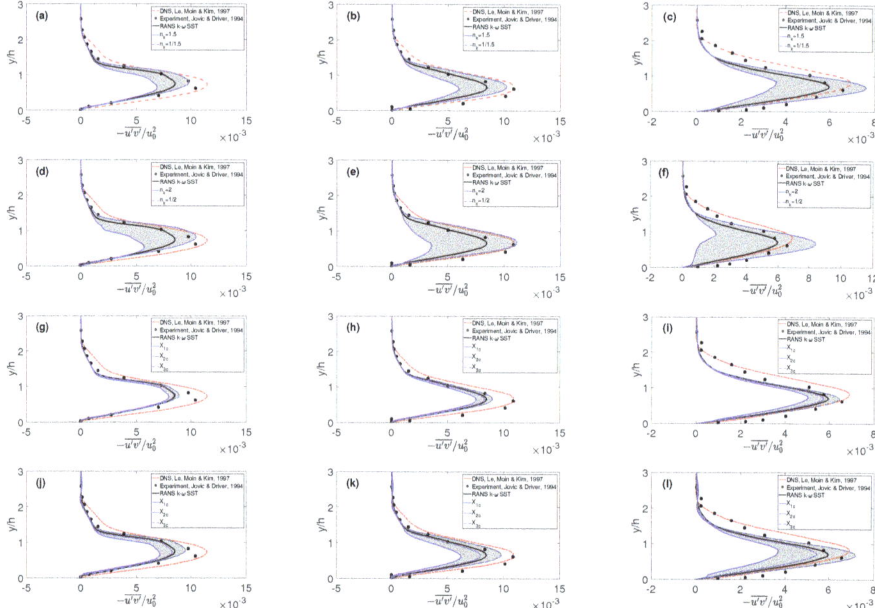

**Figure 9.** Reynolds shear stress calculated at $x/h = 4$, $x/h = 6$ and $x/h = 10$ and normalized by the inlet mean velocity $(-\overline{u'v'}/u_0^2)$. (**a**) $k$ perturbation with $n_k = 1.5$ at $x/h = 4$, (**b**) $k$ perturbation with $n_k = 1.5$ at $x/h = 6$, (**c**) $k$ perturbation with $n_k = 1.5$ at $x/h = 10$, (**d**) $k$ perturbation with $n_k = 2$ at $x/h = 4$, (**e**) $k$ perturbation with $n_k = 2$ at $x/h = 6$, (**f**) $k$ perturbation with $n_k = 2$ at $x/h = 10$, (**g**) eigenvalue perturbation with $\delta_B = 0.1$ at $x/h = 4$, (**h**) eigenvalue perturbation with $\delta_B = 0.1$ at $x/h = 6$, (**i**) eigenvalue perturbation with $\delta_B = 0.1$ at $x/h = 10$, (**j**) eigenvalue perturbation with $\delta_B = 0.25$ at $x/h = 4$, (**k**) eigenvalue perturbation with $\delta_B = 0.25$ at $x/h = 6$ and (**l**) eigenvalue perturbation with $\delta_B = 0.25$ at $x/h = 10$. $X_{1c}$, $X_{2c}$ and $X_{3c}$ represent perturbations toward the three corners of the barycentric map. DNS data [15] ($-\cdot$); experimental data [9] (●); RANS $k$-$\omega$ SST model (—). The uncertainty bounds are shown with gray areas.

## 4. Summary and Conclusions

In this article, we applied the framework, originally proposed by Emory et al. [6] for quantifying the structural uncertainties in RANS models. The quantification consisted in bounding the regions, where the most accurate results are certain to be located. This methodology focuses on the perturbation of the Reynolds stress tensor in the momentum equation. The Reynolds stress tensor is decomposed into components that represent the amplitude (turbulent kinetic energy), the shape (eigenvalues) and the orientation (eigenvectors). This investigation focused only on the influence of the amplitude and shape perturbations.

This perturbation is carried out by the turbulent kinetic energy and the eigenvalues of the Reynolds stress tensor separately and is applied to the backward-facing step case using the open-source software OpenFOAM. In this investigation, the following quantities were monitored: pressure and friction coefficients at the wall, mean streamwise velocity field and the Reynolds stress tensor. The investigation shows promising results. The results showed that for a specific amount of perturbation, the model provides a range of values where the physics of the case are well represented. An interesting feature of this research is that the perturbation of the turbulent kinetic energy shows results as good as the perturbation of the eigenvalues. The eigenvalues have well-established limits of the perturbation through the barycentric map, while the turbulent kinetic energy has not. This implies that the amount of the perturbation applied to the turbulent kinetic energy is not chosen systematically, thus it is left for future investigation.

In view of the results of this article, it is recommended for future work to focus on applying the perturbation to other parameters of the Reynolds stress such as the eigenvectors or a combination that optimizes the range that captures the most precise solution [7]. Recent studies [23,24] suggested that machine learning can be used as a powerful tool to predict discrepancies in the magnitude, anisotropy and orientation of the Reynolds stress tensor. It would be useful to continue applying this methodology to a new set of cases to monitor its behavior.

**Author Contributions:** Conceptualization, L.F.C.R., D.F.H. and M.A.; methodology, L.F.C.R., D.F.H. and M.A.; software, L.F.C.R.; validation, D.F.H., L.F.C.R. and M.A.; formal analysis, L.F.C.R., D.F.H. and M.A.; investigation, L.F.C.R., D.F.H. and M.A.; resources, D.F.H.; writing—original draft preparation, L.F.C.R.; writing—review and editing, D.F.H. and M.A.; visualization, L.F.C.R.; supervision, D.F.H. and M.A.; project administration, D.F.H. and M.A.

**Funding:** This research received no external funding.

**Acknowledgments:** This research was supported by Kamstrup A/S (Group Quality—Fluid Dynamics Center) and Aarhus University. This research is solidly based on the final Master's Thesis of Luis F. Cremades Rey [25]. The authors also acknowledge the comments by three anonymous reviewers.

**Conflicts of Interest:** The authors declare no conflict of interest. Kamstrup A/S had no role in the design of the study; in the collection, analyses, or interpretation of data; in the writing of the manuscript, or in the decision to publish the results.

## Appendix A. Implementation of the Reynolds Stress Perturbation in OpenFOAM

The application of the Reynolds stress perturbation results in the appearance of an extra term (i.e., $\Delta R_{ij}$) in the right-hand side of Equation (2) as

$$\partial_t \overline{u}_i + \overline{u}_j \partial_j \overline{u}_i = \partial_i \overline{p} + \nu \partial_j \partial_j \overline{u}_i - \partial_j (R_{ij} + \Delta R_{ij}), \tag{A1}$$

where $\Delta R_{ij}$ is the discrepancy between the perturbed Reynolds stress tensor shown in Equation (14) and the original one. This extra term is evaluated separately and added in OpenFOAM as a new line of code into the momentum equations as highlighted in Figure A1.

```
// Momentum predictor

tmp<fvVectorMatrix> UEqn(

fvm::div(phi,U)
+turbulence->ficDevReff(U)
+fvc::div(deltaR)
==
fvOptions(U)
);
```

**Figure A1.** Modification to the Momentum predictor subroutine implemented in OpenFOAM.

Figure A2 shows the mean velocity field obtained from the baseline RANS solver. To better describe the procedure of the Reynolds stress perturbation, three different locations are chosen and their positions in the barycentric map are shown in Figure A3. This figure shows how the extracted Reynolds stress tensor from the baseline case is perturbed toward different corners in the barycentric map. This procedure is done for all the points in the domain and the perturbed Reynolds stress tensor is computed following Equation (14). Next, the discrepancy between the perturbed Reynolds stress tensor and the original one from the baseline case (i.e., $\Delta R_{ij}$) is evaluated (here using a separate Matlab code) and added to the right-hand side of the momentum equations in OpenFOAM. The simulations are run again until the results are converged. The corresponding flowchart is also plotted in Figure A4.

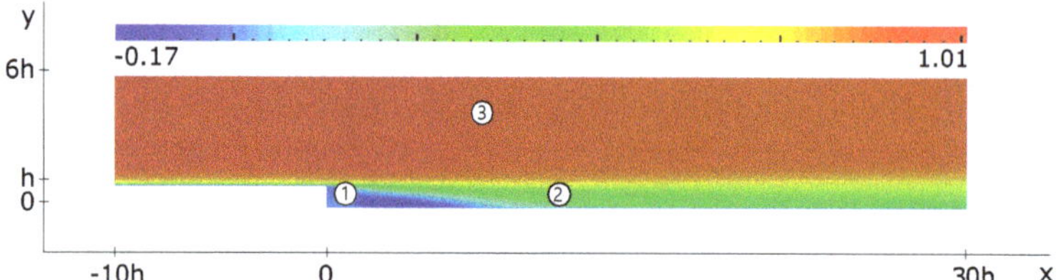


**Figure A2.** Mean velocity field in the *x*-direction, normalized with respect to the inlet velocity ($\bar{u}/u_0$) obtained from the baseline case. Three different locations have been chosen in the flow domain to study their positions in the barycentric map. The points 1, 2 and 3 are respectively located at ($h$, $0.7h$), ($10h$, $0.7h$) and ($8h$, $5h$).

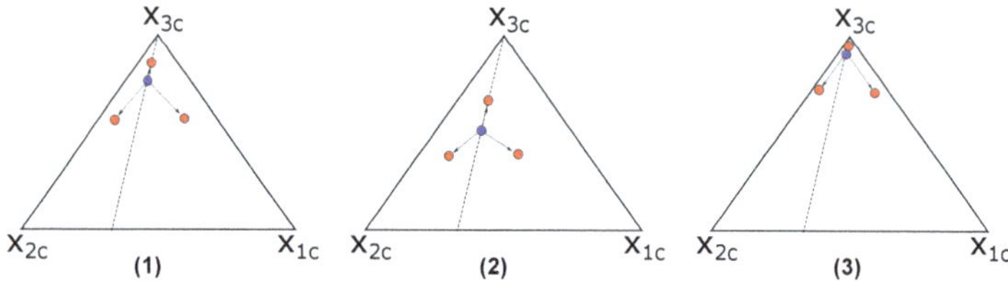


**Figure A3.** Locations of the chosen points shown in Figure A2 in the barycentric map. The points were perturbed towards the three limiting states and the amount of perturbation is $\delta_B = 0.25$.

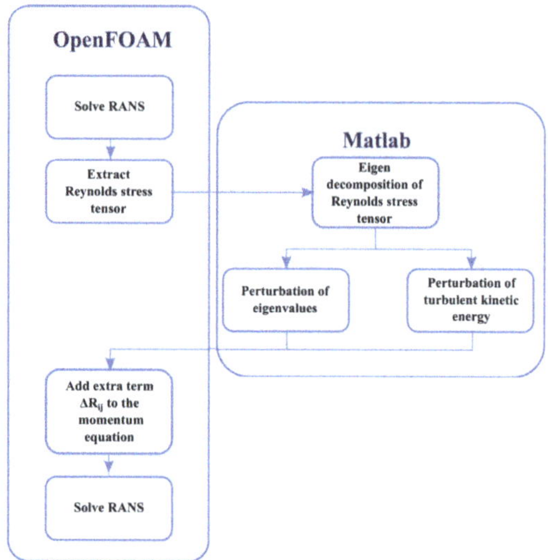


**Figure A4.** A flowchart describing the implementation of Reynolds stress perturbation in OpenFOAM.

## References


1. Emory, J.; Pecnik, R.; Iaccarino, G. Modeling Structural Uncertainties in Reynolds-Averaged Computations of Shock/Boundary Layer Interactions. In Proceedings of the AIAA Aerospace Sciences Meeting, Orlando, FL, USA, 4–7 January 2011; p. 0479.
2. Banerjee, S.; Krahl, R.; Zenger, C.H. Presentation of anisotropy properties of turbulence, invariants versus eigenvalue approaches. *J. Fluid Mech.* **2007**, *8*, N32. [CrossRef]

3.  Gorle, C.; Emory, J.; Iaccarino, G. Epistemic uncertainty quantification of RANS modeling for an underexpanded jet in a supersonic cross flow. In *Center for Turbulence Research Annual Research Briefs*; Center for Turbulence Research, Stanford University: Stanford, CA, USA, 2011; pp. 147–159.

4.  Gorle, C.; Iaccarino, G. A framework for epistemic uncertainty quantification of turbulent scalar flux models for Reynolds-averaged Navier-Stokes simulations. *Phys. Fluids* **2013**, *25*, 055105. [CrossRef]

5.  Gorle, C.; Emory, J.; Larsson, J.; Iaccarino, G. Epistemic uncertainty quantification for RANS modeling of the flow over a wavy wall. In *Center for Turbulence Research Annual Research Briefs*; Center for Turbulence Research, Stanford University: Stanford, CA, USA, 2012; pp. 81–91.

6.  Emory, J.; Larsson, J.; Iaccarino, G. Modeling of structural uncertainties in Reynolds-averaged Navier-Stokes closures. *Phys. Fluids* **2013**, *25*, 110822. [CrossRef]

7.  Iaccarino, G.; Mishra, A.A.; Ghili, S. Eigenspace perturbations for uncertainty estimation of single point turbulence closures. *Phys. Rev. Fluids* **2017**, *2*, 024605. [CrossRef]

8.  Le, H.; Moin, P.; Kim, J. Direct numerical simulation of turbulent flow over a backward-facing step. *J. Fluid Mech.* **1997**, *330*, 349–374. [CrossRef]

9.  Jovic, S.; Driver, D.M. Backward-facing step measurements at low Reynolds number, $Re_h$ = 5000. In *NASA Technical Memorandum*; Technical Report; National Aeronautics and Space Administration (NASA): Washington, DC, USA, 1994; p. 108807.

10. Patankar, S.V. *Numerical Heat Transfer and Fluid Flow*, 1st ed.; CRC Press: Boca Raton, FL, USA, 1980.

11. Pope, S.B. *Turbulent Flows*, 1st ed.; Cambridge University Press: Cambridge, UK, 2000.

12. Wilcox, D.C. *Turbulence Modeling for CFD*, 1st ed.; DCW Industries: La Canada, CA, USA, 1993.

13. Speziale, C.; Abid, R.; Durbin, P. On the realizability of Reynolds stress turbulence closures. *J. Sci. Comput.* **1994**, *9*, 369–403. [CrossRef]

14. Kreyszig, E. *Advanced Engineering Mathematics*, 10th ed.; Wiley: Hoboken, NJ, USA, 2001.

15. Le, H.; Moin, P. Direct numerical simulation of turbulent flow over a backward-facing' step. In *Center for Turbulence Research Annual Research Briefs*; Center for Turbulence Research, Stanford University: Stanford, CA, USA, 1992; pp. 161–173.

16. Spalart, P.R. Direct simulation of a turbulent boundary layer up to $Re_\Theta$ = 1410. In *NASA Technical Memorandum*; National Aeronautics and Space Administration (NASA): Washington, DC, USA, 1986; p. 89407.

17. Ferziger, J.H.; Peric, M. *Computational Methods for Fluid Dynamics*, 3rd ed.; Springer: Berlin/Heidelberg, Germany, 2001.

18. Versteeg, H.K.; Malalasekera, W. *An Introduction to Computational Fluid Dynamics: The Finite Volume Method*, 2nd ed.; Pearson Prentice Hall: Upper Saddle River, NJ, USA, 2007.

19. Rumsey, C. Turbulence Modeling Resource. Available online: https://turbmodels.larc.nasa.gov (accessed on 3 May 2019).

20. The American Society of Mechanical Engineers (ASME). *Standard for Verification and Validation in Computational Fluid Dynamics and Heat Transfer*; ASME: New York, NY, USA; 2009.

21. Roache, P.J. Perspective: A method for uniform reporting of grid refinement studies. *J. Fluids Eng.* **1994**, *116*, 405–413. [CrossRef]

22. Roache, P.J. Quantification of uncertainty in computational fluid dynamics. *Annu. Rev. Fluid Mech.* **1997**, *29*, 123–160. [CrossRef]

23. Wu, J.L.; Xiao, H.; Paterson, E. Physics-informed machine learning approach for augmenting turbulence models: A comprehensive framework. *Phys. Rev. Fluids* **2018**, *3*, 074602. [CrossRef]

24. Duraisamy, K.; Iaccarino, G.; Xiao, H. Turbulence modeling in the age of data. *Annu. Rev. Fluid Mech.* **2019**, *51*, 357–377. [CrossRef]

25. Cremades Rey, L.F. Epistemic Uncertainty Quantification of RANS Models. Master's Thesis, Aarhus University, Aarhus, Denmark, 2018.

 *fluids*

*Article*

# Shock Capturing in Large Eddy Simulations by Adaptive Filtering

**Sumit Kumar Patel * and Joseph Mathew ***

Department of Aerospace Engineering, Indian Institute of Science, Bangalore 560012, India
* Correspondence: patelsumitkumar@gmail.com (S.K.P.); joseph@iisc.ac.in (J.M.); Tel.: +91-80-2293-3027 (J.M.)

Received: 15 June 2019; Accepted: 10 July 2019; Published: 15 July 2019

**Abstract:** A method for shock capturing by adaptive filtering for use with high-resolution, high-order schemes for Large Eddy Simulations (LES) is presented. The LES method used in all the examples here employs the Explicit Filtering approach and the spatial derivatives are obtained with sixth-order, compact, finite differences. The adaptation is to drop the order of the explicit filter to two at gridpoints where a shock is detected, and to then increase the order from 2 to 10 in steps at successive gridpoints away from the shock. The method is found to be effective in a series of tests of common inviscid 1D and 2D problems of shock propagation and propagation of waves through shocks. As a prelude to LES, the 3D Taylor–Green problem for the inviscid and a finite viscosity case were simulated. An assessment of the overall performance of the method for LES was carried out by simulating an underexpanded round jet at a Reynolds number of 6.09 million, based in centerline velocity and diameter at nozzle exit plane. Very close quantitative agreement was found for the development of centerline mean pressure when compared to experiment. Simulations on several increasingly finer grids showed a monotonic extension of the computed part of the inertial range, with little change to low frequency content. Amplitudes and locations of large changes in pressure through several cells were captured accurately. A similar performance was observed for LES of an impinging jet containing normal and curved shocks.

**Keywords:** large eddy simulations (LES); shock capturing; adaptive filter; explicit filtering; jet

---

## 1. Introduction

Large eddy simulation (LES) has now become a widely-used technique. The greater expense in computing resources and time is justified when it provides accurate solutions, especially when the more common RANS (Reynolds-averaged Navier–Stokes) methods yield qualitatively incorrect solutions. From this general experience, practitioners have come to rely on their own set of procedures—combinations of numerical method and sub-grid-scale (SGS) model—that they have found to be reliable and accurate. Thus, several models co-exist, including the explicit filtering method. This paper concerns a proposal to extend explicit filtering to compressible flows with shocks by filter adaptation. Our results, from a sequence of elementary problems leading to a general turbulent flow of an underexpanded jet, reveals its excellent potential.

A recent review of methods for direct numerical simulation (DNS) and LES of compressible flows with shocks was provided by Pirozzoli [1]. There have been many efforts to devise methods that could handle shocks in inviscid flows—to incorporate procedures or terms to obtain correctly the jumps across shocks, and suppress oscillations at these jumps to avoid instability. The successes have carried over to RANS computations as well. However, specific assessments of shock-capturing methods for DNS/LES are far fewer, even though methods found to be successful for inviscid flows have been employed for DNS/LES. Pirozzoli [1] noted that such methods "exhibit excessive numerical viscosity" when combined with high-order methods designed for DNS/LES of smooth flows, and suggests that an optimal hybrid is needed for DNS/LES of flows with shocks. Typical hybrid schemes switch to a shock-capturing scheme where indicated by a shock sensor, and over a small set of neighboring gridpoints. Some examples are studies that employed a sixth-order compact scheme switching to sixth-order essentially non-oscillatory (ENO) near the shock [2], compact upwind with fifth-order ENO [3], and a fifth-order compact upwind with seventh-order weighted essentially non-oscillatory (WENO) [4]. An alternative to discontinuous switching is to apply the same discretization scheme with a continuously variable coefficient that adds artificial dissipation near shocks and nearly vanishes in smooth regions [5]. Cook and Cabot [6,7] proposed an artificial stress tensor with both shear and bulk viscosity. The generalization is to use artificial thermal conductivity and species diffusivity as well [8]. It was observed that artificial bulk viscosity provided shock capturing without affecting the vorticity field, but artificial shear viscosity damped vortical fluctuations. These artificial properties were scaled with high-order derivatives of the flow field and are therefore continuous, but rapidly changing, functions. Artificial diffusivity may also be suitable for dealing with other types of large derivatives (e.g., mixing of fluids with large density differences) [9].

A recent examination of these several approaches by Johnsen et al. [10] did not find any of the methods to be completely satisfactory: WENO provides sharp shocks but overwhelms physical dissipation; artificial viscosity works well when shocks are not too close to each other, but can be excessive when there are multiple shocks in the turbulent flow; and all methods exhibit post-shock oscillations. It would thus appear that there is still a need to study variants and other strategies that may yield reliable and accurate solutions. In this paper, we present studies with adaptive filtering. It is a natural extension for the explicit filtering approach of Mathew et al. [11] because spatial, low-pass filtering of transported fields after every time-step is an indispensable part of this LES method. The extension is to change the filter-order alone in the vicinity of shocks. An advantage is that there is no discontinuity due to switching numerical schemes since the same scheme is used. Visbal and Gaitonde [12] examined this strategy. They compared results of two proposals against a standard. In one, a 10th-order filter was applied at gridpoints away from shocks, while at the shock and at adjacent (buffer) points the filter formula was dropped to second order, but with a very high filter cutoff. At successive gridpoints away from the shock, filter order increased in steps to 10th order. In the second method, the Roe scheme was applied at cells spanning the shock and adjacent ones. Test cases included 1D and 2D inviscid flows and shock reflection from a laminar boundary layer. They found that their adaptive filter method provided a significant improvement—oscillation-free solution—over their baseline filter-stabilized compact scheme, but shocks were smeared over a few cells; the hybrid compact-Roe scheme gave sharper shocks.

To the best of our knowledge, no further studies or applications of this technique have appeared. However, the method is appealing for its simplicity and minimal overhead, thus seemed worth re-visiting. Although the study reported here also employed filter-order adaptation, the adaptation stencil is different. Here, as explained in Section 2.2, the lowest order filter was applied at only one adjacent buffer point; adaptation in a direction was selected only when the angle between the coordinate direction and the normal to the shock was not too large. Results from tests with usual problems for shock capturing, and LES of two turbulent jets with multiple shock cells show much promise. A possible criticism of this method is that the filtering seems to be ad hoc—none of filter type, shape, or filter cut-off was obtained by some kind of optimization. We argue below that optimization to arrive at the best filter is possible, but perhaps not essential.

Filter adaptation was tested in Bogey and Bailly [13] as well. Near shocks, they applied an optimized, explicit second-order filter whose filter response function lies between those of standard second- and fourth-order explicit filters. They varied the extent of the region where the lower-order filter was applied. Test cases of 1D and 2D inviscid flows were presented.

In the following, first, we discuss the basic numerical method (sixth-order, Hixon–Turkel split compact scheme for spatial derivatives, and fourth-order Runge–Kutta for time-stepping) and the proposal for filter adaptation. Then, results of a sequence of 1D and 2D inviscid test cases are presented, followed by those for the Taylor–Green problem, and LES of underexpanded, free and impinging, round jets. The five inviscid test cases reveal accurate evolution of the flow and crisp shock-capturing. Quantitative errors are at least as small as other recent strategies. The Taylor–Green problem is included to record the correctness of the code for viscous flows that develop a wide range of scales, as well as the effectiveness as LES for the inviscid case. With these elements of flows completed, we turned to LES of supersonic jets at Reynolds numbers of $O(10^6)$ (based on jet diameter and jet velocity at the nozzle exit). These are representative of applications where LES is needed since DNS is prohibitive. With the free underexpanded jet, which develops a series of shock cells, we observed close quantitative agreement with measured centerline pressure on a grid of about nine million points. By simulating on successively larger grids, up to 200 million points, we observed the extension of the computed part of the inertial range, with little change to low frequency content. The present proposal for shock capturing by adaptive filtering showed no discernible adverse effects on the development over several shock cells. The second LES of an impinging jet provides further support for the method as a suitable approach for applications.

## 2. Numerical Method

The governing equations for compressible flow were cast in the strong conservation form

$$\frac{\partial \hat{U}}{\partial t} + \frac{\partial \hat{F}}{\partial \xi} + \frac{\partial \hat{G}}{\partial \eta} + \frac{\partial \hat{H}}{\partial \zeta} = 0, \tag{1}$$

in a curvilinear coordinate system $(\xi, \eta, \zeta)$. Here, $\hat{U} = U/J$, and $U = (\rho, \rho u, \rho v, \rho w, E)$ is the vector of conserved variables with density $\rho$, Cartesian velocity components $(u, v, w)$, and energy $E = p/(\gamma-1) + \rho(u^2 + v^2 + w^2)/2$; $p$ is the pressure; and $\gamma$ is the specific heat ratio. $J = \partial(\xi, \eta, \zeta)/\partial(x, y, z)$ is the Jacobian of the transformation between Cartesian and curvilinear coordinate systems. $\hat{F}, \hat{G}$, and $\hat{H}$ include inviscid and viscous fluxes (see, for example, Tannehill et al. [14]). The fluid is Newtonian, with viscosity as per Sutherland's law, Fourier's law for conduction applies, the Prandtl number is 0.7, $\gamma = 1.4$, and specific heat $c_p = 1.005$.

### LES Model

The explicit filtering approach of Mathew et al. [11] was used for the LES presented here. The essential requirements of this approach are that the numerical method used should consist of high-resolution spatial approximations, and that a spatial, high-resolution, low-pass filter be applied to transported fields after every time step. The high-resolution numerical method should ensure that the evolution of a range of low wavenumber content is obtained accurately, while the high-resolution filter removes only a small range of the smallest computed scales. Examples of high resolution numerical methods are those that use spectral, implicit difference, or high-order explicit difference methods for derivative evaluations. Typically, implicit difference formulas, also known as compact difference formulas, have better resolution properties compared to explicit differences of the same order [15]. If explicit formulas are used instead of compact differences, they should be of higher order, and would require large stencils [16]. Here, a procedure that is approximately equivalent to a sixth-order compact scheme was used for first derivatives. Its resolution characteristics can be noted from its modified wavenumber curve (6/2 curve) plotted in Figure 1a. Filter response functions of several implicit filters of different orders and values of filter parameter are shown in Figure 3a. Either high values of the filter parameter, or high order, results in a flat response (negligible filtering) over a range of low wavenumbers. The arguments for the correctness of this approach have been given before [11,17]. LES of many different kinds of flows, by at least two other groups [18,19] who applied essentially this approach, are available. Although originally devised for high accuracy by using compact schemes [18], or unusually high-order explicit differences [16], with high wavenumber filtering for stability, the authors then demonstrated that these methods were suitable for LES. The performance of computations with our method for incompressible, variable density and compressible flows has also been documented [20–22].

## 2.1. Spatial Discretization

As noted above, an essential requirement of the LES approach adopted here is the use of high-resolution numerical schemes. While resolution quality of difference formulas increases with order, implicit difference formulas have smaller resolution error for a given order. Here, we use an extension to sixth order of the splitting proposed by Hixon and Turkel [23]. A sixth-order compact difference formula is

$$\frac{1}{5}D_{i+1} + \frac{3}{5}D_i + \frac{1}{5}D_{i-1} = \frac{7}{15}\frac{f_{i+1} - f_{i-1}}{\Delta x} + \frac{1}{60}\frac{f_{i+2} - f_{i-2}}{\Delta x}, \tag{2}$$

where $D_i$ is the first derivative of the function $f_i$. A MacCormack-like splitting

$$D_i = \frac{D_i^F + D_i^B}{2}, \tag{3}$$

leads to the two relations

$$(1-a)D_i^F + aD_{i+1}^F = \frac{1}{\Delta x}\left(-mf_{i-1} + (l+m)f_i - lf_{i+1}\right), \tag{4}$$

$$aD_{i-1}^B + (1-a)D_i^B = \frac{1}{\Delta x}\left(lf_{i-1} - (l+m)f_i + mf_{i+1}\right). \tag{5}$$

When

$$a = \frac{1}{2} - \frac{\sqrt{5}}{10}, m = \frac{1}{3(5-\sqrt{5})}, l = \frac{3\sqrt{5}-14}{3(5-\sqrt{5})},$$

$D_i^F = f_x + O(\Delta x)$, $D_i^B = f_x + O(\Delta x)$, but $(D_i^F + D_i^B)/2 = f_x + O(\Delta x)^6$. In their terminology, this would be a 6/2 scheme [23]. Equation (2) leads to a tridiagonal system, whereas Equations (4) and (5) form bidiagonal systems. Instead of finding derivatives as the mean of $D^F$ and $D^B$, we use either alone as the estimate of the derivative in successive evaluations. Tests showed that the higher-order truncation error was obtained when the time discretization error was small enough. The great advantage in splitting is that the number of operations per derivative, per time step is roughly halved, which is a tremendous saving for an LES or DNS.

Figure 1 shows the modified wavenumber $\tilde{k}$ of the 6/2 scheme. The fourth-order schemes 4/2 and 4/4 that Hixon and Turkel [23] studied are also shown. The 4/2 scheme has a first-order truncation error for each split part, while 4/4 was designed for this error to be third order. To demonstrate that higher-order behavior can be realized, we show the fall in error in solutions of the linear wave equation with unit phase speed after unit time over the region [0,5]. The initial condition was the Gaussian $u = \exp[-10(x-1)^2]$. Solutions were obtained on grids with spacing 1/5, 1/10, 1/20 and 1/40, and fourth-order Runge–Kutta time integration with CFL numbers 0.01 and 0.001. The fall in rms of the difference between numerical solutions and the exact solution is shown in Figure 1b. Solutions with the fourth-order (4/2) and sixth-order (6/2) schemes and lines with Slopes 4 and 6 are shown for reference. Clearly, the order behavior of the equivalent symmetric compact scheme has been recovered. Due to time-stepping error, the decay rate reduces at the higher CFL for the sixth-order scheme.

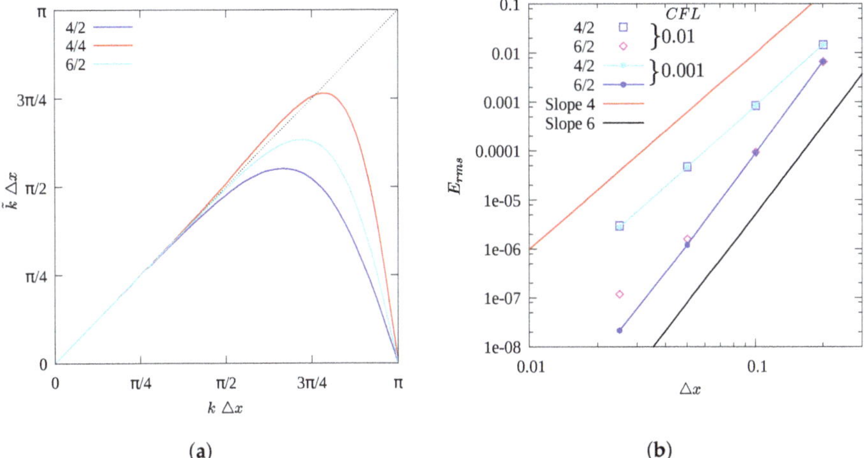

**Figure 1.** (a) Modified wavenumber characteristics of Hixon–Turkel split derivative formulas. (b) Effective orders of truncation error. (a) Modified wavenumber, (b) Order behavior.

### 2.2. Treatment of Regions with Shocks

An essential requirement of shock capturing methods is that the growth of oscillations in the vicinity of the shock must be prevented. Visbal and Gaitonde [12] proposed an adaptive filtering strategy, but do not seem to have pursued it further. It is an attractive strategy because it requires but a simple modification of coefficients of a filter that is anyway applied for the LES. A 2*N*th-order implicit filter (denoted filter F2N) is represented by the relation

$$\alpha_f \bar{f}_{i-1} + \bar{f}_i + \alpha_f \bar{f}_{i+1} = \sum_{n=0}^{N} \frac{a_n}{2}(f_{i+n} + f_{i-n}) \tag{6}$$

connecting filtered and unfiltered vectors, $\bar{f}$ and $f$, respectively. As filter order increases, the response function becomes flatter at low wavenumbers. As filter parameter $\alpha_f$ is increased the spectral content of the function that is filtered out is over a smaller range of high wavenumbers.

The explicit filtering LES sub-grid-scale (SGS) model prescribes that transported variables be filtered with a high-resolution, low-pass, spatial filter after every time-step [11]. Accordingly, filter F10 was applied to all transported fields in all three directions, after every timestep, for LES of flows without shocks. When there are shocks, filters of different orders were applied in the vicinity of the shock. The effectiveness of the adaptive filter is dependent on the shock sensor. In fact, both a review [1] and an assessment of methods [10] emphasize its importance in achieving good results. Pirozzoli [1] examined the effectiveness of four types of sensors. The Ducros sensor [24] selected only the shock, whereas other sensors selected parts of regions without shocks as well. Here, the sensor proposed by Bhagatwala and Lele [25] was used. A shock was considered to be near a gridpoint if

$$\frac{1}{2}\left(1 - \tanh\left(2.5 + \frac{10\Delta}{c}(\nabla \cdot \mathbf{u})\right)\right) \frac{(\nabla \cdot \mathbf{u})^2}{(\nabla \cdot \mathbf{u})^2 + |\nabla \times \mathbf{u}|^2 + 10^{-32}} > \epsilon. \tag{7}$$

From our numerical experiments, $\epsilon = 0.02$ was found to work well for various flows containing Mach waves to strong normal shocks. After a gridpoint was detected as being "in" a shock, the inclination of the shock was determined to ensure filter adaptation was in required directions only. The filter was adapted along a coordinate direction $i$ if $\hat{x}_i \cdot \nabla M / |\nabla M|$ or $\hat{x}_i \cdot \nabla \rho / |\nabla \rho| > 1/\sqrt{3}$ at that point, where $\hat{x}_i$ is the unit vector along the $i$th coordinate direction. The second-order filter F2 was applied at that point and at the two neighboring points. Filter order was increased progressively from 2 to 10, from shock to the smooth region, as shown in Figure 2a for a single shock and Figure 2b for multiple nearby shocks. Then, all filter stencils (F4–F10) include the point detected by the shock sensor, but do not straddle that point.

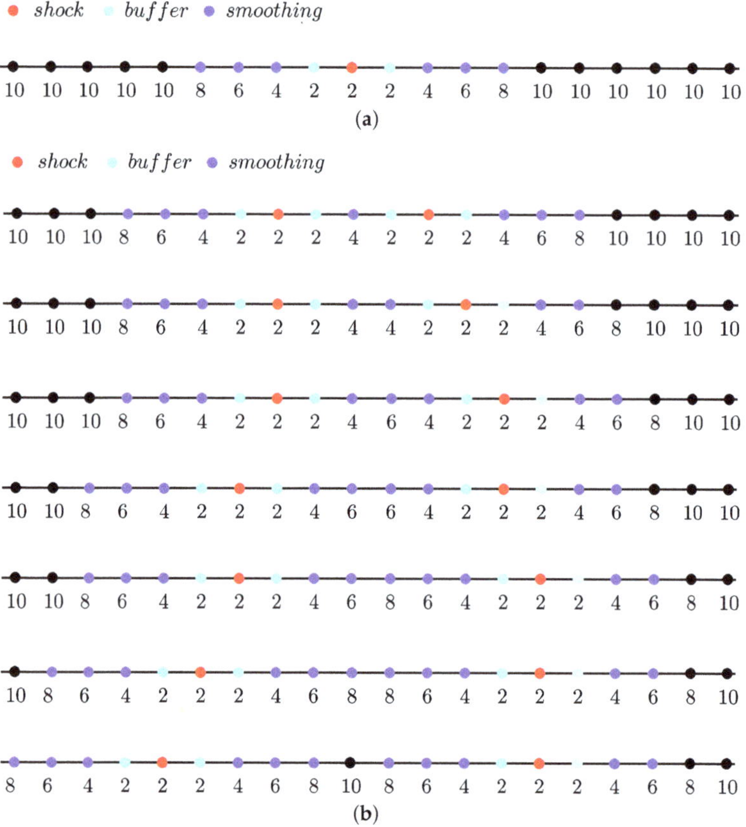

**Figure 2.** Scheme for changing filter order near isolated and proximate shocks: (**a**) isolated shock; and (**b**) proximate shocks.

By testing with a normal shock, Visbal and Gaitonde [12] determined that filter order had to be reduced to 2 to prevent the appearance of wiggles. However, the filter parameter could be set to a high value $\alpha_f = 0.498$. From a large number of our own tests, we found the following parameter-order combinations to provide wiggle-free shocks in several different simulations: filter parameter $\alpha_f = 0.47$ for filter F2; 0.48 for F4; and 0.498 for F6, F8 and F10. Filter response functions of filters of various orders are shown in Figure 3a. Higher-order filters, F6–F10, remove very little of the solution over a small range of the largest represented wavenumbers. A larger value of filter parameter for F2 may improve solutions in some cases as discussed in Section 3.1. Another comparison of filter characteristics that accentuates low wavenumber differences is obtained by plotting damping functions $D(k) = 1 - T(k)$ (Figure 3b).

The resolution error of a derivative formula is $\tilde{k}/k - 1$. To compare with filter characteristics, the function $\tilde{k}/k$ of the 6/2 derivative scheme has been included in Figure 3a, and $k - \tilde{k}$ in Figure 3b. Even when filter F2 is applied, its significant effects are over a high wavenumber range comparable to that of the sixth-order derivative formula. A formal optimization to obtain filter parameters was not carried out. It has been our practice to use the values listed above for new problems as long as stable solutions are obtained. Solution accuracy improves readily with grid refinement, thus it has not seemed worthwhile to implement formal optimization procedures, which may be problem dependent, especially when multiple interacting shocks are present.

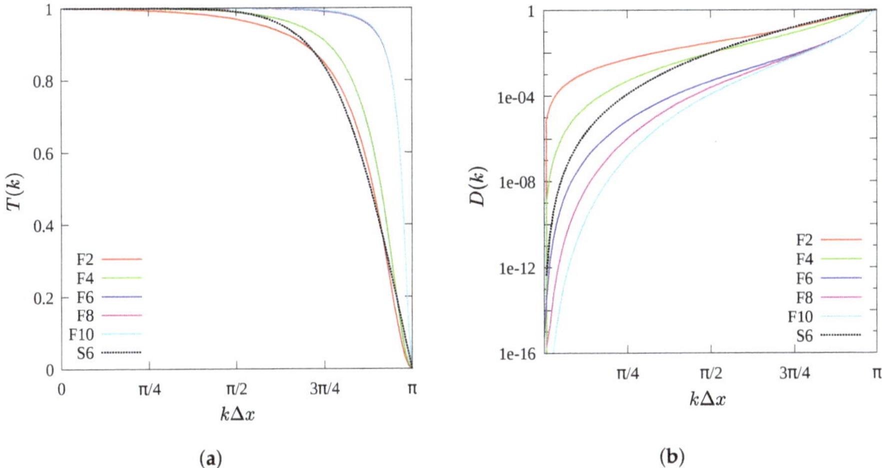

**Figure 3.** Response and damping functions of filters and scheme. F2–F10 are filters of various orders, and S6 is the sixth-order difference scheme. Filter parameters: $\alpha_f(F2) = 0.47, \alpha_f(F4) = 0.48$ and $\alpha_f(F6, F8, F10) = 0.498$. (**a**) Filter response functions; and (**b**) filter damping functions.

The second-order explicit Runge–Kutta scheme was used for the time integration. For simplicity, let us write the 1D version of Equation (1) in the form $U_t + F_\zeta^{inv} + F_\zeta^{vis} = 0$, where $F^{inv}$ and $F^{vis}$ are the inviscid and viscous fluxes, respectively. Then, the field $U^{n+1}$ at time-step $t^{n+1}$ is obtained from the field at $t^n$ as follows:

$$U^{n+1} = U^n + \frac{h^{(1)}}{2} + \frac{h^{(2)}}{2} \tag{8}$$

with,

$$h^{(1)} = -\Delta t \frac{\partial^F}{\partial \zeta} \left[ F^{inv}(U^n) + F^{vis,B}(U^n) \right]$$

$$h^{(2)} = -\Delta t \frac{\partial^B}{\partial \zeta} \left[ F^{inv}(U^n) + F^{vis,F}(U^n) \right]$$

The superscripts $F$ and $B$ signify the use of forward and backward relations (Equations (4) and (5)), respectively. The expressions above can be denoted an FBBF rule. During integration, the sequence is alternated after every time step, i.e., if FBBF is used at a time step, then at the following time step the relations are BFFB.

## 3. Basic Tests

Several tests were conducted to understand effects of adapting filtering in flows with shocks and simple fluctuations. The first three are Riemann problems and the following two are of shocks interacting with 1D and 2D waves sinusoidal incident waves.

### 3.1. Riemann Problems

Sod's conditions for a 1D Riemann problem is a common example. The initial pressure ratio is 10 and density ratio is 8 across the initial discontinuity. The fluid is air. In the simulations, viscosity $\mu = 0$, and conduction terms also vanish. Grid spacing $\Delta x = 0.005$. The numerical solution at $t = 0.2$ is displayed as the density distribution in Figure 4a, with the analytical solution for comparison. The filter order at the same gridpoints are also shown. Figure 4b shows the solution to Lax's shock problem. The initial conditions were $\rho = 0.445$, $u = 0.698$ and $p = 3.528$ for $x < 0$, and $\rho = 0.5$, $u = 0.0$ and $p = 0.571$ for $x > 0$. The solution is everywhere very close to the exact solution. For both problems, grid spacing and times at which their solutions have been presented are the same as those in Kawai and Lele [26] for direct comparison against their artificial diffusivity solutions. They found that artificial viscosity was enhanced at the shock and at the ends of the expansion, and artificial conductivity was enhanced at the shock and contact surface. Here, filter order changes are near the shock only. Small oscillations and a slight smearing at the shock are evident in the solution of the Lax problem.

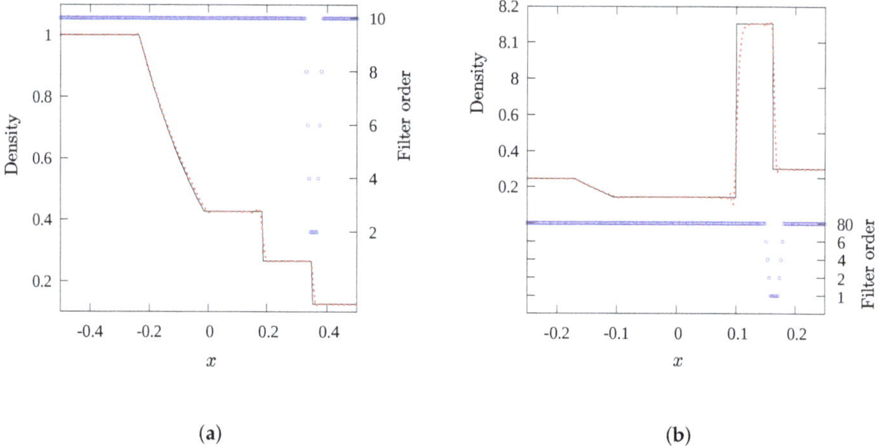

|  |  |
|:---:|:---:|
| (a) | (b) |

**Figure 4.** Solution of shock tube problems. exact (——), numerical (- - -), filter order (•); $\alpha_f(F2) = 0.47$, $\alpha_f(F4) = 0.48$ and $\alpha_f(F6, F8, F10) = 0.498$. (a) Sod's problem at t = 0.2; and (b) Lax's problem at t = 0.13.

The third basic test is with a 2D Riemann problem listed as Case 13 by Lax and Liu [27]. Initial conditions in the four quadrants are listed in Table 1. These conditions imply that there are shocks along the line $y = 0$ which move to $y < 0$ but at different speeds, and a slip line along $x = 0$. This case was used by Pirozzoli [1] to illustrate the relative performance of standard and hybrid WENO schemes. Figure 5a shows the density distribution at time $t = 0.3$ on a grid with $\Delta x = \Delta y = 1/1200$. Gridpoints where filter order has been lowered for treatment of shocks are marked in Figure 5b; where shocks are nearly perpendicular to the $y$-axis, adaptation is along $y$ only. The shocks are crisp. The vortex sheet is unstable to perturbations of any wavelength. Here, the rolled-up vortices seen along the slip line are selected by the grid spacing, which selects the smallest wavelength. Owing to the high-resolution derivative formula and restriction of low-order filtering to the vicinity of shocks, there is no discernible smearing of the vortices along the slip line. Flow features, especially shock locations, agree closely with the solution given in Lax and Liu [27] using their scheme. Shocks are sharper in the present solution. In their solution, the slip line is thicker and, understandably for an inviscid scheme, does not have any of the vortices shown here.

**Table 1.** Initial condition for 2D Riemann problem—Case 13 by Lax and Liu [27].

| Quadrant | $\rho$ | $p$ | $v$ |
|---|---|---|---|
| I | 1.0 | 1.0 | −0.3 |
| II | 2.0 | 1.0 | 0.3 |
| III | 1.0625 | 0.4 | 0.8145 |
| IV | 0.5313 | 0.4 | 0.4276 |

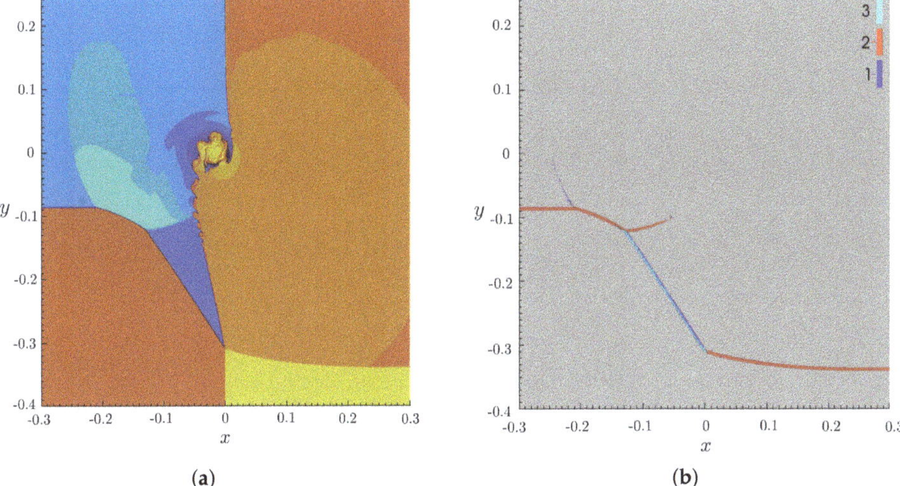

(a)　　　　(b)

**Figure 5.** Solution at $t = 0.3$ for Case 13 by Lax and Liu [27], and filter adaptation: (**a**) Density; and (**b**) Filter adaptation ((3) (**cyan**) adapted along $x$ and $y$; (2) (**red**) along $y$ only; and (1) (**blue**) along $x$ only).

### 3.2. Interaction of Plane Waves with Shocks

In view of the intended application of the proposed method to LES, it is useful to examine its performance on flows in which fluctuations pass through a shock wave. The Shu–Osher problem [28] is initialized with a Mach 3 shock wave travelling into a quiescent region with sinusoidal density variations. The initial conditions are

$$(\rho, p, u) = \begin{cases} (3.857143, 10.33333, 2.629369) & (-5 < x < -4), \\ (1 + 0.2 \sin 5x, 1, 0) & (-4 \leq x < 5). \end{cases}$$

A solution on a grid with spacing $\Delta x = 0.05$ at $t = 1.8$ is shown in Figure 6. The reference solution is that on a finer grid with spacing $\Delta x = 6.25 \times 10^{-3}$. These are the spacings used in Johnsen et al. [10]. The solutions are identical on the finer grid. At this time, acoustic waves have traveled farther into the post-shock region than the entropy waves, and have themselves steepened into weak shocks (in $-2.5 < x < 0$). The most noticeable differences between solutions on the two grids are in density and entropy immediately downstream of the shock. Entropy fluctuations are damped at the shock only, visible as a drop in amplitude at the peak adjacent to the one at the shock (Figure 6c). There is no further loss to this wave as it propagates away from the shock. The present solution is comparable to that with the Hybrid scheme (sixth-order central switching to fifth-order WENO over discontinuous regions) in Johnsen et al. [10]. It is slightly worse than the artificial diffusivity methods, but better than WENO methods; entropy continues to decay along the wave for WENO methods.

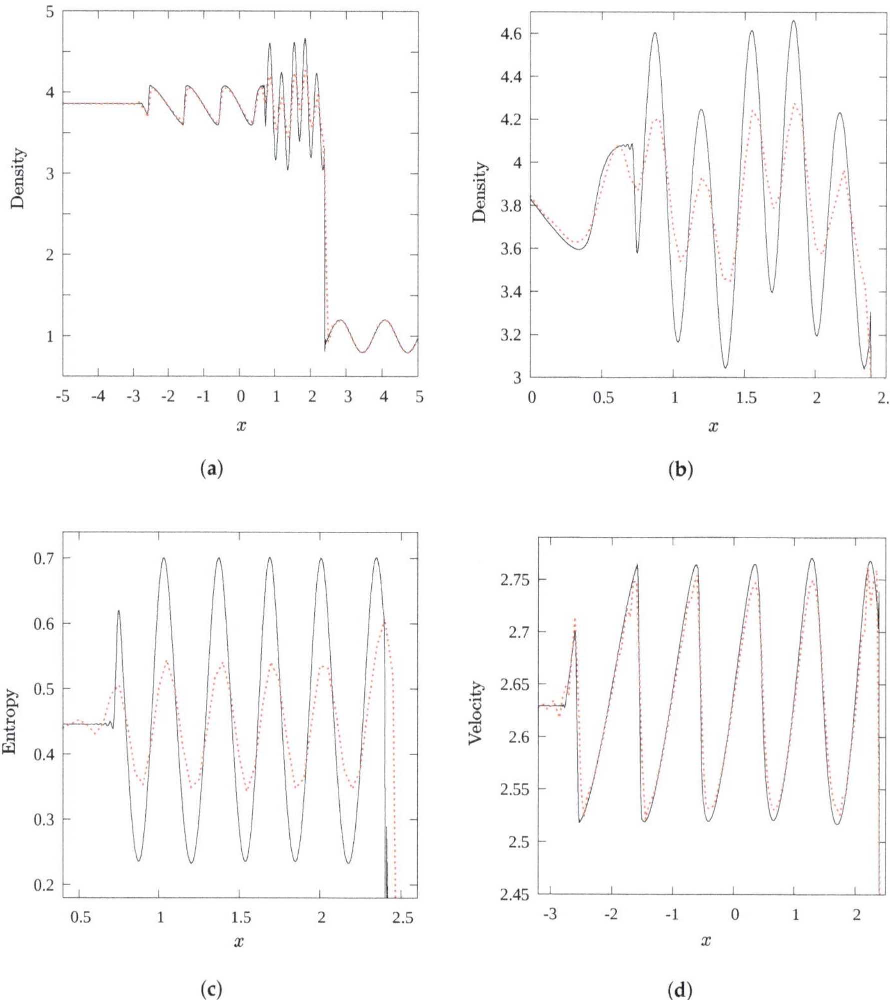

**(a)**

**(b)**

**(c)**

**(d)**

**Figure 6.** Shu–Osher problem at t = 1.8 (red dotted curve: $\Delta x = 0.05$; black curve: $\Delta x = 6.25 \times 10^{-3}$).

A 2D version of the Shu–Osher problem can be posed as the interaction between a plane shock and an inclined plane wave with sinusoidal vorticity and entropy variations. Johnsen et al. [10] examined the performance of several schemes on this problem (considered earlier by Mahesh [29]). The region of interest is $0 < x < 4\pi$ and $-\pi < y < \pi$. The solution is periodic in $y$ with period $2\pi$. There is a supersonic inflow at $x = 0$, and a subsonic outflow. A buffer region over $4\pi < x < 5.2\pi$ ensured that there were no reflections. An initial field was setup, representing steady flow separated by a plane shock at Mach 1.5, given by left and right states

$$(\rho_0, p_0, u_0) = \begin{cases} (\rho_L, p_L, u_L) & = (1, 0.714286, 1.5) \quad (0 \leq x \leq 3\pi/2), \\ (\rho_R, p_R, u_R) & = (1.862069, 1.755952, 0.8055556) \ (3\pi/2 \leq x \leq 4\pi). \end{cases}$$

To this, fluctuations

$$\rho' = \rho_L A_e \cos(k_x x + k_y y), \qquad\qquad p' = 0,$$
$$u' = u_L A_v \sin\psi \cos(k_x x + k_y y), \qquad\qquad v' = -u_L A_v \cos\psi \cos(k_x x + k_y y),$$

representing a plane wave incident at angle $\psi = \tan^{-1} k_y/k_x$ were added. This wave enters at $x = 0$ from the unsteady boundary conditions

$$\rho = \rho_L + \rho_L A_e \cos(k_y y - k_x u_L t), \qquad\qquad p = p_L,$$
$$u = u_L + u_L A_v \sin\psi \cos(k_y y - k_x u_L t), \qquad\qquad v = -u_L A_v \cos\psi \cos(k_y y - k_x u_L t).$$

Grid spacings of $\Delta x = \pi/50$, $\Delta y = \pi/32$ are identical to those of Johnsen et al. [10]. The change due to refinement along $x$ alone ($\Delta x = \pi/100$, $\Delta y = \pi/32$) and two levels of refinement in both directions are presented below. Taking small values for amplitudes ($A_e = A_v = 0.025$) allow quantitative comparisons against linear theory (linearization of Rankine–Hugoniot jump conditions) as in [29].

Contours of vorticity $\omega_z$ for a wave train incident at angle $\psi = 45°$ are shown in Figure 7a. The wave refracts and its amplitude changes. The solution along a line ($y = 0$) is found to be free of oscillations. On halving $\Delta x$ alone, a sharper jump is obtained at the shock. Oscillations appear in the vicinity of the shock when the wave is incident at a higher angle (Figure 7c), and are carried into the post-shock region. The grid refinement has confined significant oscillations to a smaller region, and made the post-shock solution smoother. The differences in solution structure arise because the post-shock wave is evanescent at $\psi = 75°$ [29].

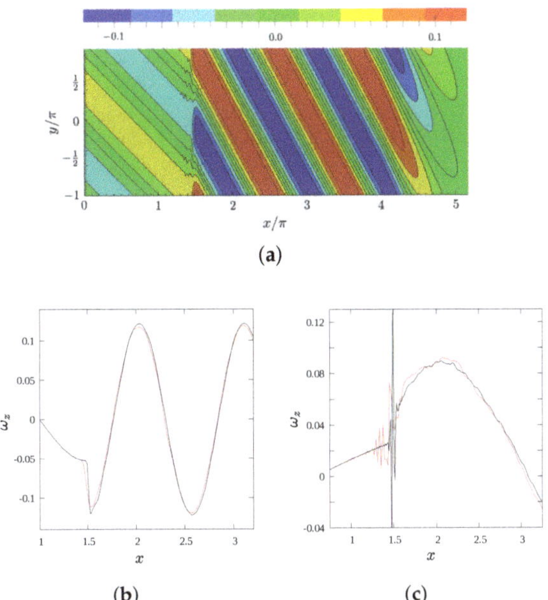

**Figure 7.** Shock-vorticity/entropy wave interaction. $\Delta y = \pi/32$; Red curve: $\Delta x = \pi/50$; black curve (finer grid): $\Delta x = \pi/100$. (**a**) Vorticity $\omega_z$ contours at $t = 25$ for $\psi = 45°$, $k_y = 1$; (**b**) $\omega_z$ along $y = 0$ for $\psi = 45°$, $k_y = 1$ at $t = 25$; and (**c**) $\omega_z$ along $y = 0$ for $\psi = 75°$, $k_y = 1$ at $t = 32$.

Quantitative comparisons against linear theory are shown in Figure 8. The amplification of the mean enstrophy $< \omega_z^2 >$ across the shock is shown for $\psi = 45°, 75°$, and $k_y = 1, 2$. Averaging is over the time period $2\pi/(k_x u_L)$. The amplification factor obtained in the simulation is slightly smaller than that of linear analysis at $\psi = 45°$, and slightly larger for $\psi = 75°$. The differences increase as wavenumber is increased. On halving grid spacing, in every case, there is a significant improvement as solution comes closer to that of linear theory. In addition, the amplitude of the post-shock oscillations and the extent of the region where these occur decrease significantly. With further refinement, the numerical solution is essentially the same as that of linear theory for $\psi = 45°$. When $\psi = 75°$, there is further improvement for $k_y = 2$ although not significantly for $k_y = 1$.

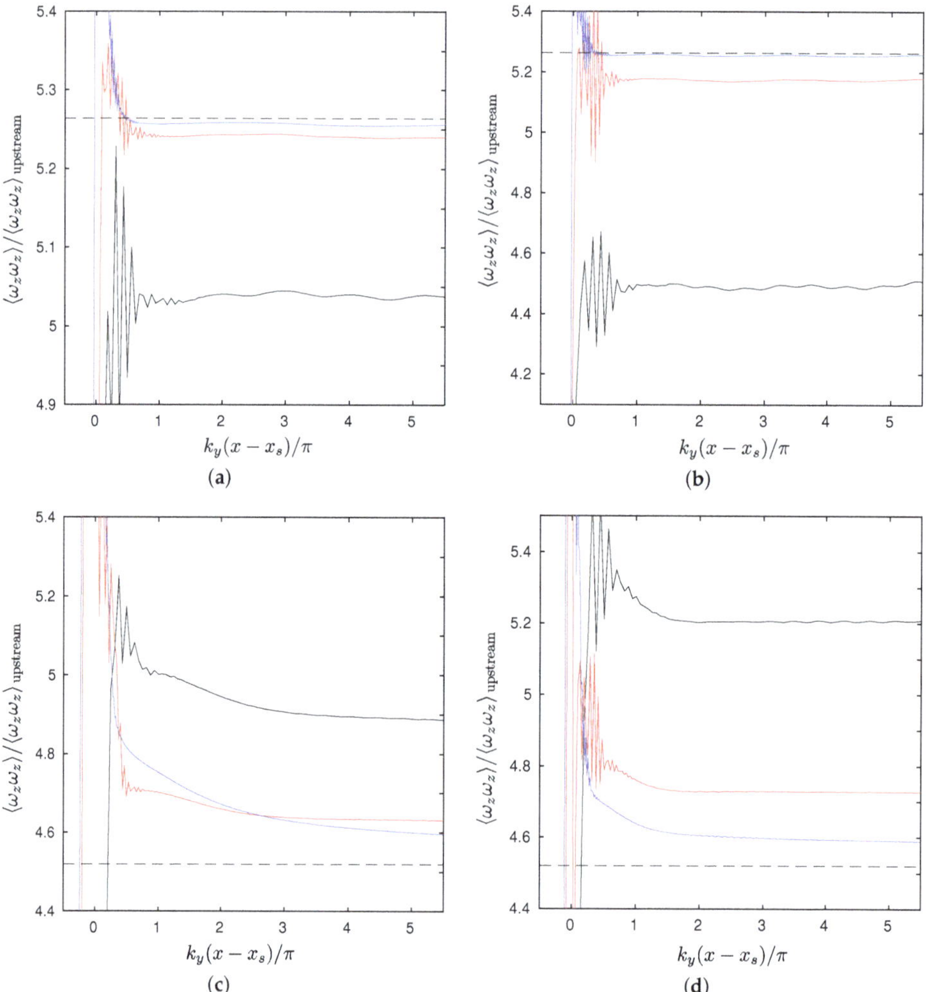

**Figure 8.** Shock-vorticity/entropy wave interaction. Black dotted line: linear solution; black: grid as in Johnsen et al. [10]; red: medium grid, $\Delta x$, $\Delta y$ halved; blue: fine grid $\Delta x$, $\Delta y$ halved again. (**a**) $\psi = 45°$, $k_y = 1$; (**b**) $\psi = 45°$, $k_y = 2$; (**c**) $\psi = 75°$, $k_y = 1$; and (**d**) $\psi = 75°$, $k_y = 2$.

The solutions can be compared with the results of Johnsen et al. [10]. On the same grid as theirs, the amplification factor in the present solutions differ more from the linear analysis solution than some other methods (Hybrid WENO/central differences, WENO and ADPDIS3D). However, when the grid was refined, the differences fell in all cases, and post shock oscillations also reduced in amplitude and occurred over a smaller region. They supposed (Section 4.2.1 in Johnsen et al. [10]) that these are post-shock oscillations owing to, generally, slowly moving shocks in shock-capturing schemes; these oscillations would "not disappear under grid or time step refinement" [10]. In the present simulations, the differences between numerical solutions and linear theory generally diminish with grid refinement.

In all the tests reported above, the same adaptation of filter order as in Figure 2, and same set of filter parameters as in Figure 3 was used. For a given problem, it is always possible to find a set of filter parameters that result in a "better" solution on a given grid. Not only would the solutions look more attractive because of reduced oscillation amplitudes, even accuracy can improve. For example, in the 2D extension of the Shu–Osher problem, with slightly less filtering, the amplification factor comes closer to that of linear theory. However, our proposal is not to seek a new, optimal set of filter parameters for every problem, but to use, e.g., the set given here. Since the computations remained stable for all the problems reported here with this set of parameters, and solutions tend to the reference one (or DNS/experiment), this appears to be a reliable method for obtaining accurate solutions.

### 3.3. 3D Evolution

The Taylor–Green vortex problem evolves from a periodic initial field and develops smaller scales due to nonlinearity. We consider both inviscid and viscous solutions. No shocks form in this problem. It is included to demonstrate accuracy of the numerical scheme, and convergence with grid refinement by comparisons with reference solutions.

For the inviscid case, there is no lower limit on the scale. An LES model and/or numerical method must stabilize the computation beyond the time at which structures of the order of the grid spacing appear. Excessive damping can be revealed by this test problem. The initial condition is

$$\rho = 1, \qquad p = 100 + \frac{(\cos 2z + 2)(\cos 2x + \cos 2y) - 2}{16},$$

$$u = \sin x \cos y \cos z, \qquad v = -\cos x \sin y \cos z, \qquad w = 0.$$

Simulations were carried out in a cube of edge length $2\pi$ on uniformly-spaced grids of $512^3$, $128^3$ and $64^3$ points. The growth of enstrophy in the box is shown in Figure 9a. A semi-analytical solution is included for comparison [30]. All simulations remained stable but departed from each other when $t > 4$, approximately. As the grid is refined, the numerical solution agrees with enstrophy of the reference solution to later times. The kinetic energy in the box should be preserved because there is no viscous dissipation. Figure 9b shows the kinetic energy to begin to fall at $t \approx 4$ on the $64^3$ grid, and later on the finer grids. Comparisons at two times are shown in Table 2. The monotonic convergence with grid refinement is evident. Owing to the flat filter characteristics, dissipation is active only after small scales have grown.

The viscous Taylor–Green solution was found for Re $= 1600$ based on the initial length scale (length of cube edge, $2\pi$), and maximum velocity. The reference data for comparison are from DeBonis [31] (available at https://eprints.soton.ac.uk/401892/1/512.dat). The simulation was carried out for $0 \le t \le 20$. Differences are evident on coarse grids, but diminishes with refinement.

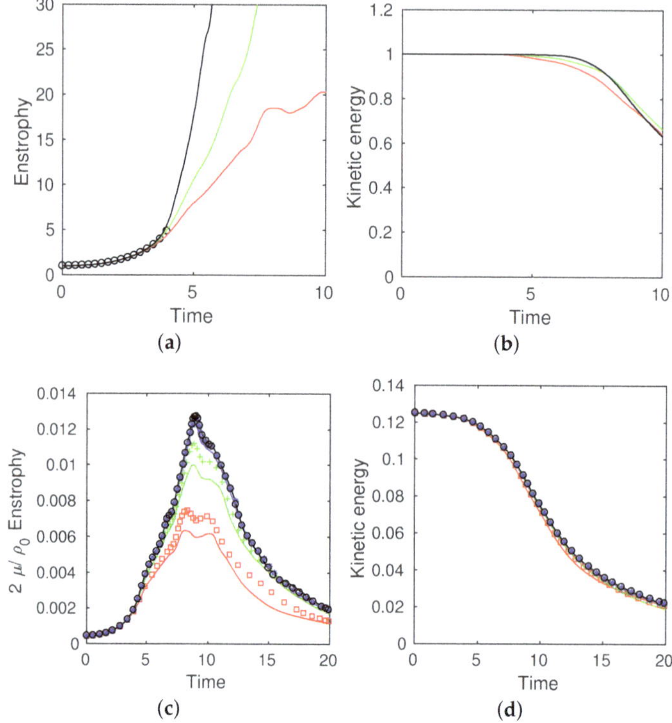

**Figure 9.** Evolution of Taylor–Green vortex. Curves are from simulations on grids of $512^3$ (**black**), $256^3$ (**blue**), $128^3$ (**green**), and $64^3$ (**red**) points. (**a**,**b**) Inviscid case with symbols from [30]; and (**c**,**d**) viscous case with solid line, present LES and symbols from [31].

**Table 2.** Measures of Taylor–Green vortex solution.

|  | $64^3$ | $128^3$ | $512^3$ | Brachet et al. [30] |
|---|---|---|---|---|
| Inviscid |  |  |  |  |
| Energy ($t = 5$) | 0.9846 | 0.9943 | 0.9992 | 1.00 |
| Enstrophy ($t = 3.5$) | 3.276 | 3.402 | 3.458 | 3.459 |
| Viscous |  |  |  |  |
| Energy ($t = 5$) | 0.9423 | 0.9468 | 0.9476 |  |
| Enstrophy ($t = 3.5$) | 3.080 | 3.145 | 3.154 |  |

## 4. Jet LES

Now that the performance of the proposed method has been evaluated carefully in canonical problems of isolated shocks, wave refraction by a shock, and the evolution of a turbulent flow at a moderate Reynolds number accessible to DNS, we turn to simulations of high Reynolds number, turbulent, supersonic jets. Imperfectly expanded jets exhibit a train of shocks. As turbulent fluctuations travelling through these shocks are distorted, turbulent dissipation and production can get affected. Excessive numerical dissipation can then alter the jet's development. The following two test cases are model turbulent flows with stationary shocks that can indicate the accuracy that can be expected from LES for compressible flow applications. Comparisons with experiments, and another LES for the impinging case are shown. The Reynolds numbers are quite large, $O(10^6)$. The intent is to examine the usefulness of this method for practical applications. It is shown that, as the grid is refined, the energy spectrum broadens by extending the computed part of the inertial range.

### 4.1. Free, Underexpanded Round Jet

Large eddy simulations were performed on the supersonic round jet experiments of Norum and Seiner [32]. A contoured C-D nozzle designed for exit plane Mach number $M_E = 2$ was used. The nozzle pressure ratio (NPR), which is the ratio of reservoir pressure to nozzle exit pressure, was 9.187. Because the jet was slightly underexpanded, the fully expanded Mach number would be $M_j = 2.103$. The Reynolds number based on these nozzle exit plane conditions, centerline velocity $U_j$, and nozzle inner diameter $d$, was $6.09 \times 10^6$. The simulation was set up in a rectangular domain that was $7d$ along jet axis, and $4d$ in cross stream directions. Simulations were performed on a sequence of grids. Each was distinguished by the number of uniformly spaced gridpoints across the jet diameter. Outside the jet, the grid was stretched—spacing increased in geometric progression at 1% in cross stream directions. A much milder stretching of 0.1% was used in the streamwise direction. Buffer zones, with twenty points at the lateral and thirty points at the outflow boundaries, were added with a stretching of 10%. On the coarsest grid there were 20 points across the jet diameter leading to $2.23 \times 10^6$ gridpoints in total. A refined grid had 40 points across the jet and $9.3 \times 10^6$ points in total (medium grid). Finer grids had 23 million (fine), 43 million (finer), and 200 million (finest) points.

Boundary conditions at the inflow plane ($x = 0$) were obtained from radial profiles

$$
\begin{aligned}
\frac{u}{U_j} &= \frac{1}{2} + \frac{1}{2}\tanh\left(\frac{r_j - r}{2\delta_\theta}\right), \\
\frac{p}{p_\infty} &= 1 + \left(\frac{p_j}{p_\infty} - 1\right)\left[\frac{1}{2} + \frac{1}{2}\tanh\left(\frac{r_j - r}{2\delta_\theta}\right)\right], \\
\frac{T}{T_j} &= \frac{T_\infty}{T_j}\left(1 - \frac{u}{U_j}\right) + \frac{u}{U_j} + \left(\frac{\gamma - 1}{2}\right)M_j^2\frac{u}{U_j}\left(1 - \frac{u}{U_j}\right).
\end{aligned}
\tag{9}
$$

Here, $r_j = d/2$ is the mean jet radius and $\delta_\theta$ the momentum thickness of the jet (taken to be $r_j/20$). Subscripts $j$ and $\infty$ denote jet centerline and ambient conditions, respectively. To anchor the inflow profile, the following procedure proposed by Bogey and Bailly [19] was employed at every time step. Flow variables $U = \{\rho, u, v, w, p\}$ at the inflow plane, obtained after applying non-reflecting conditions, were corrected to $U^c = (1 - \sigma_r)U + \sigma_r U_{ref}$. Here, $U_{ref}$ is the prescribed inflow profile (Equation (9)), and $\sigma_r = 5 \times 10^{-2}$ for $r \le 2r_0$ and $5 \times 10^{-3}$ for $r > 2r_0$. Within the jet shear layer alone, small perturbations,

$$
\frac{u'}{U_j} = 10^{-1}\left[\frac{4u}{U_j}\left(1 - \frac{u}{U_j}\right)\right]\sum_{i=1}^{6}\cos\left(St_i\frac{U_j}{d} + \phi_i\right),
\tag{10}
$$

were added to the inflow axial velocity to represent inflow turbulence of the nozzle boundary. Strouhal numbers $St_i$ and phase $\phi_i$ varied from 0.1 to 0.6 and 0 to $5\pi/6$, respectively. The quantity within square brackets ensures that fluctuations are added within the shear layer only.

Figure 10a is a visualization of the jet in terms of the mean pressure on a plane containing the jet axis, showing the expected cell structures. Static pressure along the centerline in the experiment was reported (page 27 [32]). Figure 10b shows solutions from the coarse, medium and finest grids along with data from the experiment. There is little change in mean pressure between the medium (9 million points) and the finest grid (200 million points). We can observe that both the locations and strengths of the pressure variations through the cells have been captured accurately. On refining from coarse to medium, pressure variations occur at the same places, but peaks are slightly sharper. The pressure distribution at an instant is also shown. Significant fluctuations on the centerline appear only beyond $x/d > 3$. Where such fluctuations are significant, we can observe that gradients at an instant are slightly larger than in the mean (see pressure rise near $x/d = 4$ and 6). Contours of pressure on a longitudinal plane also show that significant unsteadiness sets in the 3rd shock cell.

The effect of grids finer than the medium one can be seen in energy spectra. Figure 10b shows spectra from the five simulations, calculated from time series at a point near the jet boundary at

$x/d = 5.7$. As the grid spacing is reduced, a clear inertial range develops, and extends to smaller scales. The magnitude of low frequency content changes little. At the Reynolds number of the flow, the dynamically significant range of scales is still larger, but quantities such as the centerline mean pressure can be obtained accurately in an LES, on a grid of moderate size. Especially, shock positions and strengths change little with refinement. This serves as strong support for the adaptive filtering method for shock capturing used in these simulations. The quality of the simulation was similar for an overexpanded jet [33].

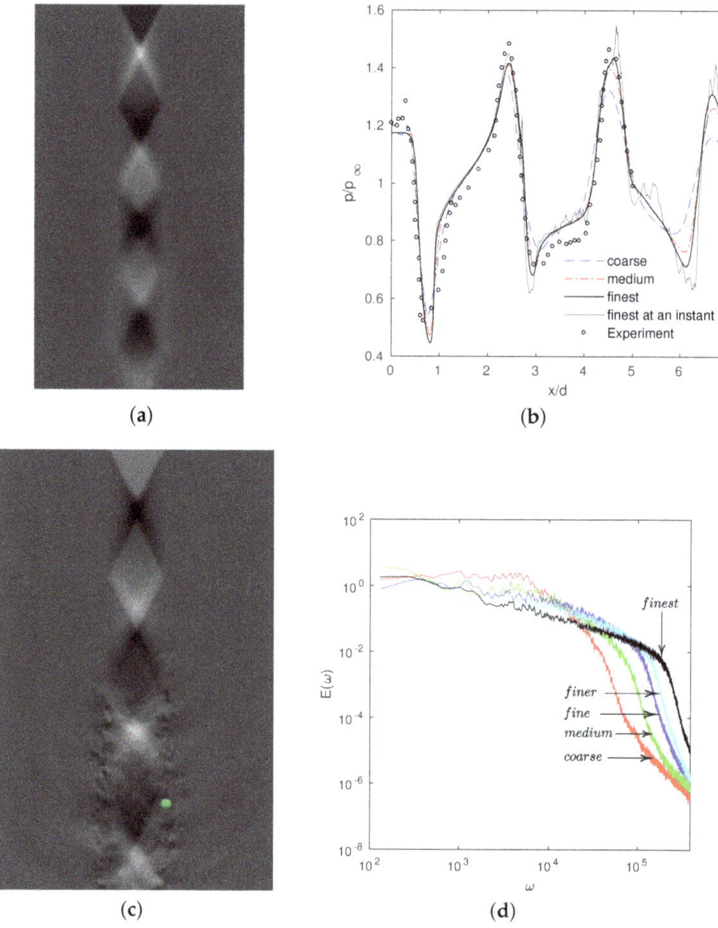

**Figure 10.** LES of underexpanded jet for $M_j = 2.103$, $p_0/p_a = 9.187$. Experiment by Norum and Seiner [32] (case on page 27 of their report). (**a**) Mean pressure on longitudinal plane. (**b**) Pressure along jet axis. Curves for simulation data, symbol for experiment. (**c**) Instantaneous pressure on longitudinal plane (Medium grid). (**d**) Energy spectra (at green dot in the left figure).

## 4.2. Impinging Round Jet

The second case, representative of flows of interest in applications, is the impinging jet experiment of Henderson et al. [34]. Velocity data from PIV were reported. This is a sonic jet with NPR = 4.03 impinging on a flat wall placed at a distance of $4.16d$ from the nozzle exit. Dauptain et al. [35] reported an LES of this problem. They used 7.5, 16 and 22 million tetrahedral volumes including a region within the nozzle as well. Here, two grids with 30 and 60 points across the jet at the inflow plane were taken, with 5% stretching in the lateral direction. Only lateral buffer zones were used in the current simulation. The total number of points were then $2.38 \times 10^6$ for the coarse grid and $7.70 \times 10^6$ for the fine grid. The Reynolds number was $1.5 \times 10^6$. A comparison of the mean streamwise velocity with the experiment along the jet axis is shown in Figure 11a. The simulations are consistent among themselves: on the finer grid, the gradients are sharper everywhere. Surprisingly, Dauptain et al. [35] found their coarsest grid to be closest to the experiment. Especially in the vicinity of $x/d \approx 1$, the velocity dropped to about 150 m/s whereas in the experiment it is about 300 m/s. Smaller but significant differences were found where flow decelerates again ($x/d \approx 3$). They supposed that, in these strongly decelerating regions, the measurement could be in error. Our simulations do not show such large differences with experiment. Figure 11b compares present simulations with data from Dauptain et al. [35] of the pressure along the centerline (experimental data of pressure were not available). Pressure data agree quite closely with the other simulation. Since the pressure jump near $x/d \approx 1$ is about the same, it is surprising that the velocity jump could be different.

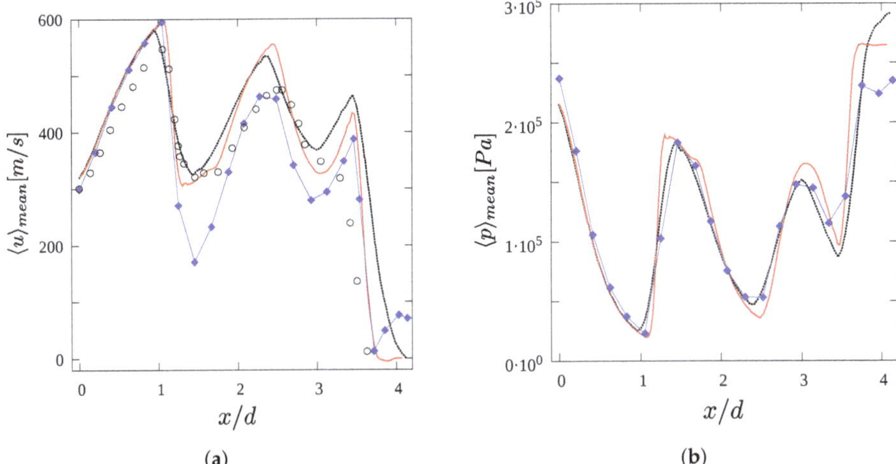

**Figure 11.** Mean axial velocity and pressure along centerline of impinging jet. Coarse grid (**black curve**); fine grid (**red**), experiment of Henderson et al. [34] (**black circles**), and simulation of Dauptain et al. [35] (blue curve with filled symbol). (**a**) Velocity; and (**b**) Pressure.

Since the objective of this test was to determine the usefulness of the proposed shock-capturing scheme, and not the other implications of this flow, such as impingement acoustics, the simulation domain did not include the flow within the nozzle. There has not been any adverse effect on the quantities that we could compare with experiment and another LES, such as any shift in locations of Mach disk or extent of the cells. Indeed, the very close agreement between our simulations and the one which included the nozzle flow [35] in the initial region ($0 < x/d < 1$) affirms this.

## 5. Conclusions

An explicit filtering approach to LES of flows has been extended to flows with shocks by filter order adaptation. Although the usual treatment of flows with shock capturing is to reduce the order of the spatial discretization in the vicinity of shocks, here, the discretization remains the same everywhere, and was sixth-order. Instead, the order of the spatial filter was reduced from 10 in smooth regions to 2 at gridpoints where a shock was detected. Filter order was increased in steps at neighboring gridpoints. However, filtering is an essential part of this method for LES. Thus, the adaptation is a convenience. Adaptive filtering had been proposed by Visbal and Gaitonde [12] earlier, but does not appear to have been taken forward either in method assessment or in applications. They lowered filter order to 2 at a band of seven points spanning shocks. Here, it sufficed to use a band of three points. It is not surprising that Bogey and Bailly [13], who also performed LES with an explicit filter, examined filter adaptation briefly. We can expect that the studies presented here may invite its further use.

Although, generally, the results of adaptive filtering are very encouraging, quantitative comparisons with the results in Johnsen et al. [10], for the 2D Shu–Osher problem (Section 3.2), revealed that, on the same grid, some other methods have better performance. Nevertheless, close quantitative agreement with the reference solution was readily obtained on grid refinement. Indeed, it is a common experience in the problems considered here that solutions tend to the reference ones as the grid is refined (explicitly included above for the 2D Shu–Osher, Taylor–Green and free jet problems). For the 2D Shu–Osher problem, not only does the amplification factor approach the linear analysis result, the post-shock oscillations diminish in amplitude and extent. Since computations were stable for all these problems with the same set of parameters, and tend monotonically to reference or exact solutions, we find this to be a reliable method. No search for an optimal set of parameters was considered, since such an optimal set may offer small improvements in computation economy but may be problem-dependent.

Although filter adaptation is but a small change in the LES method used in the examples presented here, because there is always a filtering step, the adaptation strategy can also be used with any other LES code. For example, it can be an alternative to other strategies for shock capturing such as adding artificial diffusion in the vicinity of shocks.

**Author Contributions:** Software, S.K.P.; all other aspects S.K.P. and J.M.

**Funding:** This research received no external funding.

**Conflicts of Interest:** The authors declare no conflict of interest.

## References

1. Pirozzoli, S. Numerical Methods for High-Speed Flows. *Annu. Rev. Fluid Mech.* **2011**, *43*, 163–194. [CrossRef]
2. Sangsan Lee, S.K.L.; Moin, P. Interaction of isotropic turbulence with shock waves: Effect of shock strength. *J. Fluid Mech.* **1997**, *340*, 225–247.
3. Adams, N.; Shariff, K. A High-Resolution Hybrid Compact-ENO Scheme for Shock-Turbulence Interaction Problems. *J. Comput. Phys.* **1996**, *127*, 27–51. [CrossRef]
4. Pirozzoli, S. Conservative Hybrid Compact-WENO Schemes for Shock-Turbulence Interaction. *J. Comput. Phys.* **2002**, *178*, 81–117. [CrossRef]
5. Yee, H.C.; Sandham, N.D.; Djomehri, M.J. Low-Dissipative High-Order Shock-Capturing Methods Using Characteristic-Based Filters 1. *J. Comput. Phys.* **1999**, *238*, 199–238. [CrossRef]
6. Cook, A.W.; Cabot, W.H. A high-wavenumber viscosity for high-resolution numerical methods. *J. Comput. Phys.* **2004**, *195*, 594–601. [CrossRef]
7. Cook, A.W.; Cabot, W.H. Hyperviscosity for shock-turbulence interactions. *J. Comput. Phys.* **2005**, *203*, 379–385. [CrossRef]
8. Cook, A.W. Artificial fluid properties for large-eddy simulation of compressible turbulent mixing. *Phys. Fluids* **2007**, *19*, 055103. [CrossRef]
9. Bonelli, F.; Viggiano, A.; Magi, V. How does a high density ratio affect the near-and intermediate-field of high-Re hydrogen jets? *Int. J. Hydrog. Energy* **2016**, *41*, 15007–15025. [CrossRef]
10. Johnsen, E.; Larsson, J.; Bhagatwala, A.V.; Cabot, W.H.; Moin, P.; Olson, B.J.; Rawat, P.S.; Shankar, S.K.; Sjögreen, B.; Yee, H.; et al. Assessment of high-resolution methods for numerical simulations of compressible turbulence with shock waves. *J. Comput. Phys.* **2010**, *229*, 1213–1237. [CrossRef]
11. Mathew, J.; Foysi, H.; Sesterhenn, J.; Friedrich, R. An explicit filtering method for large eddy simulation of compressible flows. *Phys. Fluids* **2003**, *15*, 2279–2289. [CrossRef]
12. Visbal, M.R.; Gaitonde, D.V. Shock capturing using compact-differencing-based methods. In Proceedings of the 43rd AIAA Aerospace Sciences Meeting & Exhibit, Reno, NV, USA, 10–13 January 2005; AIAA 2005-1265.

13. Bogey, C.; de Cacqueray, N.; Bailly, C. A shock-capturing methodology based on adaptative spatial filtering for high-order non-linear computations. *J. Comput. Phys.* **2009**, *228*, 1447–1465. [CrossRef]
14. Tannehill, J.C.; Anderson, D.A.; Pletcher, R.H. *Computational Fluid Mechanics and Heat Transfer*, 2nd ed.; Taylor & Francis: Abingdon, UK, 1984.
15. Lele, S.K. Compact Finite Difference Schemes with Spectral-like. *J. Comput. Phys.* **1992**, *103*, 16–42. [CrossRef]
16. Bogey, C.; Bailly, C. A family of low dispersive and low dissipative explicit schemes for flow and noise computations. *J. Comput. Phys.* **2004**, *194*, 194–214. [CrossRef]
17. Mathew, J.; Foysi, H.; Friedrich, R. A new approach to LES based on explicit filtering. *Int. J. Heat Fluid Flow* **2006**, *27*, 594–602. [CrossRef]
18. Visbal, M.R.; Rizzetta, D.P. Large-Eddy Simulation on Curvilinear Grids Using Compact Differencing and Filtering Schemes. *ASME J. Fluids Eng.* **2014**, *124*, 836–847. [CrossRef]
19. Bogey, C.; Bailly, C. Computation of a high Reynolds number jet and its radiated noise using large eddy simulation based on explicit filtering. *Comput. Fluids* **2006**, *35*, 1344–1358. [CrossRef]
20. Chakravorty, S.; Mathew, J. A high-resolution scheme for low Mach number flows. *Int. J. Numer. Methods Fluids* **2004**, *46*, 245–261. [CrossRef]
21. Chakravorty, S. On Large Eddy Simulation of Reacting Flows Using the Explicit Filtering Method with a Filtered Mass Density. Ph.D. Thesis, Indian Institute of Science, Bengaluru, India, 2010.
22. Ganesh, S. Flow and Sound Field of Free and Impinging Jets from LES Using Explicit Filtering Method. Ph.D. Thesis, Indian Institute of Science, Bengaluru, India, 2013.
23. Hixon, R.; Turkel, E. Compact Implicit MacCormack-Type Schemes with High Accuracy. *J. Comput. Phys.* **2000**, *158*, 51–70. [CrossRef]
24. Ducros, F.; Ferrand, V.; Nicoud, F.; Weber, C.; Darracq, D.; Gacherieu, C.; Poinsot, T. Large-Eddy Simulation of the Shock/Turbulence Interaction. *J. Comput. Phys.* **1999**, *152*, 517–549. [CrossRef]
25. Bhagatwala, A.; Lele, S.K. A modified artificial viscosity approach for compressible turbulence simulations. *J. Comput. Phys.* **2009**, *228*, 4965–4969. [CrossRef]
26. Kawai, S.; Lele, S. Localized artificial diffusivity scheme for discontinuity capturing on curvilinear meshes. *J. Comput. Phys.* **2008**, *227*, 9498–9526. [CrossRef]
27. Lax, P.D.; Liu, X.D. Solution of two-dimensional Riemann problems of gas dynamics by positive schemes. *SIAM J. Sci. Comput.* **1998**, *19*, 319–340. [CrossRef]
28. Shu, C.W.; Osher, S. Efficient implementation of essentially non-oscillatory shock capturing schemes. *J. Comput. Phys.* **1988**, *77*, 439–471. [CrossRef]
29. Mahesh, K. The Interaction of a Shock Wave with a Turbulent Shear Flow. Ph.D. Thesis, Stanford University, Stanford, CA, USA, 1996.
30. Brachet, M.E.; Meiron, D.I.; Orszag, S.A.; Nickel, B.G.; Morf, R.H.; Frisch, U. Small-scale structure of the Taylor–Green vortex. *J. Fluid Mech.* **1983**, *130*, 411–452. [CrossRef]
31. DeBonis, J. Solutions of the Taylor-Green vortex problem using high-resolution explicit finite difference methods. In Proceedings of the 51st AIAA Aerospace Sciences Meeting including the New Horizons Forum and Aerospace Exposition, Grapevine, TX, USA, 7–10 January 2013; AIAA 2013-0382. [CrossRef]
32. Norum, T.D.; Seiner, J.M. *Measurements of Mean Static Pressure and Far-Field Acoustics of Shock-Containing Supersonic Jets*; NASA Technical Memorandum 84521; NASA: Washington, DC, USA, 1982.
33. Patel, S.K.; Mathew, J. Large Eddy Simulation of Supersonic Impinging Jets by adaptive, explicit filtering. In Proceedings of the 22nd AIAA Computational Fluid Dynamics Conference, Dallas, TX, USA, 22–26 June 2015; Paper No. AIAA-2015-2298.
34. Henderson, A. An experimental study of the oscillatory flow structure of tone-producing supersonic impinging jets. *J. Fluid Mech.* **2005**, *542*, 115–137. [CrossRef]
35. Dauptain, A.; Cuenot, B.; Gicquel, L.Y.M. Large Eddy Simulation of Stable Supersonic Jet Impinging on Flat Plate. *AIAA J.* **2010**, *48*, 2325–2338. [CrossRef]

*Article*

# Cross-Correlation of POD Spatial Modes for the Separation of Stochastic Turbulence and Coherent Structures

Daniel Butcher * and Adrian Spencer

Department of Aeronautical and Automotive Engineering, Loughborough University, Loughborough LE11 3TU, Leicestershire, UK
* Correspondence: D.Butcher@lboro.ac.uk

Received: 5 June 2019; Accepted: 11 July 2019; Published: 16 July 2019

**Abstract:** This article describes a proper-orthogonal-decomposition (POD) based methodology proposed for the identification and separation of coherent and turbulent velocity fluctuations. Typically, POD filtering requires assumptions to be made on the cumulative energy content of coherent modes and can therefore exclude smaller, but important contributions from lower energy modes. This work introduces a suggested new metric to consider in the selection of POD modes to be included in a reconstruction of coherent and turbulent features. Cross-correlation of POD spatial modes derived from independent samples is used to identify modes descriptive of either coherent (high-correlation) or incoherent (low-correlation) features. The technique is demonstrated through application to a cylinder in cross-flow allowing appropriate analysis to be carried out on the coherent and turbulent velocity fields separately. This approach allows identification of coherent motions associated with cross-flow transport and vortex shedding, such as integral length scales. Turbulent flow characteristics may be analysed independently from the coherent motions, allowing for the extraction of properties such as turbulent length scale.

**Keywords:** proper orthogonal decomposition; coherent structures; turbulence; vector flow fields; PIV

---

## 1. Introduction

In the field of fluid mechanics, it is possible to obtain detailed, highly-resolved time-space velocity information from both experimental and computational approaches. Contained within these datasets are coherent structures and stochastic turbulence, often indistinguishable in raw velocity vector fields, the understanding of which is crucial in research and development. Coherent structures, often considered as vortices, are responsible for many flow phenomena, across many applications—lift, drag, mixing and heat transfer, for example. Similarly, an understanding of the turbulent characteristics of an aerodynamic system is vital in the study of many processes such as turbulent combustion. It is therefore important that these distinct structure classes be effectively identified and separated before further analysis may be carried out on each.

There have been multiple approaches suggested in literature to achieve velocity field decomposition. Ölçmen et al. [1] present a comparative analysis of suggested approaches—Reynolds decomposition (referred to as ensemble averaging in that study), proper orthogonal decomposition (POD) and frequency filtering amongst others.

Of these, Reynolds decomposition is the least computationally expensive to calculate and is often a useful first step in the analysis of a flow field. However, cyclic or oscillatory structures would be considered together with the turbulent field and therefore for many applications further decomposition is required. This is demonstrated by Stone [2] and Butcher [3] in the field of internal combustion engines, where there is a cyclic variation and a relatively increased frequency turbulent fluctuation both included within the fluctuating term which should be separated as in Equation (1):

$$U_i = \overline{U} + U_i^* + U_i' \ (= \ U_{true} + \varepsilon) \tag{1}$$

where $U_i$ is velocity measurement, $\overline{U}$ is ensemble mean velocity and $U^*$ and $U\prime$ are coherent and turbulent fluctuations, respectively.

Frequency filtering, such as LES decomposition [4], applies a homogeneous filter to identify all structures above or below a cut-off scale allowing this separation. Using a priori knowledge of the flow field and expected structure sizes, the cut-off may be set such that coherent fluctuations, $U^*$, be filtering in (or out) as required.

POD is a commonly used technique in the fluid mechanics field for this purpose and will be the focus of this paper. The following subsection provides a background to the technique.

*Proper Orthogonal Decomposition*

Since its introduction to fluid mechanics by Lumley [5], the proper orthogonal decomposition (POD) technique has been widely adopted to identify coherent structures in flow fields. In the context of fluid mechanics, POD decomposes a velocity field into spatiotemporal modes according to Equation (2):

$$(x, t_i) = \sum_{k=1}^{M} a_k(t_i)\varphi_k(x) \tag{2}$$

where $u$ is the velocity, $M$ is the number of modes, $a_k$ is the temporal coefficient and $\varphi_k$ is the eigenmode (or spatial mode). In the determination of spatial modes, $\varphi_k$, it is commonly accepted that the classical method [6,7] is inefficient for typical PIV datasets—where the number of spatial locations is significantly greater than the temporal steps—and therefore the 'snap-shot' method is used [8]. The modes are then typically arranged according to their energy content, i.e., the associated eigenvalue.

For the reconstruction of velocity fields, it is possible to consider a reduced order representation of velocity if the number of POD modes; $M$ is less than the total number of snapshots. This is a commonly employed technique for filtering velocity field data. If the number of POD modes for the reconstruction is equal to the number of snapshots in the original dataset, then the reconstructions would be exactly equal to the original snap-shots. Therefore, the methodology for the selection of $M$ is integral to the analysis process—as it defines how much information is represented.

The most commonly used criteria for determining a cut-off value, $M$, are based on the cumulative energy threshold [9], $E$, defined in Equation (3):

$$\frac{\sum_{k=1}^{M} \lambda_k}{\sum_{k=1}^{T} \lambda_k} > E \tag{3}$$

where $T$ is the total number of snap-shots in the original data and $\lambda$ is the mode eigenvalue.

Defining the cut-off mode in this manner allows a total fraction of the energy desired in the reconstructions to be considered—meaning that only the most energetic structures and fluctuations are considered. Typically, 90% is used for the cumulative energy threshold [8,9], but the subjective user-defined, arbitrary nature of the criterion has led to a range of values being defined in literature between 75 and 90%. Physically, by limiting the amount of energy included in the reconstructions, desired information may be inadvertently excluded, particularly in cases with high levels of turbulence and/or when that energy is similar to periodic or cyclic fluctuations as is the case, for example, in internal combustion engines in-cylinder flow [3,10]. In addition, defining the cut-off based on energy may increase the susceptibility to corruption by erroneous velocity vectors with large magnitude and therefore large energy value, if these have not been successfully removed prior [11].

The corruption of higher energy modes by erroneous vectors is highlighted by Epps and Techet [11] who suggest the consideration of the root mean square (RMS) stating that only those modes whose RMS velocity exceeds that of the underlying PIV measurement RMS should be considered. Raiola et al. [12] propose that the POD spectrum instead be optimised based on the comparison between the POD eigenvalue decay rate and the turbulent spectrum in Fourier space, defining the cut-off mode, $M$ according to Equation (4):

$$M = \left(\frac{Tq}{\sigma^2 \xi(5/3)}\right)^{\frac{3}{5}} \tag{4}$$

where $q$ is the average turbulent kinetic energy (TKE) over all spatiotemporal data and $\xi(s)$ is the Euler–Riemann zeta function. Brindise and Vlachos [9] propose the combination of discrete cosine transform (DCT) and Shannon entropy to define which modes should be considered. A significant difference in this approach is that the modes are re-ordered by entropy before a cut-off is defined. This allows for some low order modes which may be corrupted by erroneous vectors to be disregarded, irrespective of their surrounding energy-ordered POD modes.

A novel approach to defining the cut-off is employed by Doosttalab et al. [13]. In the study of a bio-inspired coating on an airfoil, the cut-off is determined by considering the streamwise auto-correlation. An iterative approach is used, increasing the cut-off and comparing the zero-crossing point of the streamwise auto-correlation from the pre-cutoff estimation and the original data. When this occurs at the same location, this mode number is used. It is worth noting that, using this approach, the authors observed a cutoff of 532 from a possible 4000 modes. This is significantly larger than the cutoff values typically referenced elsewhere in literature.

## 2. Methodology

The literature presented thus far highlights the criteria commonly used for the selection of POD modes. The methodology presented in this paper provides an alternative approach to defining which POD spatial modes are associated with the coherent motions, and which represent flow features representative of turbulence.

The methodology is underpinned by considering a pair of repeated experiments under similar conditions. In this situation, it would be expected that when considering flow statistical measures, the two sets of results would be highly similar, albeit with some stochastic variation evident in the comparison of instantaneous flow fields. Further, performing decomposition, POD, independently on each set of results would result in similar spatial modes representing the coherent motions; with a similar number of POD modes comprising the turbulent energy content of the flow for each experiment. Therefore, by the comparison of these independently calculated POD spatial modes from each result set, it may be deduced that those spatial modes that appear in both sets contain coherent flow information. In addition, the stochastic nature of turbulence suggests that there should be no similarity in the spatial modes, other than the number of modes and the total flow energy proportion that they represent.

Building on this concept, the presented identification technique allows a new definition of coherent and stochastic turbulent parts of a single set of velocity vector field to be separated for further analysis as required. This is achieved by generating two sets of POD modes each based on half of the original set of velocity fields. The process will be described in detail by application to an experimentally obtained dataset. The details of the experimental setup are given in Section 2.1, and the identification technique is described in Section 2.2.

### 2.1. Experimental Setup

The dataset to be used for demonstration of the identification methodology is an example of the cross-flow behind a cylinder—a case which is well studied in literature [14,15]. For this study, flow fields are experimentally obtained using planar image velocimetry (PIV) in the Loughborough University (LU) horizontal water flow tunnel facility, depicted in Figure 1. The test piece is a single

1″ (25.4 mm) diameter, D, brass cylinder which is placed within the 0.3 m × 0.37 m working section, spanning the full height. The blockage co-efficient (diameter width) was 0.069. Further design details of the tunnel may be found in literature [16]. Upstream velocity, $U_0$ = 0.302. m/s, giving a Reynolds number based on cylinder diameter, $Re$ = 8600, i.e., turbulent flow.

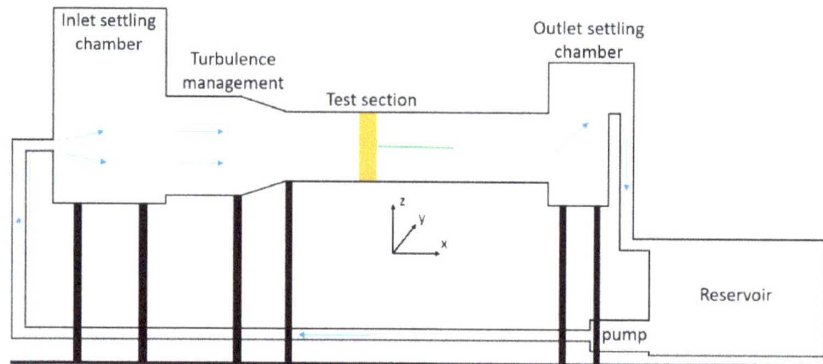


**Figure 1.** Schematic of LU horizontal water flow tunnel. Indicative measurement plane represented with horizontal green line, downstream of test piece.

Water PIV measurements were made in the mid-height plane, 0.15 m, (as indicated in Figure 1) immediately downstream of the test piece in a region covering 0.12 m × 0.12 m using well-mixed 20 µm polyamide seeding particles. A new Wave Pegasus 527 nm laser provided illumination of 10 mJ/pulse at 1 kHz. A single Photron APX 1MP camera, operating at 1 kHz, was used to capture PIV images. All timing, synchronisation and vector field processing was controlled and carried out using LaVision hardware, PTU8 and software, DaVis v7. A Multi-pass, reducing size interrogation window strategy was used on the image pairs to produce 2-component, 2-dimentional (2C2D) vector fields. Interrogation windows for initial passes were 64 × 64 with 50% overlap, with two subsequent passes at 32 × 32, 50% overlap. The resulting temporal and spatial resolutions were 1 kHz and 1.905 mm × 1.905 mm respectively. Flow was set according to an upstream, streamwise velocity of 0.3 m/s. Measurements show that the average velocity was 0.302 m/s in the streamwise direction. Upstream RMS velocity was also assessed for both streamwise and cross components and was 0.004 m/s and 0.003 m/s respectively.

It should be noted that in the experimental testing, an optical aberration was present around the center line, in the region of X = 45–60 mm. This was a result of an imperfection on the tunnel wall which caused a lens/blurring effect over a small, localized area. This is not visible in ensemble average (Figure 2b), instantaneous or decomposed figures. However, it is considered that the data in this area would not be reliable in the calculation of higher-order statistics and frequency content. Therefore, this has been masked from all subsequent presented figures, and the region has not been included for any of the presented analysis.

*2.2. Flow Characteristics*

An example of instantaneous flow and the ensemble average is presented in Figure 2a,b respectively, with the cylinder placed at (0,0). Throughout the paper, distance is non-dimensionalised by the cylinder diameter.

The flow exhibits a counter-rotating vortex pair immediately behind the cylinder, visible in the ensemble average. Free-stream regions exist either side of a central turbulent region, with a slight bias in the negative y-direction, as further evident in the profiles presented in Figure 3.

There is little fluctuation in the free-stream area, as evident in the velocity RMS presented in Figure 4, indicating low turbulence in this region. As the ensemble average is removed prior to POD analysis as is convention, it is expected that many of the POD modes will contain information relating to the central region.

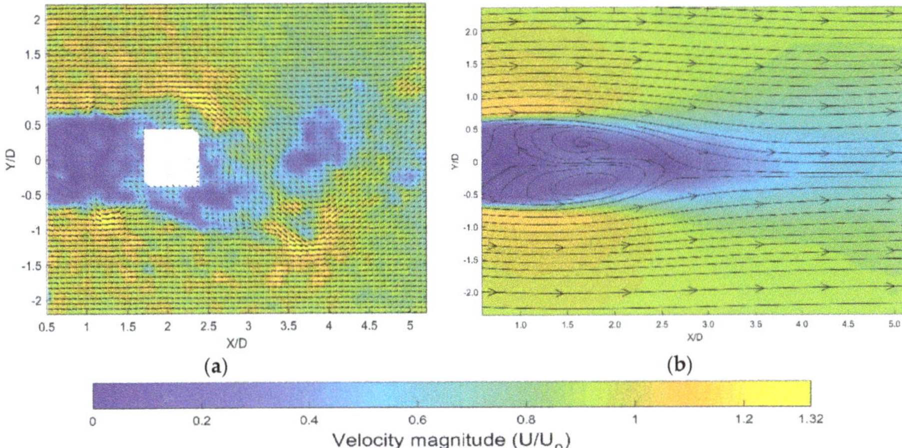

**Figure 2.** Experimentally obtained flow fields. (**a**) Instantaneous vector field, (**b**) Ensemble average stream lines. Colour represents velocity magnitude, normalized by free-stream velocity, $U_0$.

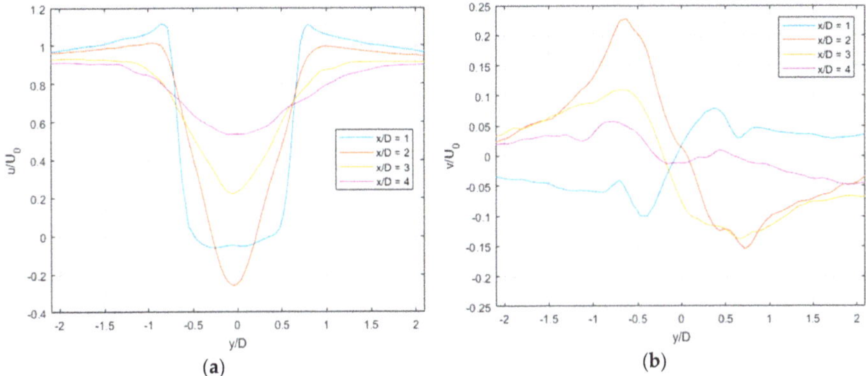

**Figure 3.** Velocity profiles (y-direction) at selected points downstream. (**a**) x-component, (**b**) y-component.

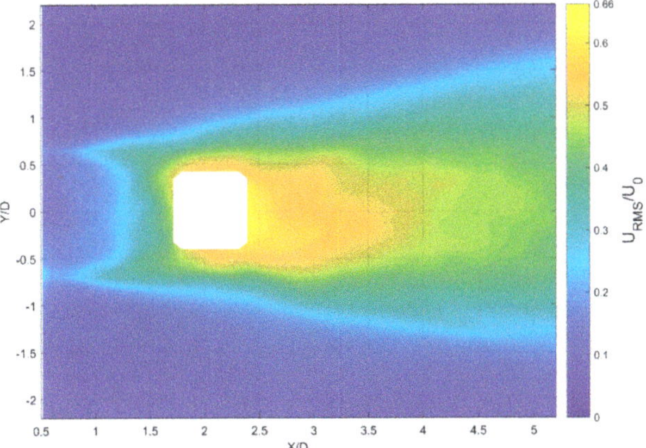

**Figure 4.** Velocity RMS distribution for full dataset.

Analysis of the velocity frequency content, presented in Figure 5, reveals a peak at between 1.953 Hz and 2.279 Hz. The true peak appears to be located between these values, with the resolution limited by the duration of the measurement.

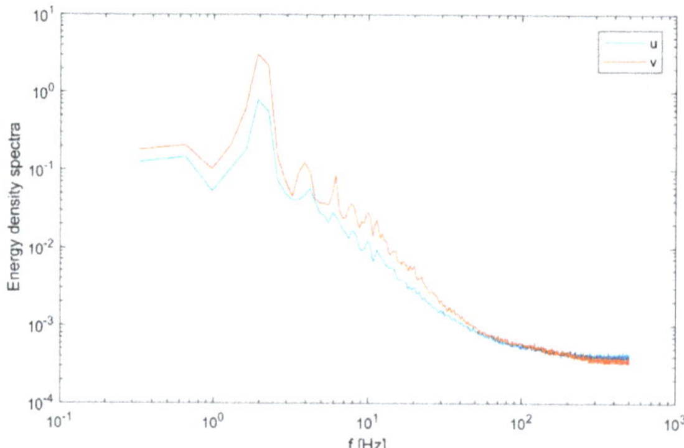

**Figure 5.** Field averaged spectral energy density for measured velocity components.

The frequencies correspond to a non-dimensional Strouhal number in the range 0.164–0.192 calculated according to Equation (5). This is related to the vortex shedding that occurs in this flow case, agreeing with reported studies in literature, for example, Perrin et al. [17,18]. In that work, flow past a cylinder in a wind tunnel is studied, with a peak frequency of 22.5 Hz observed (in air flow), corresponding to a similar Strouhal number of 0.21.

$$St = \frac{fD}{U} \tag{5}$$

*2.3. Identification Methodology*

Firstly, it should be determined that the dataset is of sufficient size (N) to adequately represent the flow features using half of the time instances captured (N/2), ordered randomly. This may be effectively assessed by considering the convergence of statistics between the two half-datasets (set A and set B) and the full dataset such as a comparison of the rms velocity component distributions. Figure 6 demonstrates parameters that may be used to assess the suitability of the number of snapshots considered. Here, the order has been randomised and the mean velocity distribution for a given number of snapshots is compared with the population mean and the velocity error, $\varepsilon$, the difference between population mean velocity at each point and the mean velocity at each point for a given number of snapshots, is evaluated by a PDF. Figure 6a presents the PDF for 1536 (i.e., half of the dataset) snapshots error from which the standard deviation may be extracted.

Further, if this process is repeated from 1 up to the number of available snapshots, the trend may be plotted as in Figure 6b. It can be seen that after a modest number of snapshots, the error is less than 0.01 magnitude, and by 1536 it is approximately of magnitude 0.001 which in this case is deemed to be acceptable. In addition to this check, the velocity RMS Figure 6c for a given number of snapshots can be seen to converge after approximately 1000 snapshots. Therefore, it is deemed that using half of the dataset, 1536 snapshots may sufficiently represent the full dataset.

Whilst here it has been determined that N/2 vector fields are sufficient, this may not always be the case; this would depend on many factors including the frequency of both the large-scale flow structures and the measurement.

Once it is determined that both set A and set B are statistically similar to each other, and to the full dataset, POD analysis may be performed on each set *independently.* Following decomposition according to Equation (2), two sets of POD spatial modes should exist, allowing a comparison between them.

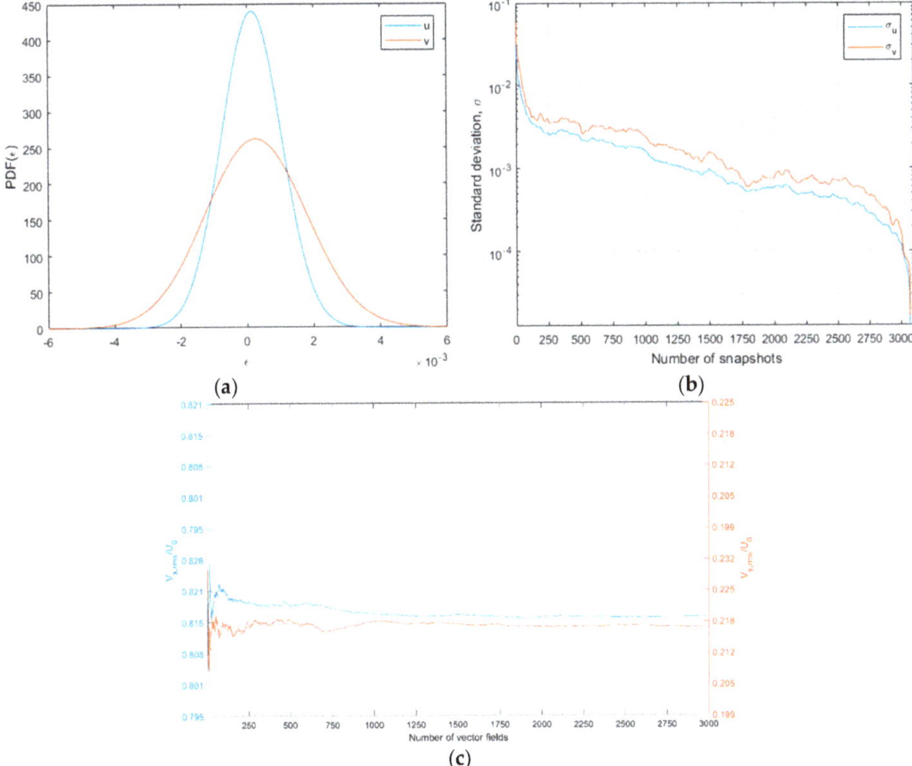

**Figure 6.** Demonstration of velocity convergence with number of samples. (**a**) PDF of error in u velocity, $\varepsilon_u$ for 1536 snapshots; (**b**) Trend of standard deviation of error, $\sigma$ with increasing samples calculated from PDFs; (**c**) Field average velocity RMS convergence.

Following conventional POD processes, the energy contained within each mode may be calculated according to Equation (3) and examined. This is calculated for set A and B and presented in Figure 7. Both sets provide a similar distribution of energy with the first two modes for each containing the majority of flow information. This indicates that much of the large-scale, dominant flow structures are captured by these modes.

It should be noted that during the application of the POD approach, the ensemble average is first subtracted. This is because it would otherwise appear as the first, most energetic mode, sometimes referred to as mode 0 in cases where it is not removed prior to POD analysis. Therefore, the decomposition is only carried out on the fluctuating velocity component. The mean is later recombined in the reconstruction.

As previously discussed, one method for determining the cut-off mode between coherent and turbulent motions is to consider the energy knee-point, which using Figure 7 would suggest mode 3. However, there may be subsequent modes which still contain coherent motions, but at lower energy. Considering the first three modes only, Figure 8 presents the associated spatial modes obtained from each of the two sets. Firstly, there is clear evidence from Figure 8e,f that visually similar spatial modes are obtained for the third mode in both sets, which would indicate the likelihood of good correlation and therefore imply that these modes are representative of coherent motions and therefore should be considered as such.

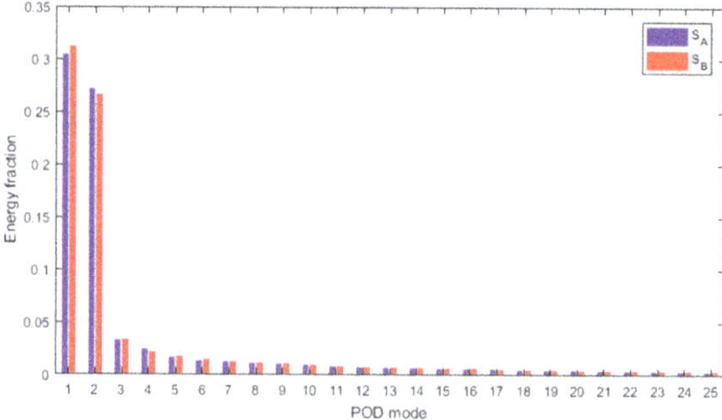

**Figure 7.** Comparison of energy fraction for each proper-orthogonal-decomposition (POD) mode of Set A and B. Curtailed at mode 25 for clarity.

Examination of both the energy content, Figure 7, and the spatial mode distributions, Figure 8, for modes 1 and 2 suggests that there is a pairing between them. Pairs of modes often exhibit very similar energy as is the case here, when their pairing is descriptive of oscillatory behaviour. By exploring the temporal coefficient information, this pairing can be further characterised. Firstly, Figure 9a presents the first two modes plotted against each other, revealing the pairing dependence of these two modes. Considering the frequency content of the temporal coefficients in Figure 9b reveals strong evidence that the oscillatory pairing is actually that which is responsible for the same frequency peak at approximately 2 Hz identified earlier. For comparison, the frequency content of mode 3, $a_3$ is also plotted, exhibiting similar frequency energy distribution across the frequency range, outside of the 2 Hz peak.

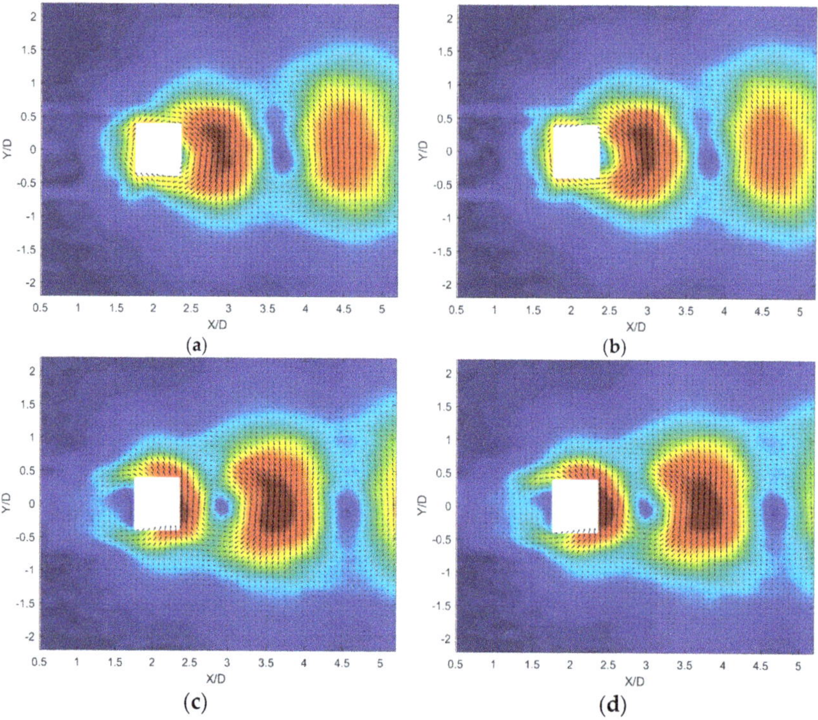

**Figure 8.** *Cont.*

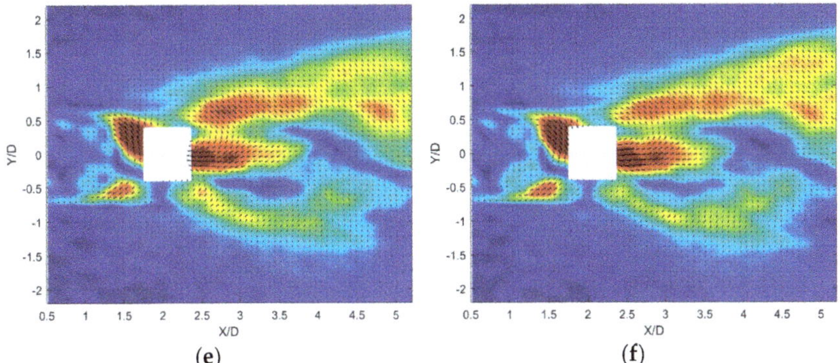

Figure 8. POD spatial modes: (a) $S_A = 1$, (b) $S_B = 1$, (c) $S_A = 2$, (d) $S_B = 2$, (e) $S_A = 3$, (f) $S_B = 3$. Color scale represents magnitude and is normalized to a common factor across all subfigures.

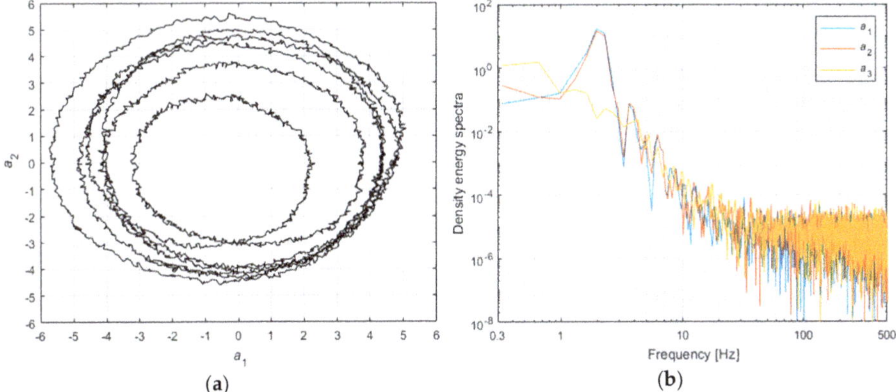

**Figure 9.** POD temporal coefficient analysis for early modes. (a) First, $a_1$ and second, $a_2$ temporal coefficients; (b) Spectral energy content for the first three sets of temporal modes.

Now, considering higher order modes, which are not expected to be representative of coherent motions, Figure 10 presents POD spatial mode 30 for each set. In contrast to those modes presented in Figure 8, there is no clear visual similarity between the two POD spatial modes. This indicates that the two presented modes are not representative of coherent motions. However, if the size of the structures represented in each mode is considered, these are of similar character and of an expected magnitude for turbulent length scales in this flow. Further, the region in which the POD spatial mode structures are contained is within the central, turbulence containing region, with insignificant contributions in the free-stream area.

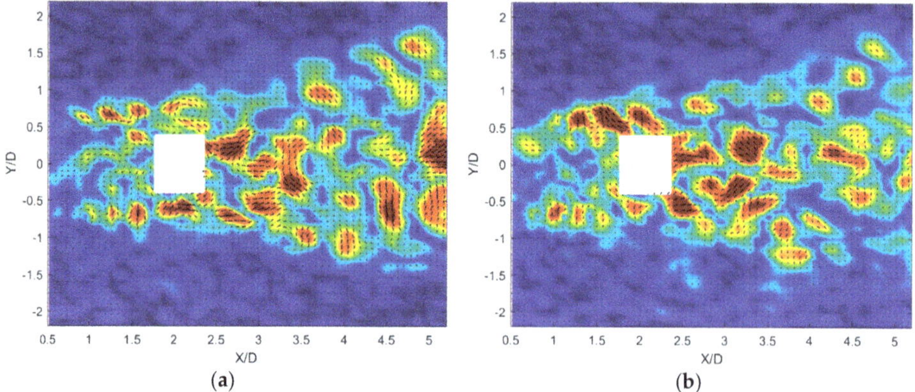

**Figure 10.** POD spatial modes for higher order: (**a**) $S_A = 30$, (**b**) $S_B = 30$. Color scale represents magnitude and is normalized to a common factor.

Similarity between two flow fields—or in this case spatial modes represented as velocity vectors—may be achieved by considering two-point statistics, specifically the cross-correlation function.

The generalized equation for the cross-correlation function between any two-points separated by the vector, *r* and time $\tau$ may be defined as:

$$R_{ij}(x, r, \tau) \;=\; \frac{\langle u_i(x,t)u_j(x+r,t+\tau)\rangle}{\sqrt{\langle u_i^2(x,t)\rangle}\,\sqrt{\langle u_j^2(x+r,t)\rangle}} \tag{6}$$

where $u_i$ and $u_j$ are the velocity components for correlation. It is common to use a simplified form of Equation (6) for correlations at the same point in time, spatial velocity correlation (SVC, $\tau = 0$) or correlations performed at the same point in space, autocorrelation function (ACF, $r = 0$). For the application presented, it is the cross-correlation of POD spatial modes that is to be considered, thus no true time association exists. However, for the purposes of the comparison, it is convenient to consider the two desired POD modes for comparison as two time-steps at the same spatial location, leading to the following equation:

$$R_{ii,AB} \;=\; \left| \frac{\langle u_i(x)_A u_i(x)_B\rangle}{\sqrt{\langle u_i^2(x)_A\rangle}\,\sqrt{\langle u_i^2(x)_B\rangle}} \right| \tag{7}$$

where subscripts A and B define the POD mode from set A and its corresponding (or other as appropriate) POD mode from set B respectively.

Each mode number is considered in turn from mode 1 through to the total number of spatial modes available in each of the two datasets. For each spatial mode, the surrounding +/−5 modes may also be compared. For example, $S_A = 10$ will be compared using Equation (7) to modes $S_B = 5{:}15$. This is to account for any minor re-ordering of POD modes which can occur when two or more consecutive POD modes have similar energy and therefore may be slightly re-ordered during the calculation of eigenvalues. This leads to a set of (11) correlation values for each mode considered, as represented in Figure 11. It should be noted that the absolute correlation should be considered as it is possible for two POD analyses of nominally the same velocity fields to produce an equal but opposite velocity spatial mode, with the corresponding reversed temporal co-efficient.

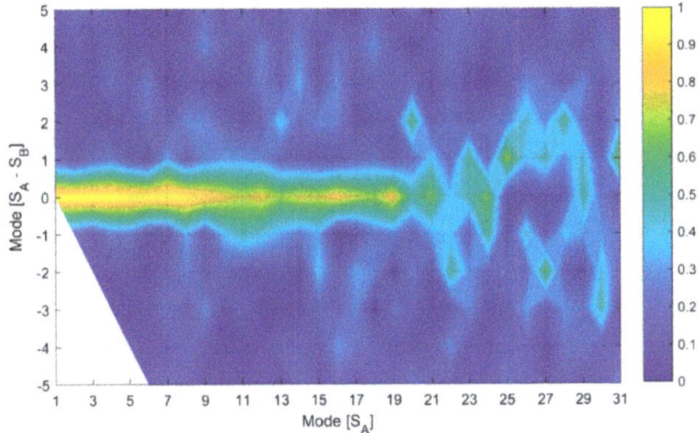

**Figure 11.** Absolute cross-correlation of POD spatial modes from set A and B. Curtailed at 31 for clarity.

It can be seen in Figure 11 that for the majority of modes in set A, the strongest correlation occurs with the corresponding mode in set B as one would expect. However, there is evidence of some mode re-ordering, for example, for the modes 18 and 22 where the strongest correlation occurs with a mode +2 and −2 respectively, showing evidence of a pair with similar energy whose order has been swapped. Following this approach, the most significant correlation value is extracted for each mode in set A to obtain Figure 12. For the purposes of selecting a cut-off mode, it has been deemed appropriate to smooth the trend to easier identify the point at which POD spatial modes have poor self-correlation. The reasoning is that in turbulent fields, with similar sized features, spikes of increased correlation may be observed for a single spatial mode. In addition, in consideration of correlation in Figure 12, it is necessary to define a value below which it is considered poor correlation. The introduction of this new parameter, $R$, allows tuning of the coherent structures to be included in the low order model. In this case, 0.6 is selected as the threshold to poor correlation.

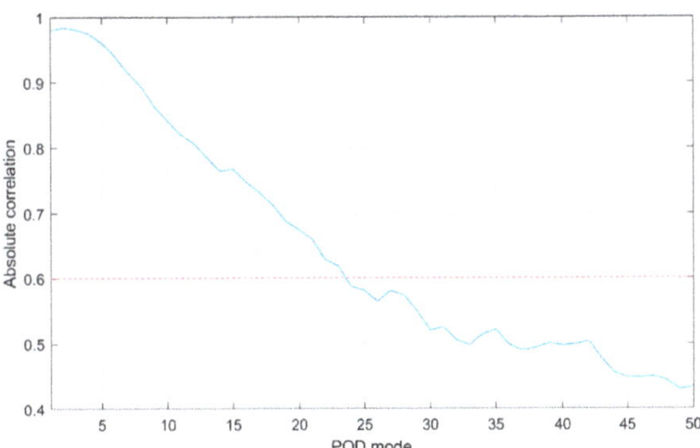

**Figure 12.** Maximum absolute correlation for each POD spatial mode (curtailed at mode 50). Smoothed data is used to produce trend. A cut-off at 60% correlation is indicated.

From Figure 12, it is apparent that poor correlation (defined here as less than 60%) occurs at greater than 23 POD modes for the datasets presented. Therefore, it follows that the first 23 modes contain high levels of coherent information and should therefore be considered when POD filtering is applied to these velocity fields. Subsequent analysis of coherent structures and their characteristics may be carried out on a reconstruction using the first 23 POD modes. Analysis on stochastic turbulence characteristics of the flow should be carried out on a reconstruction using the remaining 24-1536 POD modes (otherwise known as the residual fields). At this point, it would be possible to instead only consider the modes which have a significant correlation.

The final step of the technique is to recombine the POD spatial and respective temporal co-efficients to obtain coherent and turbulent velocity fields. The cut-off POD mode should be used (in this case 23) for the reconstruction of coherent fields of each dataset, $S_A$ and $S_B$. The two datasets may then be recombined, taking note of the randomization carried out earlier in the presented methodology. This is particularly important if subsequent frequency analysis of coherent structures is to be carried out.

The resulting decomposed flow field provides a coherent and incoherent part of each instantaneous velocity field as presented in Figure 13a,b respectively. The original, raw velocity field is also presented in Figure 13c. The effect of reconstructing the velocity field using only the POD modes associated with coherence is a smoothed velocity distribution which still contains the larger, bulk features without the small-scale structures—this is visually evident in a direct comparison between Figure 13a,c.

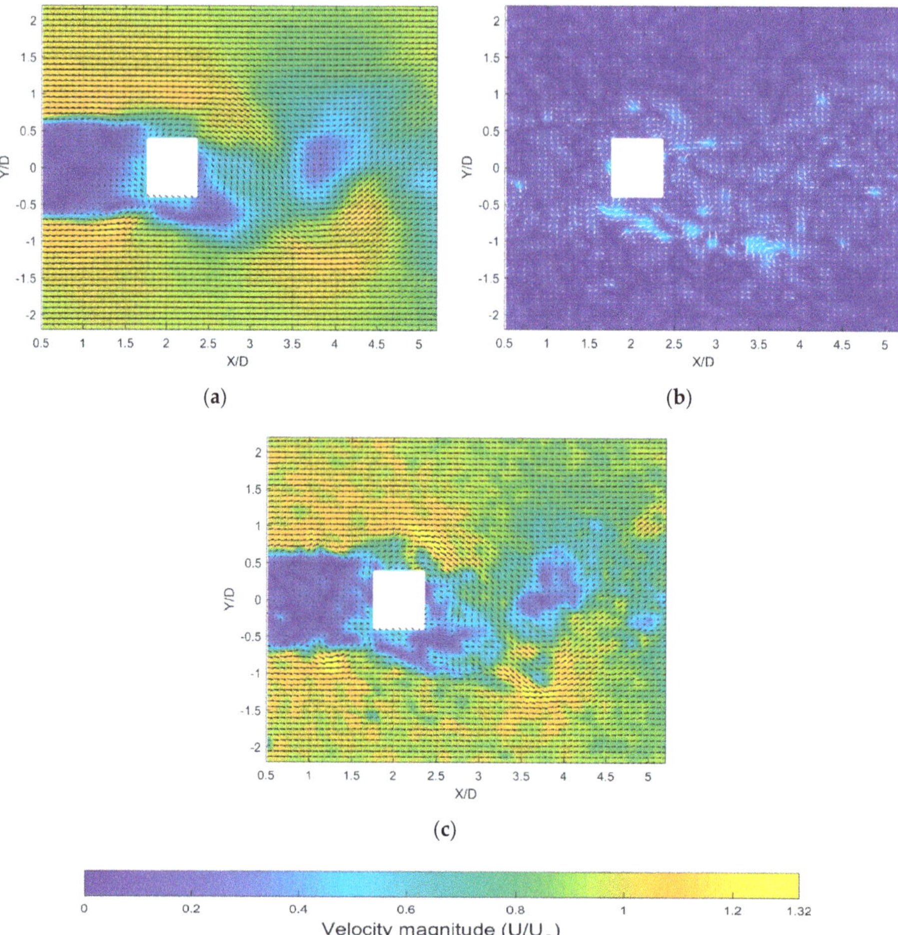

**Figure 13.** Example of a decomposed instantaneous vector field. (**a**) Coherent part; (**b**) incoherent part (vectors shown white for clarity); (**c**) raw velocity field. Colour scale represents velocity magnitude normalized with free-stream velocity, $U_0$.

The entire process is summarised as a flowchart in Figure 14 which should be followed for implementation with any new dataset. It is possible to omit some stages such as the check for convergence, if desired.

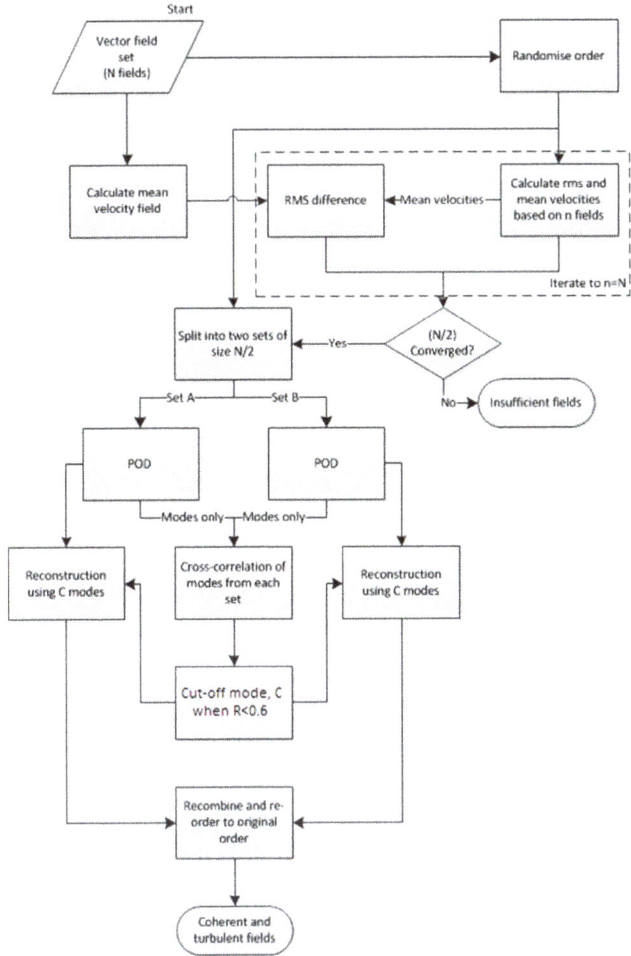


**Figure 14.** Diagrammatic overview of methodology.

## 3. Analysis of Separated Velocity Fields

Once the velocity vector distributions have been reconstructed and re-ordered in the original order, analysis of coherent structures and turbulent parts may be carried out separately, as appropriate. Figure 15 presents an example instantaneous flow field, with each of the possible decomposed velocity fields. Figure 15c,d are the results of the presented methodology, representing coherent and turbulent parts respectively.

For reference, Figure 15b presents the Reynolds decomposed fluctuations, which are equal to the sum of the fields in (c) and (d). However, the purpose of presenting the Reynolds decomposed field is to illustrate that visibly, the large-scale motions are evident, but it is difficult to make quantitative evaluation due to the inclusion of turbulence. Considering the coherent fluctuating motions only, Figure 15c, these may be characterised by large vortices responsible for the bulk transport in the y-axis direction. The structures lead to relatively large integral length scales associated with the coherent part, as is evident in Figure 16a,b. The inhomogeneous distribution of length scales is influenced by the presence of these structures with values in the region of $^1L_{11} = 22$ mm and $^2L_{22} = 50$ mm in the surrounding region (approximately 1–2 D), compared to a field average length scales of 15 mm and 42 mm respectively. Further, there is a clear link between these structures, and the first two POD modes presented in Figure 8a,c (or Figure 8b,d), underlining the dominance of these modes as expected from the significant energy contained within, as demonstrated in Figure 7. It should be noted that having a masked region within the flow has an impact on the calculated length scales, particularly where these are particularly large. This is most evident in Figure 16b; here, it is not possible to calculate the length scale immediately below the masked region as it is expected to be larger than the distance between the masked region and the edge of the domain. This strip of data should not be regarded any further in analysis. However, in the case where the integral length scale is much smaller, for example in the turbulent field, Figure 16d, there is no evidence of this artifact.

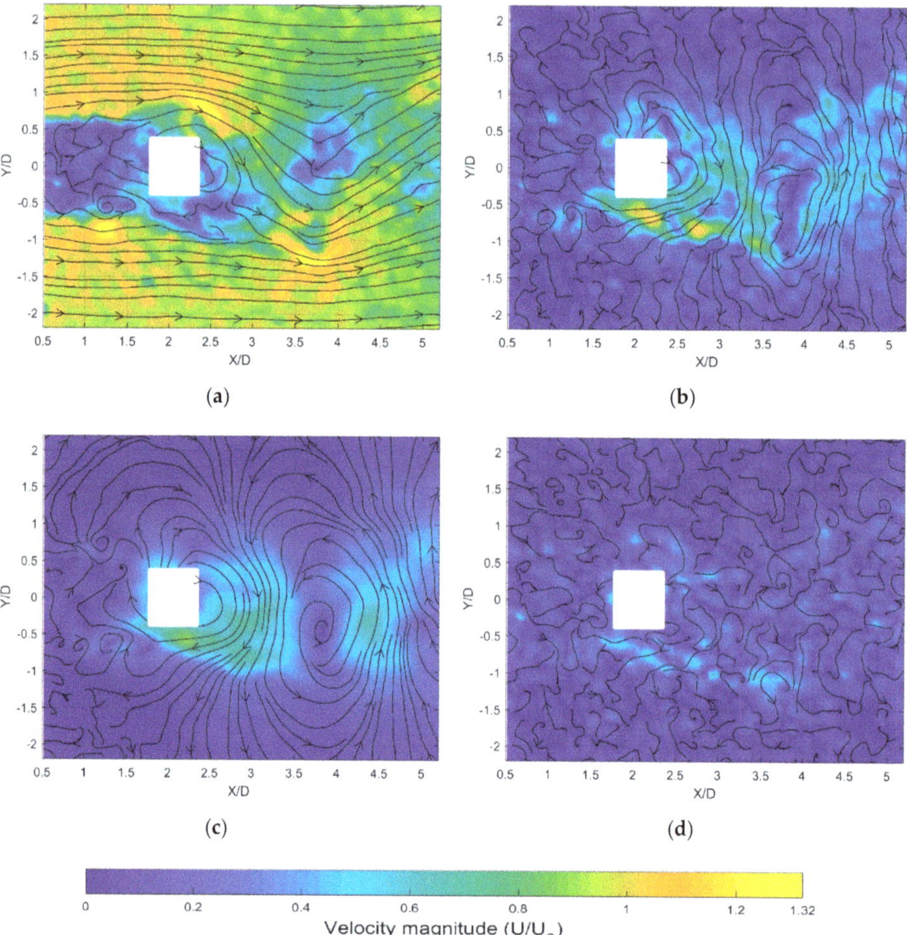

**Figure 15.** Comparison of fluctuating flow fields for an example flow field. (**a**) Original flow field; (**b**) All fluctuations (Reynolds decomposition); (**c**) Coherent part; (**d**) incoherent part. Colour scale represents velocity magnitude (common scale).

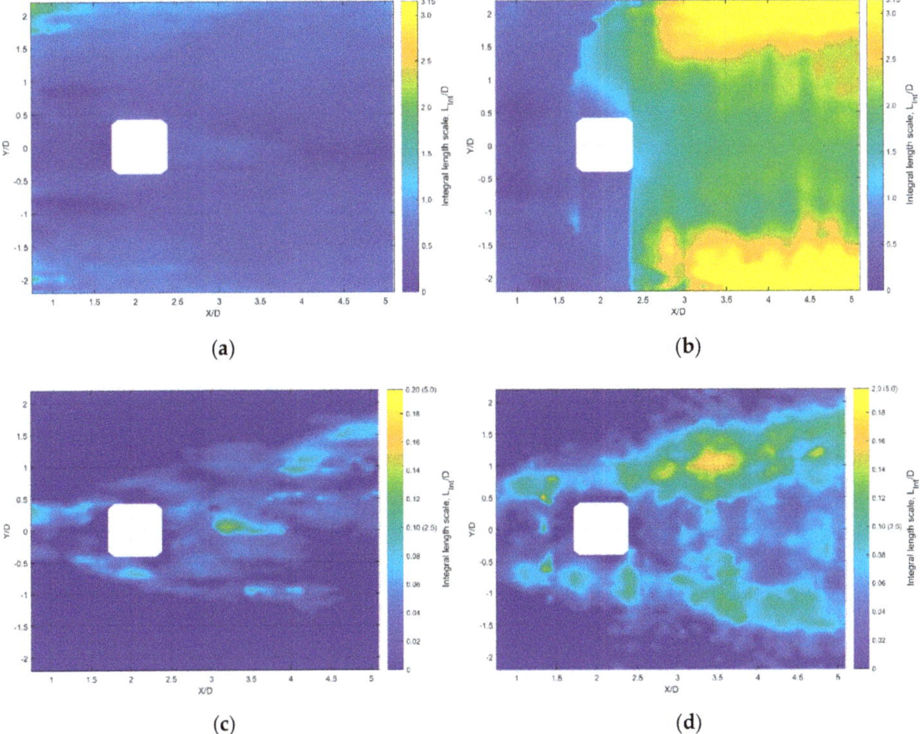

**Figure 16.** Integral length scales associated with (**a,b**) coherent fields, $^1L_{11}$ and $^2L_{22}$ respectively; (**c,d**) turbulent fields, $^1L_{11}$ and $^2L_{22}$ respectively. Note colour scale changes between subfigures. Parentheses values in (**c**) and (**d**) are mm.

Similar analysis of integral length scales of the turbulent field, presented in Figure 16c,d reveals a distribution with substantially smaller associated length scales of 1.5–3.5 mm (approximately 0.1 D). The length scale distributions show how only the wake region contains significant turbulence, as one may expect, with little evident in the freestream area.

It is commonly acknowledged that the velocity and velocity gradient distributions should conform to a Gaussian distribution in the case of stochastic turbulence. It is therefore expected that a study of these parameters in the incoherent decomposed velocity fields presented will exhibit the characteristics of turbulent flow. Figure 17 presents the bivariate distributions for u and v velocity components for each of the considered velocity sets.

Firstly, if the original velocity data is considered in Figure 17a, there is evidence of a strong bias as would be expected in the direction of the freestream velocity (i.e., positive u), peaking at close to $u/U_0 = 1$, the mean upstream velocity. Performing Reynolds decomposition effectively removes this bias, as demonstrated in Figure 17b. While the peak is now centered close to (0,0), the spread is much greater in the cross-flow direction, v, compared to streamwise due to the shape of the large (coherent) structures that can be seen to dominate, for example, in Figure 15b. This is further evidenced when considering the distribution associated with the decomposed coherent velocity data in Figure 17c which has a similar distribution, showing the domination of these motions in the Reynolds decomposed flow fields.

In contrast, Figure 17d presents the distribution associated only with the incoherent decomposed velocity fields. In this case, the velocities appear to adhere to a Gaussian distribution in both u and v components, centered on (0,0). However, in both the u and v components, calculation of the fourth standardized moment, kurtosis, shows a slight leptokurtic tendency, i.e., the distribution exhibits a more asymptotic tail than would be expected from a true Gaussian distribution. This may be attributed to experimental error noise inherent in the incoherent velocity data.

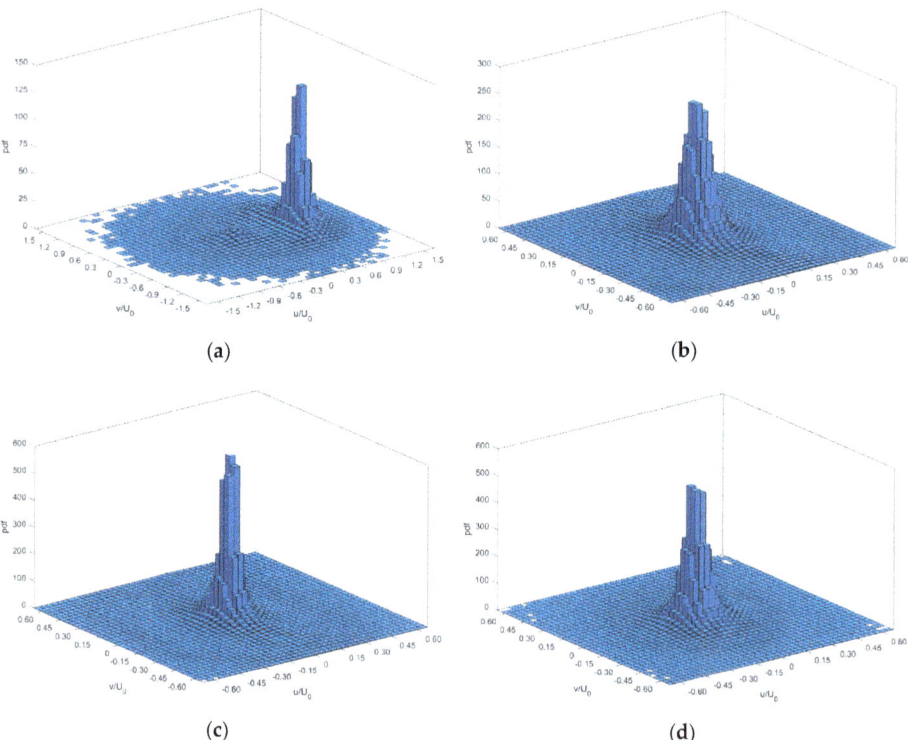

**Figure 17.** Probability density functions (pdf) of velocity components. (**a**) Original data; (**b**) all fluctuations—Reynolds decomposed; (**c**) coherent part; (**d**) turbulent part.

Further turbulent characteristics may be identified in the incoherent velocity data by consideration of the spectral energy content. Figure 18 presents the density energy spectra for both u and v components of the decomposed incoherent velocity dataset. In the figure, a line representing the −5/3 law is plotted, showing that the inertial range exhibits expected behavior up to 100 Hz, with the higher frequency content expected to be attributed to experimental error and noise.

This may be directly compared to the spectra obtained from the coherent velocity fields, presented in Figure 19. The behavior of this velocity data does not fit with the spectra expected for turbulent content. Furthermore, the peak at approximately 2 Hz is maintained from the original dataset, presented earlier in Figure 5. To aid interpretation of the presented spectra, the five-thirds law is plotted as a dashed line to contextualise the gradient.

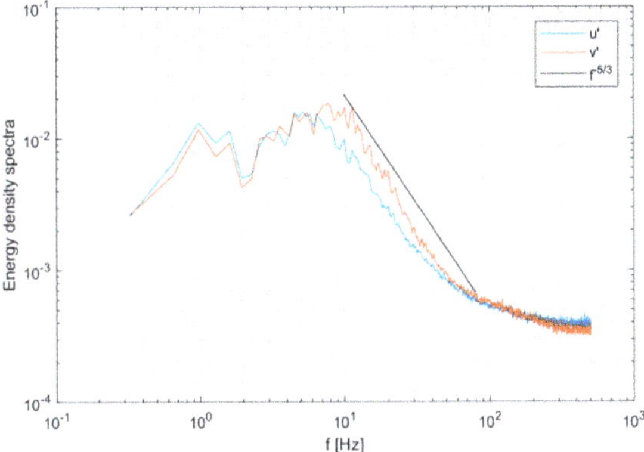

**Figure 18.** Spectral energy density for turbulent fluctuations, u′. Five-thirds law plotted for reference.

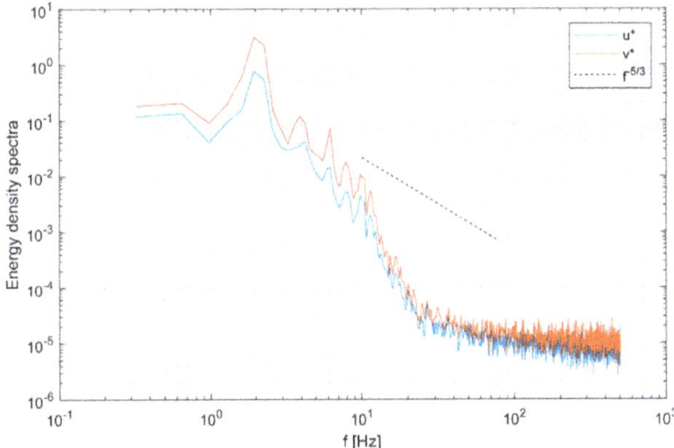

**Figure 19.** Spectral energy density for coherent fluctuation velocities, u*. Five-thirds law plotted for reference.

In a similar fashion to the length scales discussed, it is possible to use the time resolved point-based information to calculate the integral time scales associated with the flow. Figure 20a,b present the integral time scales associated with the decomposed turbulent fields for u-component and v-component respectively. It can be seen that the calculated values within the wake region are comparable for both components, indicating significant isotropy.

Outside of the wake area, much lower values are calculated, of the order 0.001 and below; this appears to be a consequence of the flow in this area having negligible turbulence, as shown in Figure 4. Therefore, once the coherent content is removed, the calculation is not reliable in this area due to low auto-correlation. The average time scale may be obtained from the two distributions in Figure 20, allowing the equivalent frequencies to be calculated, presented in Figure 21. The upper limit for colour scale is chosen to cover the range of frequencies within the wake region (high RMS). It is evident that the majority of the content in the wake region is below 150 Hz, suggesting that the tail in Figures 18 and 19 is associated with experimental noise, evident when considering the freestream region. This noise may be filtered by an appropriate method such as low-pass frequency filter, as required.

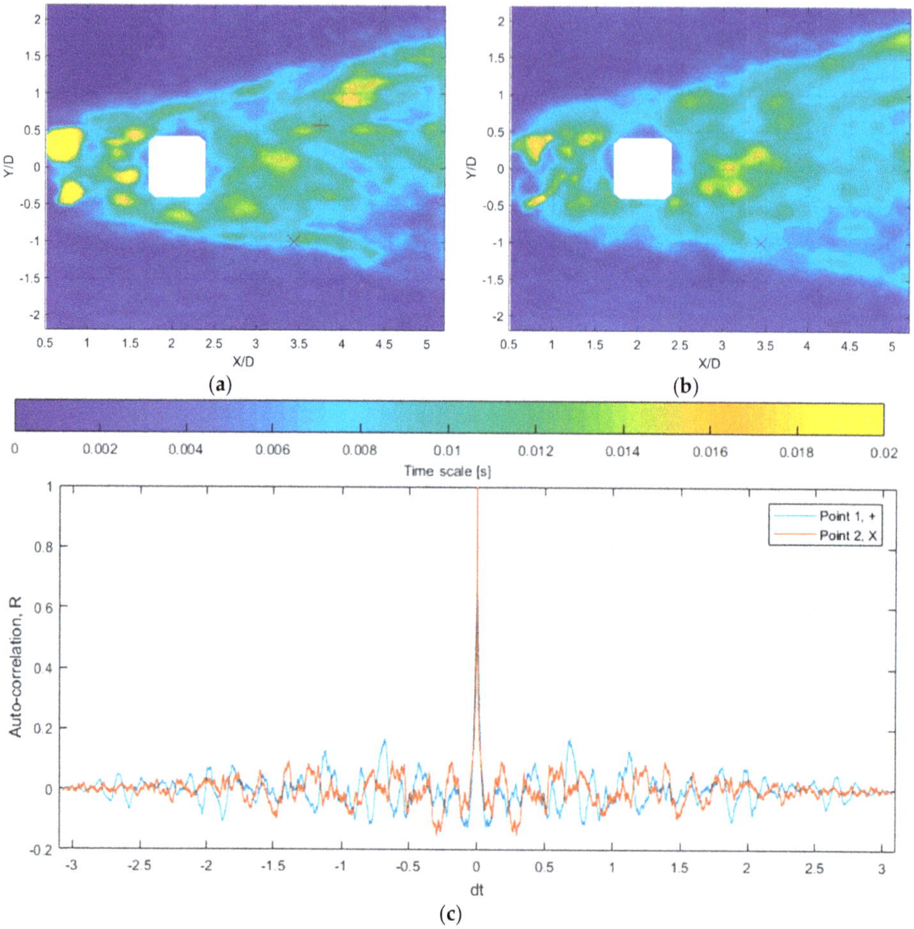

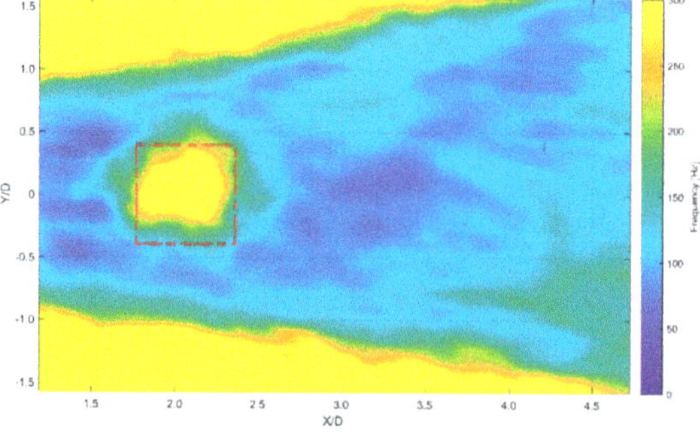

**Figure 20.** Integral time scale distributions associated with turbulent parts. (**a**) uu; (**b**) vv; (**c**) Auto-correlation function, uu about two example points, locations indicated in (**a**) and (**b**).

**Figure 21.** Spatial distribution of frequency content associated with integral time scales in turbulent field $(1/\tau)$, scaled appropriately to the wake region only.

Evident in the wake region is an anomalous area which has been omitted from analysis throughout the paper and masked in all figures except Figure 2a. An optical aberration, as mentioned in the method section was responsible for this. In all but the time and length scale analysis, the data in this region has looked contiguous and credible within the rest of the field (up to and including Figure 15). However, the frequency and lengthscale analysis identified that there is a strong noise content to the velocity signal in this region, hence it has not been included within the analysis. This is highlighted here to show how only considering first and second order statistical analysis of data may not always indicate when there could be underlying quality issues with the measured data.

## 4. Conclusions

In this paper, a methodology for the separation of coherent motions and turbulent fluctuations from spatially distributed experimentally obtained datasets is suggested and described. The methodology uses the well-established proper orthogonal decomposition (POD) technique to derive spatial eigenmodes associated with the flow which are then cross-correlated between two independent datasets, introducing a new metric, $R$. This value presents the correlation matrix between spatial modes of the independent datasets, allowing definition of those which relate to coherent motions, through high correlation, and those related to incoherent motions, through low correlation.

The methodology has been demonstrated on the commonly studied condition, a cylinder in cross-flow, revealing reported phenomena in addition to information that would otherwise not be possible to extract. Distributions of turbulent length and time scales reveal the characterisation of the structures within the wake region that would overwise be distorted by including coherent fluctuations within the analysis, for example, considerably overestimating the integral lengthscale.

**Author Contributions:** Conceptualization, A.S. and D.B.; Methodology, D.B. and A.S.; Programming, D.B.; Manuscript preparation, D.B.; Manuscript review and editing D.B. and A.S.

**Funding:** This research received no external funding.

**Conflicts of Interest:** The authors declare no conflict of interest.

## References

1. Ölçmen, S.; Ashford, M.; Schinestsky, P.; Drabo, M. Comparative Analysis of Velocity Decomposition Methods for Internal Combustion Engines. *Open J. Fluid Dyn.* **2012**, *2*, 70–90. [CrossRef]
2. Stone, R. *Introduction to Internal Combustion Engines*, 4th ed.; Palgrave Mscmillan: Basingstoke, UK, 2012; ISBN1 9780230576636 (hbk.). ISBN2 023057663X.
3. Butcher, D.S.A. *Influence of Asymmetric Valve Timing Strategy on In-Cylinder Flow of the Internal Combustion Engine*; Loughborough University: Loughborough, UK, 2016.
4. Adrian, R.J.; Christensen, K.T.; Liu, Z.-C. Analysis and interpretation of instantaneous turbulent velocity fields. *Exp. Fluids* **2000**, *29*, 275–290. [CrossRef]
5. Lumley, J.L. The Structure of Inhomogeneous Turbulent Flows. In *Atmospheric Turbulence and Wave Propagation*; Publishing House Nauka: Moscow, Russia, 1967; pp. 166–178.
6. Chatterjee, A. An introduction to the proper orthogonal decomposition. *Curr. Sci.* **2000**, *78*, 808–817. [CrossRef]
7. Holmes, P.; Lumley, J.L.; Berkooz, G. *Turbulence, Coherent Structures, Dynamical Systems and Symmetry*; Cambridge University Press: Cambridge, UK, 1996.
8. Sirovich, L. Turbulence and Dynamics of Coherent Structues. Part I: Coherent Structures. *Q. Appl. Math.* **1987**, *45*, 561–571. [CrossRef]
9. Brindise, M.C.; Vlachos, P.P. Proper orthogonal decomposition truncation method for data denoising and order reduction. *Exp. Fluids* **2017**, *58*, 28. [CrossRef]
10. Butcher, D.; Spencer, A.; Chen, R. Influence of asymmetric valve strategy on large-scale and turbulent in-cylinder flows. *Int. J. Engine Res.* **2018**, *19*, 631–642. [CrossRef]
11. Epps, B.P.; Techet, A.H. An error threshold criterion for singular value decomposition modes extracted from PIV data. *Exp. Fluids* **2010**, *48*, 355–367. [CrossRef]

12. Raiola, M.; Discetti, S.; Ianiro, A. On PIV random error minimization with optimal POD-based low-order reconstruction. *Exp. Fluids* **2015**, *56*. [CrossRef]

13. Doosttalab, A.; Dharmarathne, S.; Bocanegra Evans, H.; Hamed, A.M.; Gorumlu, S.; Aksak, B.; Chamorro, L.P.; Tutkun, M.; Castillo, L. Flow modulation by a mushroom-like coating around the separation region of a wind-turbine airfoil section. *J. Renew. Sustain. Energy* **2018**, *10*. [CrossRef]

14. Norberg, C.; Sunden, B. Turbulence and reynolds number effects on the flow and fluid forces on a single cylinder in cross flow. *J. Fluids Struct.* **1987**, *1*, 337–357. [CrossRef]

15. Park, J.; Choi, H. Numerical solutions of flow past a circular cylinder at reynolds numbers up to 160. *KSME Int. J.* **1998**, *12*, 1200–1205. [CrossRef]

16. Behrouzi, P.; McGuirk, J. Experimental data for CFD validation of impinging jets in cross-flow with application to ASTOVL flow problems. In Proceedings of the AGARD Conference Proceedings 534, Computational and Experimental Assessment of Jets in Cross-Flow, Winchester, UK, 19–22 April 1993.

17. Perrin, R.; Cid, E.; Cazin, S.; Sevrain, A.; Braza, M.; Moradei, F.; Harran, G. Phase-averaged measurements of the turbulence properties in the near wake of a circular cylinder at high Reynolds number by 2C-PIV and 3C-PIV. *Exp. Fluids* **2007**, *42*, 93–109. [CrossRef]

18. Perrin, R.; Braza, M.; Cid, E.; Cazin, S.; Barthet, A.; Sevrain, A.; Mockett, C.; Thiele, F. Obtaining phase averaged turbulence properties in the near wake of a circular cylinder at high Reynolds number using POD. *Exp. Fluids* **2007**, *43*, 341–355. [CrossRef]

*Article*

# Numerical Modelling of Air Pollutant Dispersion in Complex Urban Areas: Investigation of City Parts from Downtowns Hanover and Frankfurt

**Mohamed Salah Idrissi [1], Nabil Ben Salah [2] and Mouldi Chrigui [1,3,*]**

[1]  Research Unit "Mechanical modeling, Materials and Energy", National school of engineer of Gabes, University of Gabes, Gabes 6029, Tunisia
[2]  CADFEM AN, Sousse Technopole, Sousse 4000, Tunisia
[3]  Institute for energy and power-plant Technology, Darmstadt University of Technology, Otto-Berndt-Str.3, 64287 Darmstadt, Germany
*  Correspondence: mouldi.cherigui@gmx.de; Tel.: +216-753-92-100

Received: 7 May 2019; Accepted: 16 July 2019; Published: 18 July 2019

**Abstract:** Hazardous gas dispersion within a complex urban environment in 1:1 scaled geometry of German cities, Hanover and Frankfurt, is predicted using an advanced turbulence model. The investigation involves a large group of real buildings with a high level of details. For this purpose, Computer Aided Design (CAD) of two configurations are cleaned, then fine grids meshed in. Weather conditions are introduced using power law velocity profiles at inlets boundary. The investigation focused on the effects of release locations and material properties of the contaminants (e.g., densities) on the convection/diffusion of pollutants within complex urban area. Two geometries demonstrating different topologies and boundaries conditions are investigated. Pollutants are introduced into the computational domain through chimney and/or pipe leakages in various locations. Simulations are carried out using Large Eddy Simulation (LES) turbulence model and species transport for the pollutants. The weather conditions are accounted for using a logarithmic velocity profile at inlets. $CH_4$ and $CO_2$ distributions, as well as turbulence quantities and velocity profiles, show important influences on the dispersion behavior of the hazardous gas.

**Keywords:** buildings; urban area; pollution dispersion; turbulence; Large Eddy Simulation (LES)

---

## 1. Introduction

A growing concern about air pollution in urban environments results in stringent legislation for devices certification and energy utilization [1]. Several studies have been interested in the pollutant emissions reduction and the optimization of combustion systems by the use of new burners [2,3] or by capturing target species [4]. To assess the problem in detail, accurate prediction of the transport and dispersion of airborne contaminants in urban environments is needed. Several methods have been used to investigate air pollution [5,6]. Wind tunnels only provide an understanding of the dynamics and scaled data. Such approaches are able to perform parameters sampling points and to deliver well-documented data sets for the evaluation of numerical models [7]. Yet, due to scale limitations, it is a challenge for such models to fully be tested in large scale atmospheric turbulent flow conditions [8]. Numerical approaches have numerous advantages, e.g., high controllability, lower cost, three-dimensional (3D) data output, etc. [9].

Many preceding studies showed that Computational Fluid Dynamics (CFD) simulations, based on the Reynolds averaged Navier-Stokes (RANS) equations, delivered unaccrued results in reproducing the wind-flow pollutant dispersion around buildings [10–14]. Britter and Hanna [11] pointed out that the RANS turbulence model can be used in simple geometry (isolated building) to produce reasonable qualitative results for mean flows. When compared with laboratory or field experiments, this model showed a better performance than that of simple operational models. Blocken et al. [13] tested three cases of tracer gas dispersion in a neutrally stable atmospheric boundary layer and compared their results against wind tunnel full-scale measurements. For all three cases, they concluded that RANS simulations significantly underestimate the cross-wind (lateral) dispersion of the pollutants. Tominaga and Stathopoulos [14] numerically studied the dispersion around an isolated cube. They found that RANS models underpredict horizontal concentrations at the roof of simulated buildings.

To reach more accurate results, several researchers have turned to the Large-Eddy Simulation (LES) approach, which demonstrated its capability and potential in microscale atmospheric dispersion modelling [15–25]. Liu and Barth [15] studied the flow in 3D streets using scalar transport. Their results showed the limitations of eddies developments with sizes larger than 0.5H (H height of the building) in the so-called free surface layer and produce less turbulent kinetic energy (TKE) above the rooftop. Sada and Sato [18] simulated the stack-gas diffusion around a cubical building in order to predict the instantaneous concentration fluctuation of a plume. Their results were compared with data from wind tunnel experiments and showed a good agreement. Liu et al. [25] used the one-equation subgrid-scale (SGS) parameterization and the LES model to study the flow and pollutant transport in 2D urban street canyons. Their results showed that the combined model is an appropriate method for predicting wind field and pollutant dispersion in the crowded urban area.

Recent works [26–30] focused on the study of more complex geometries. Gausseau et al. [26] used LES to investigate the effect of two wind directions on near-field pollutant dispersion from stacks located at a low-rise building in downtown Montreal Canada. Their results showed some similarities with the case of dispersion around an isolated building. Van Hoof and Blocken [28] studied the $CO_2$ dispersion in a semi-enclosed stadium and validated their results using onsite measurements. The validated model was used to evaluate the carbon dioxide concentration in the stadium. Amorim et al. [29] chose two European urban areas (Lisbon and Aveiro, Portugal) to evaluate the effect of trees over the dispersion of carbon monoxide (CO) emitted by road traffic. They found that trees are responsible for an increase of 12% CO concentration in the first configuration (nonaligned wind 45°). In the second configuration (aligned wind), they observed a decrease of 16% of CO concentration.

In light of presented studies, there are many qualitative, as well as quantitative parameters, which need to be studied when investigating pollutants dispersion in an urban area, especially for complex geometries. Hence, the present study aims at providing more insight into the dispersion process by analyzing the effect of buildings arrangement and the material properties (such as density) on the dispersion of hazardous gases. For this purpose, two different configurations, which exhibit complex geometries of Hanover and Frankfurt cities, Germany, are investigated. The most challenging point is the high density of the surrounding buildings, which makes the carrier phase flow and pollutants concentration field complex, anisotropic, and difficult for prediction. Different scenarios of pollutant injection into the computational domain are tested. Dense buildings are included up to a distance of 600 m in diameter and a high-resolution refined grid with over 40 million cells was used. LES simulations are performed, then results are compared against published data from Leitl, B. and Schatzmann. [30] in order to validate the numerical models [31].

## 2. Methodology

### 2.1. Turbulence Model

In LES, the flow variables are decomposed into the resolved-scale components and the (SGS) components using a filtering operation [32]. The filtering process effectively filters out eddies whose scales are smaller than the filter width or grid spacing used in the computations. Resulting in filtered, continuity and momentum equations, the incompressible Navier–Stokes equations are given as follows:

$$\frac{\partial \overline{\rho}}{\partial t} + \frac{\partial \overline{\rho \widetilde{u}_i}}{\partial x_i} = 0 \tag{1}$$

$$\frac{\partial}{\partial t}(\overline{\rho}\widetilde{u}_i) + \frac{\partial}{\partial x_j}(\overline{\rho}\widetilde{u}_i\widetilde{u}_j) = -\frac{\partial \overline{p}}{\partial x_i} + \overline{\rho}g_i + \frac{\partial}{\partial x_j}\left[\overline{\rho v}\left[\frac{\partial \widetilde{u}_i}{\partial x_j} + \frac{\partial \widetilde{u}_j}{\partial x_i}\right] - \frac{2}{3}\overline{\rho v}\frac{\partial \widetilde{u}_k}{\partial x_k}\delta_{ij} - \overline{\rho}\tau_{ij}^{SGS}\right] + \overline{S} \tag{2}$$

where bars and over-bars express mean and filtered quantities. The variables $u$, $\rho$, $p$, $g$, and $\delta_{ij}$ denote the velocity, the density, the hydrostatic pressure, the gravitational force, and the kronecker delta, respectively.

The quantity $\widetilde{v}$ represents the molecular viscosity and $\overline{S}$ is the source term. The Smagorinsky SGS model is applied to close the system of equations and determine the SGS stresses $\tau_{ij}^{SGS}$ using the Equation (3)

$$\tau_{ij}^{SGS} = \widetilde{u_i u_j} - \widetilde{u}_i\widetilde{u}_j = -2v_{SGS}\widetilde{S}_{ij} + \frac{1}{3}\delta_{ij}\tau_{kk}^{SGS} \tag{3}$$

### 2.2. Dispersion Model

Numerous models are accessible to predict the dispersion of airborne material. In this study, pollution dispersion was modelled using the species transport approach by calculating convection-diffusion as described by Equation (4). This equation is solved for N-1 species, where N is the total number of species in the fluid flow. The sum of the mass fractions of all species must be equal to unity, the Nth mass fraction is given from the sum of the N-1 solved mass fractions.

The conservation equation takes the following general formula:

$$\frac{\partial}{\partial t}(\rho\widetilde{Y}_i) + \frac{\partial}{\partial x_j}(\rho\widetilde{u}_j\widetilde{Y}_i) = \frac{\partial}{\partial x_j}\left(\rho D_i\frac{\partial \widetilde{Y}_i}{x_j}\right) + \widetilde{R}_i + \widetilde{S}_i \tag{4}$$

where $\rho$ is the density, $\widetilde{R}_i$ is the net rate of production of species by chemical reaction, $\widetilde{S}_i$ is the rate of creation by addition from the dispersed phase, $D_i$ is the diffusion flux of species, $\widetilde{Y}_i$ is the mass fraction of specie $I$, and $\widetilde{u}_j$ is the flow velocity.

### 2.3. Numerical Procedure

The commercial CFD software ANSYS FLUENT 17.0 is employed to simulate the airflow and pollutants dispersion. For all configurations, the dispersion of the pollutant is simulated using LES and dynamic Smagorinsky, SGS, along with the species transport model. The bounded central-differencing scheme is used to discretize the momentum equation. A second order upwind scheme is applied for the pollutant concentration and energy equation. Time integration is bounded to second order implicit and pressure interpolation is second order. For both cases, the time step advancement is set to $\Delta t = 10^{-3}s$, which produces a Courant number "C" less than one within the computational domain. The presented results are averaged over a period of 220 s, which corresponds to five flow-through times.

### 2.4. Validation of the Numerical Model

In order to validate the numerical model, the (CEDVAL B1-1) configuration [30] (Figure 1) is investigated using the same boundaries conditions and physical models. This study was conducted in the BLASIUS wind tunnel at the Meteorological Institute of the University of Hamburg [30]. A modified Standing-Spires and uniform LEGO-roughness was used to generate a neutral atmospheric boundary layer ABL at a scale of 1:200 with $Re = 37{,}252$. First, the boundary layer flow was validated based on detailed flow measurements and comparison with full-scale data. Then, the building models were placed in the test section. Inside the wind tunnel, a 3 $\times$ 7 array of buildings composed of rectangular blocks with the same dimensions H = 0.125 m are used to model the dispersion of carbon dioxide ($CO_2$). The $CO_2$ was emitted by four sources, located at the leeward of building source, with exhaust velocity $we = 0.033$ m/s. Validation is achieved by comparing LES results against the published measurement [30]. The horizontal plane at z = 1.5 m (full-scale) height was used as a measurement plane in which pollutants data were acquired. The same experiment is also simulated by Longo et al. [31]. Identical computational domain, as well as boundary conditions, are used for the present work. The inlet boundary is set to 1 m upstream of the buildings, in which the ABL profiles are measured within the wind tunnel. The outlet boundary is located at 4 m downstream of the last array of the buildings. The width, depth and height of the domain are 1 m, 5 m, and 1 m, respectively, which correspond to the wind tunnel size.

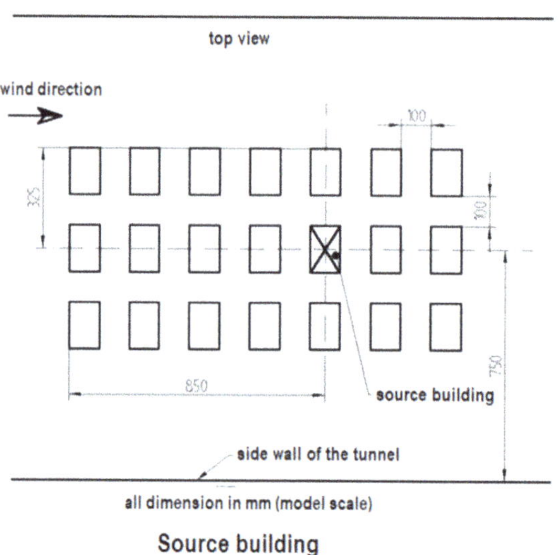


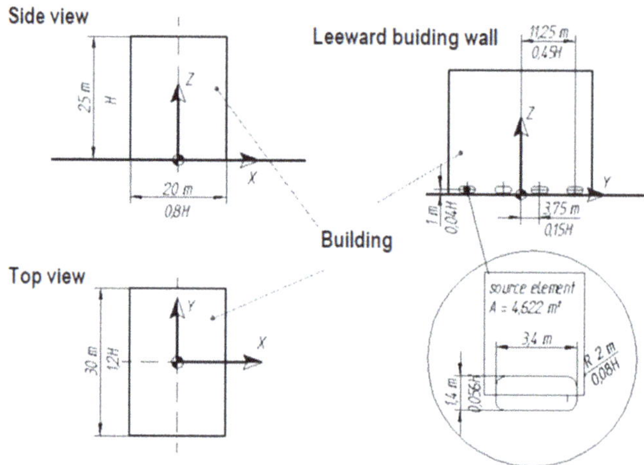


**Figure 1.** Dimensions of the building models, the source building, and the source [30].

In the simulation, a power law velocity profile is used at the inlet, a zero-pressure gradient is used at the outlet and top, no-slip velocity is used at the ground and walls, and symmetry is used for lateral boundary conditions. All the boundaries and initial conditions are summarized in Table 1. Although the temperature is isothermal (300 k), due to thermal stratification, the buoyancy is not present. Yet, we do have stratification due to density difference between carbon dioxide and air.

**Table 1.** Initial and boundaries conditions used for CEDVAL configuration.

| CEDVAL B1-1 | | | |
| --- | --- | --- | --- |
| **Boundary** | **Velocity** | **Temperature** | **Species** |
| Inlet | Profile: Experimental Data | 300 K | air |
| Pollutant source | 0.033 m/s | 300 K | $CO_2$ |
| Outlet | Neuman BC for the pressure | - | - |
| Top | Neuman BC for the pressure | - | - |
| Walls | No slip | - | - |

An unsteady, 3D, double precision, pressure-based solver with second order schemes for momentum, turbulence quantities, and concentrations are used. The coupled algorithm is selected for pressure-velocity coupling. A transport equation approach is used for the dispersion study. A grid-independent performance of the numerical simulation results is verified by comparing the results of test cases having different grid numbers. Three meshes with different levels of resolution are investigated. A fine resolved mesh of 10 million elements is used for all computations. Simulations are performed using parallels computation of 40 CPUs. The total flow period is set to 12.0 s, which allows the flow to sweep the domain five times.

Figure 2 shows a comparison between the dimensionless turbulent kinetic energy, dimensionless velocity components, and the dimensionless concentration of $CO_2$ obtained by numerical simulation against the wind tunnel experimental data. This figure demonstrates a good agreement between LES numerical results and measurements. The values of turbulent kinetic energy, velocity components, and $CO_2$ concentration are well-predicted and follow the experimental measurements, demonstrating the ability of LES Smagorinsky, SGS, and species transport models in predicting the pollutants dispersion in atmospheric conditions.

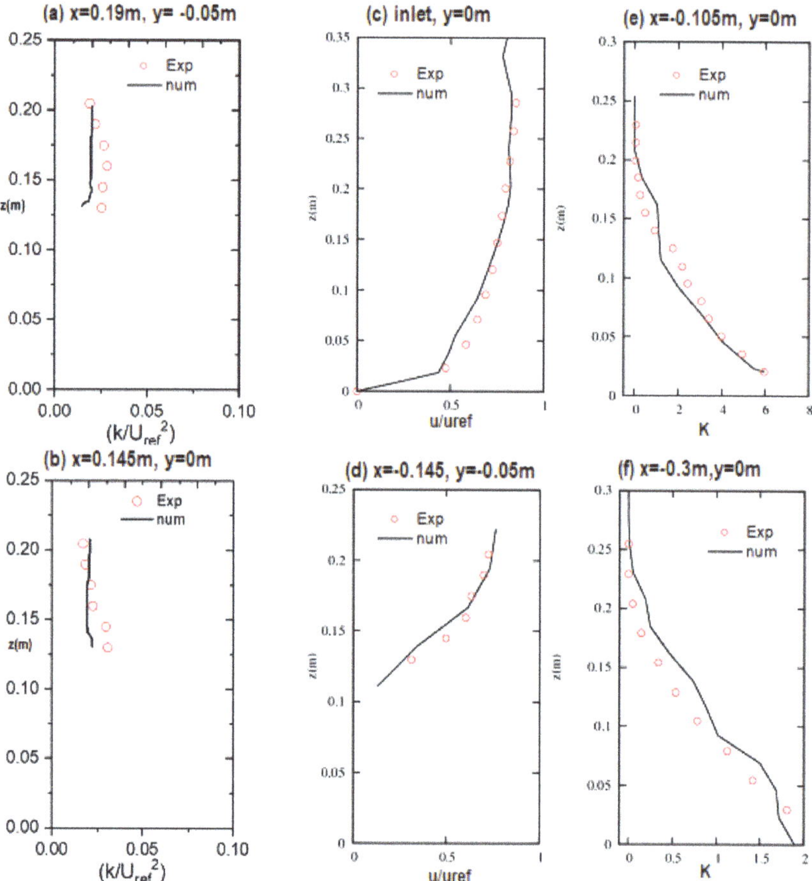

**Figure 2.** Comparison between the numerical and experimental results of the normalized turbulent kinetic energy (**a,b**), non-dimensional velocity (**c,d**), and dimensionless concentration (**e,f**).

## 3. Computational Domain and Boundaries Conditions

### 3.1. Geometry and Boundaries Conditions

In order to represent real case scenarios, two computational domains are generated, one for each case (Figure 3a,b). In case 1 (Hanover city), the focus is placed on the effect of buildings arrangement, whereas in case 2 (Frankfurt city), the behaviors of two hazardous gases with different densities are studied. At the inlet, flow is perpendicular to the injection plane. The streamwise, spanwise, and vertical coordinates are represented by x, y, and z, respectively. The buildings are modeled based on the existing full-scale data. To limit the number of cells, vegetation is not considered. Figure 3a depicts the computational domain used to simulate the configuration of Hanover city sited at the Latitude and Longitude coordinates of 52°22′18.3″N and 9°44′23.3″E, respectively. The dimensions of the computational domain are L = 600 m in length, W = 580 m in width, and H = 200 m in height, inside of which various buildings having different dimensions and showing multiple features are included. In the case of Frankfurt city, shown in Figure 3b, the configuration is located at the coordinates 50°6′49.19″N and 8°39′51.13″E. The dimensions of the computational domain are L = 500 m in length, W = 500 m in width, and H = 300 m in height.

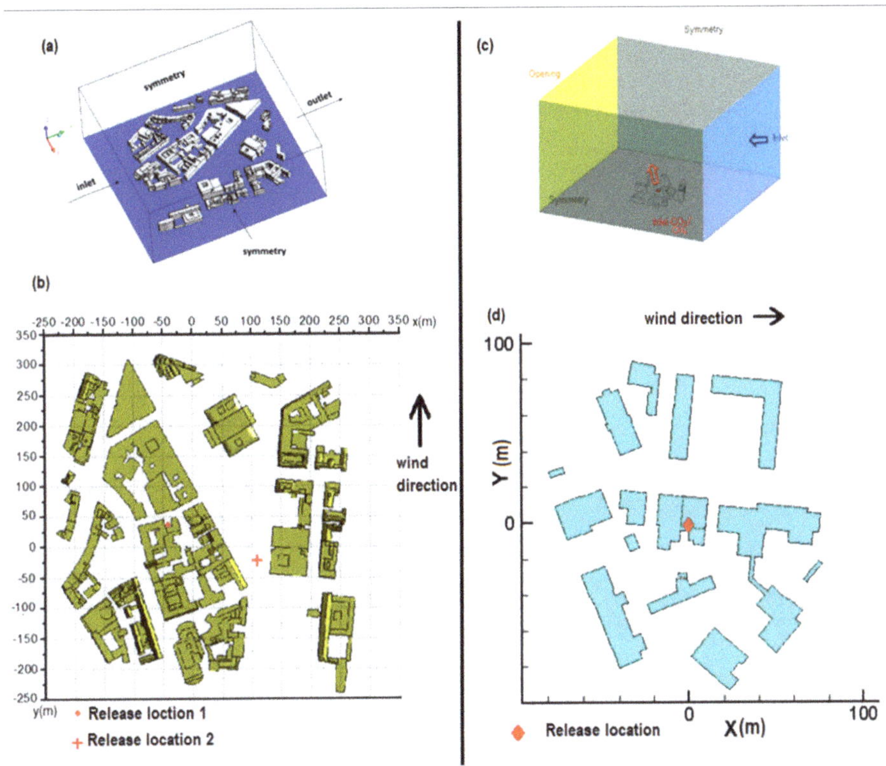


**Figure 3.** Geometry, boundaries conditions and release locations: (**a**) Geometry and boundaries conditions used for Hanover city, (**b**) Release locations used for Hanover city, (**c**) Geometry and boundaries conditions used for Frankfurt city, (**d**) Release locations used for Frankfurt city.

In case 1, shown in Figure 3a, results highlight the effect of buildings arrangement. For this purpose, two different scenarios dealing with different release locations are investigated. The first one is an accidental leakage of methane. The hazardous gas (methane) is introduced through an opening located at a confined geometry ($x = -27$ m, $y = 26$ m, $z = 1$ m). The choice of this location is based on its closeness to the buildings and/or distance to the exit. $CH_4$ is injected toward the wall of a building from an inlet having 1 m$^2$, a velocity of 8 m/s, and a temperature of 310 K. The second leakage scenario focuses on the dispersion of methane within an open surface, i.e., along the street and far from buildings ($x = 110$ m, $y = -29$ m, $z = 1$ m). Identical boundary conditions for methane are applied at the location 2. For both leakages, $CH_4$ is injected with 100% mass-fraction. Figure 3b shows the effect of gas density for case 2. Two gases having different densities, i.e., methane and carbon dioxide, are used for the investigation. The choice of these gases is based on the density ratio to the ambient air. $CO_2$ is heavier than ambient air ($\rho$ ($CO_2$) = 1.84 kg/m$^3$), while the methane is lighter than ambient air ($\rho$ ($CH_4$) = 0.71 kg/m$^3$). The $CO_2$ and $CH_4$ gases are injected from a chimney sited at ($x = 0$ m, $y = 0$ m, $z = 20$ m) by applying an inlet velocity of 2 m/s and a temperature equal to 310 K.

For the carrier phase boundary conditions, a power law profile is used to describe the variation of the inlet wind speed as a function of the height given in Equation (5) [33]

$$\frac{U(z)}{U_{ref}} = \left(\frac{z}{z_{ref}}\right)^a \tag{5}$$

where $U_{ref} = 9$ m/s is the velocity reference at $z_{ref} = 10$ m and $a = 0.21$.

The turbulence was generated using the inlet boundary condition. For LES simulation, the vortex method was used with a number of vortices Nv = 190 in order to establish a time-dependent inlet profile for the inlet velocity. In this method, a number of vortices Nv are generated and convected randomly at each time step.

With this approach, a perturbation is added on a specified mean velocity profile via a fluctuating vorticity field (i.e., two-dimensional in the plane normal to the streamwise direction). The vortex method is based on the Lagrangian form of the 2D evolution equation of the vorticity and the Biot-Savart law. A particle discretization is used to solve this equation. These particles, or "vortex points'," are convected randomly and carry information about the vorticity field. If $N$ is the number of vortex points and A is the area of the inlet section, the amount of vorticity carried by a given particle $i$ is represented by the circulation $\Gamma_i$ and an assumed spatial distribution $\eta$:

$$\Gamma_i(x,y) = \sqrt[4]{\frac{\pi A k(x,y)}{3N[2\ln(3) - 3\ln(2)]}} \tag{6}$$

$$\eta(\vec{x}) = \frac{1}{2\pi\sigma^2}\left(2e^{-|x|^2/2\sigma^2} - 1\right)2e^{-|x|^2/2\sigma^2} \tag{7}$$

where $k$ is the turbulence kinetic energy. The parameter $\sigma$ provides control over the size of a vortex particle. The resulting discretization for the velocity field is given by

$$\vec{u}(\vec{x}) = \frac{1}{2\pi}\sum_{i=1}^{N}\Gamma_i\frac{((\vec{x}_i - \vec{x}) \times \vec{z})(1 - e^{|\vec{x} - \vec{x}'|^2/2\sigma^2})}{|\vec{x} - \vec{x}'|^2} \tag{8}$$

where $\vec{z}$ is the unit vector in the streamwise direction. Originally, the size of the vortex was fixed by an adhoc value of $\sigma$. To make the vortex method generally applicable, a local vortex size is specified through a turbulent mixing length hypothesis.

The turbulent kinetic energy, $k$, and the turbulence dissipation rate, $\varepsilon$, used for RANS simulations are determined using the Equations (9) and (10) [26]:

$I_u$ is the streamwise turbulence intensity ($I_u = \sigma_u/U$, with $\sigma_u$ the standard deviation of $u$)

$$k(z) = (I_u U)^2 \tag{9}$$

$$\varepsilon(z) = \frac{(u^*)^3}{\kappa(z + z_0)} \tag{10}$$

The variable $u^*$ is the friction velocity, $\kappa = 0.41$ represents the von Karman constant and $z_0 = 1$ m (full scale) is the aerodynamic roughness length. All the boundaries and initial conditions are summarized in Table 2.

**Table 2.** Initial and boundaries conditions used for Hanover and Frankfurt city.

| Boundary | Hanover City Case | | | Frankfurt City Case | | |
|---|---|---|---|---|---|---|
| | Velocity | Temperature | Species | Velocity | Temperature | Species |
| Inlet | Profile: Equation (5) | 300 K | air | Profile: Equation (5) | 300 K | air |
| Pollutant source | 8 m/s | 310 K | $CH_4$ | 2 m/s | 310 K | $CO_2$ $CH_4$ |
| Outlet | Neuman BC for the pressure | - | - | Neuman BC for the pressure | - | - |
| Top | Neuman BC for the pressure | - | - | Neuman BC for the pressure | - | - |
| Walls | No slip | - | - | No slip | - | - |

### 3.2. Computational Grid

Figure 4a exhibits the mesh used for Hanover city simulations. A mesh having 40 million elements is generated by the patch independent algorithm. This algorithm is based on the top-down approach, in which volumes are meshed first and cells are projected to faces and edges. It is mainly used for complex geometries that demonstrate various features having different tolerances. In order to well-capture the boundary layer and well=predict the continuous phase physics, an appropriate refinement near the source of $CH_4$ and close to the ground was performed. The grid quality is improved by increasing the cell number. Figure 4b shows the grid used for the Frankfurt city case. In the building zones, the high-resolution computational grids are composed of prism and tetra elements only. Faraway from the building areas, a hexahedral mesh is applied. For both meshes, the resolution is determined using a grid-sensitivity analysis. It is worth mentioning that grid quality varies between 0.05 and 0.95. The metrics used to quantify the quality is the orthogonality, which relates the angles between adjacent elements. The optimum angle is 90° for quadrilateral faced cells and 60° for triangular faces elements. The mesh quality plays a significant impact on the accuracy of the numerical prediction, as well as the stability of the simulation.

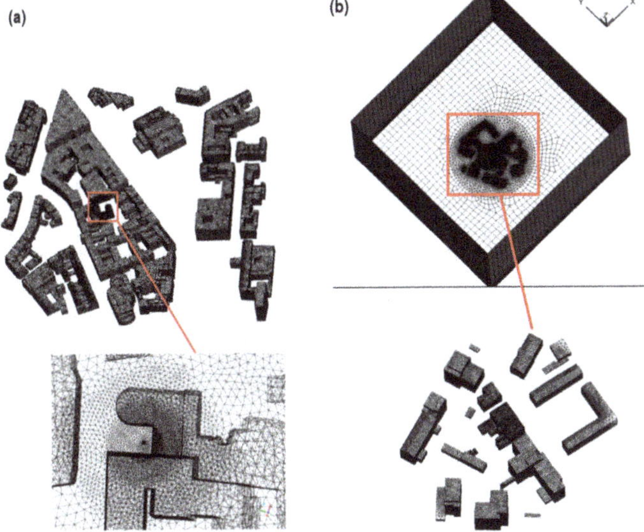


**Figure 4.** (**a**) Mesh used for Hanover city case 40 million cells, (**b**) mesh used for Frankfurt city case 10 million cells.

## 4. Results and Discussion

### 4.1. Hanover City Results

Figure 5 shows the characteristics of the flow at the section $z = 3$ m (Figure 5a) and at vertical section $y = -250$ m, (Figure 5b,c). The maximum velocity magnitude, 7 m/s, is at the central location section, which is the narrowest passage along the street. This is explained by the conservation of mass, i.e., if a fluid flow having a constant flow rate passes through smaller section, it will be accelerated. As the flow is incompressible, the transport and dispersion of pollutant species is governed by the wind velocity. Behind buildings, zones of recirculation are noticed. Figure 5b,c depict the resulting vertical profiles of the mean streamwise velocity <$u$> and the turbulence intensity ($I$), obtained at ($y = -250$ m) using LES model. The low power profile is well-reproduced and clearly shows the frictional effect of the ground. For the turbulence intensity, one notices a low variation from 4% ($z = 4$ m) to 6% ($z = 200$ m), which are smaller values compared to the literature [23] that predicted 12%. Consequently, one concludes that inlet turbulence has a relatively low influence on the dispersion of pollutants. In fact, the turbulence at the release location is mainly governed by the presence of the surrounding buildings, which generate velocity gradients, in turn enhancing the turbulence.

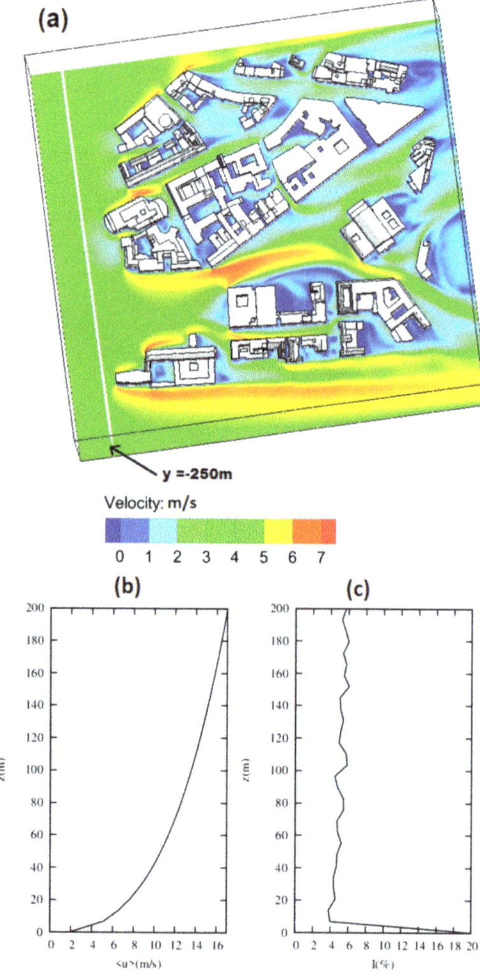


**Figure 5.** Characteristics of approaching flow: (**a**) Velocity at $z = 3$ m, (**b**) mean streamwise velocity <u> at $y = -250$ m, (**c**) turbulent intensity at $y = -250$ m.

Figure 6 shows qualitative results of methane dispersion. Two different injection locations are investigated. The release location significantly affects the dispersion of pollutants. When methane is injected for a time period of 220 s, the burnable methane-air mixture, represented by an iso-surface equal 4% of $CH_4$-mass fraction, reaches the wall of the nearby building (Figure 6b). The burnable mixture shows a well-developed vertical plume. Parts of this mixture are in touch with the ground. In this zone, the buildings interfere with the wind velocity, delivering a flow that is mainly characterized by the urban canopy and loosely dependent on atmospheric wind flow. As a result, the methane concentration increases in the wakes of the building where wind speed is considerably reduced and the recirculation zone becomes remarkably larger. Figure 6a depicts the concentration on buildings surfaces and demonstrates a critical value of 4% close to the wall of the nearby building. From a practical point of view, if a leakage occurs at this location, the ventilation intakes of these buildings should preferably not be located on this face. Indeed, pollutants can contaminate the indoor air of the nearby building if the windows of the leeward faces are open.

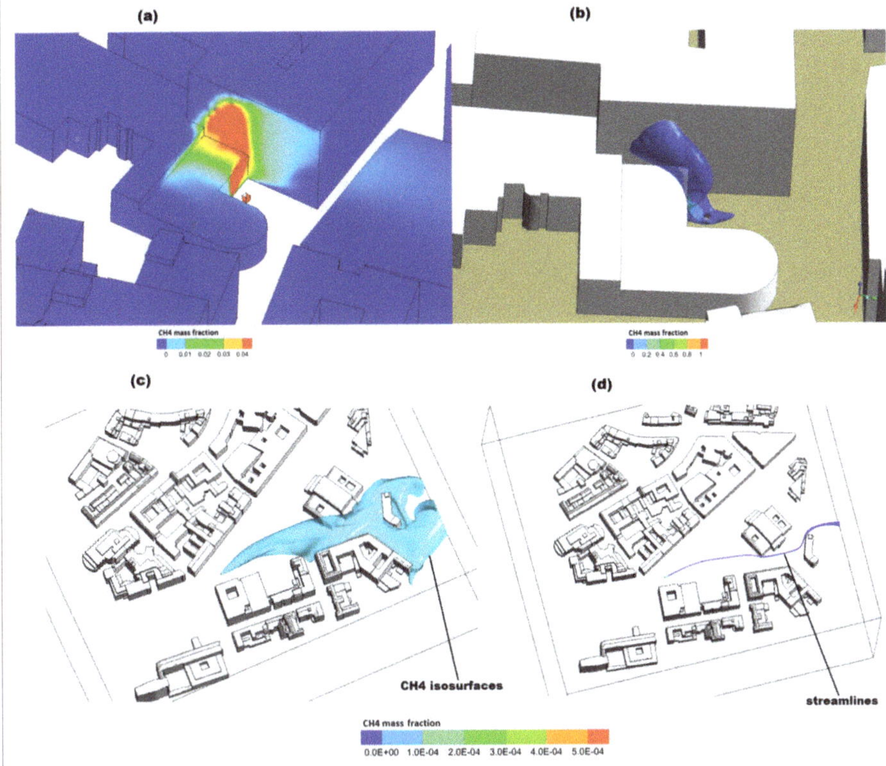


**Figure 6.** Qualitative results showing the dispersion of methane $CH_4$: (**a**,**b**) From location 1 sited at ($x = -27$ m, $y = 26$ m, $z = 1$ m) (**c**,**d**) from location 2 sited at ($x = 110$ m, $y = -29$ m, $z = 1$ m).

Figure 6c,d show a different behavior of methane dispersion due to the new location of the pollutant source, which is placed in an open area in which convective transport prevails. Thus, a change in the stack location results in a very different plume behavior. The pollutant species are subjected to a carrier phase flow having anisotropic convective fluctuations. The turbulence intensity along the y-direction is more important than that in all other directions, making the species iso-surfaces show a narrowly opened cone (Figure 6c).

Figure 6d shows the streamlines starting from the $CH_4$ source (location 2), in which the velocity of the carrier phase is the highest and exceeds 7 m/s. $CH_4$ mass fraction shows different pattern compared to location 1. Pollutants are transported horizontally more than that in the vertical direction. The important carrier phase momentum demonstrates a dominant effect. The horizontal time scale of the mixture momentum is clearly smaller than that in the vertical direction. Despite the importance of buoyancy forces, it appears that the dynamic/momentum of the carrier phase is the parameter that controls the motion and dispersion of $CH_4$.

Figure 7a,b depicts the quantitative result of the $CH_4$ dispersion obtained in location 1-simulation. The hazardous zone of an explosion and/or toxicity (mass fraction between 0.04 and 0.14) is bounded between (29 m < x < −28 m), (26 m < y < 27 m) and (2 m < z < 10 m). At the level z = 5 m, the curve bandwidth is less important in "y-direction" than that in "x-direction". The flow diffusion (dominated by the turbulence) is clearly not isentropic. This can be explained by the vortex effect behind buildings, which transports methane in the positive x-direction. Faraway from the source (z = 10 m), the concentration of methane decreases significantly to reach 0.1. Yet, it is strong enough to cause an explosion. Figure 7d,c shows the quantitative distribution of $CH_4$ dispersion produced in the simulation of location 2-case. The hazardous zone is bounded between (109 m < x < 111 m), (−29 m < y < 24 m) and (0 m < z < 2 m). Moving toward the outlet in the positive y-direction, the concentration decreases considerably and falls below 10% after 5 m. This demonstrates that the wind flow transports rapidly $CH_4$ and prevents a pollutants accumulation.

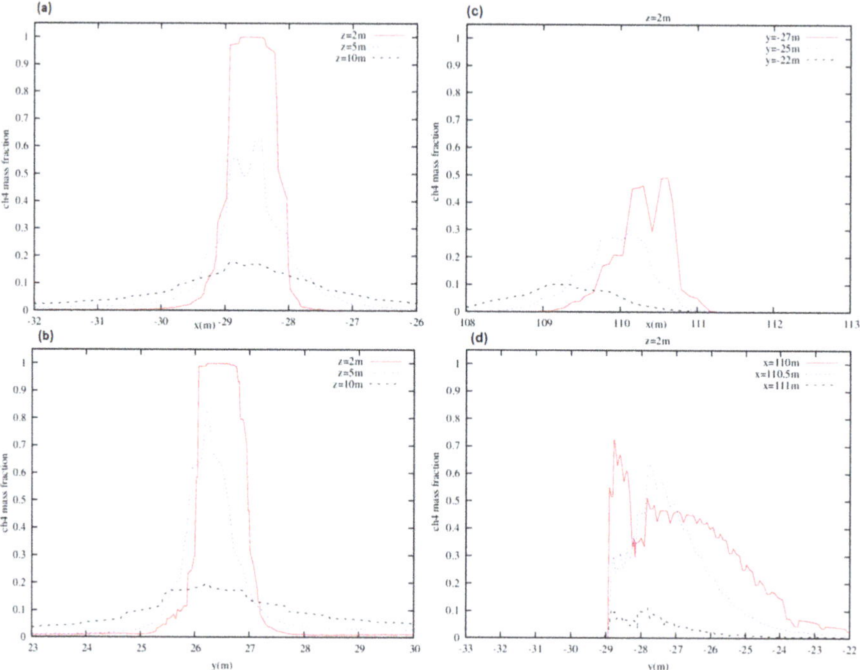

**Figure 7.** Methane ($CH_4$) mass fraction distribution in x and y directions: (**a,b**) For location 1 (at z = 2, 5, 10 m); and (**c,d**) for location 2 (at z = 2 m).

### 4.2. Frankfurt City Results

Figure 8a shows an instantaneous velocity contour of air at level z = 10 m. The velocity magnitude varies between zero and 9 m/s. Close to the buildings, obvious stagnation zones are registered. Velocity gradients are also observed downstream the buildings. Figure 8b depicts the streamlines obtained at level z = 10 m. The streamlines demonstrate a separation in front of the first raw of buildings and an acceleration of the carrier phase. The form of streamlines changes as we move toward the center of the geometry, and recirculation zones are observed behind buildings. Figure 8c,d display an instantaneous concentration of $CH_4$ and $CO_2$ at the ground, respectively. The $CO_2$ concentration is one order of magnitude larger than that of $CH_4$. This concentration difference demonstrates that the deposition of pollutants is mainly controlled by the density ratio and not by the air-velocity.

The maximum concentration of both gases is observed behind the source building. A significant pollutant concentration is observed in the vicinity of computational domain outlet. This is due to the horizontal convective transport of the pollutant toward the outlet due to high wind velocity and negligible diffusion.

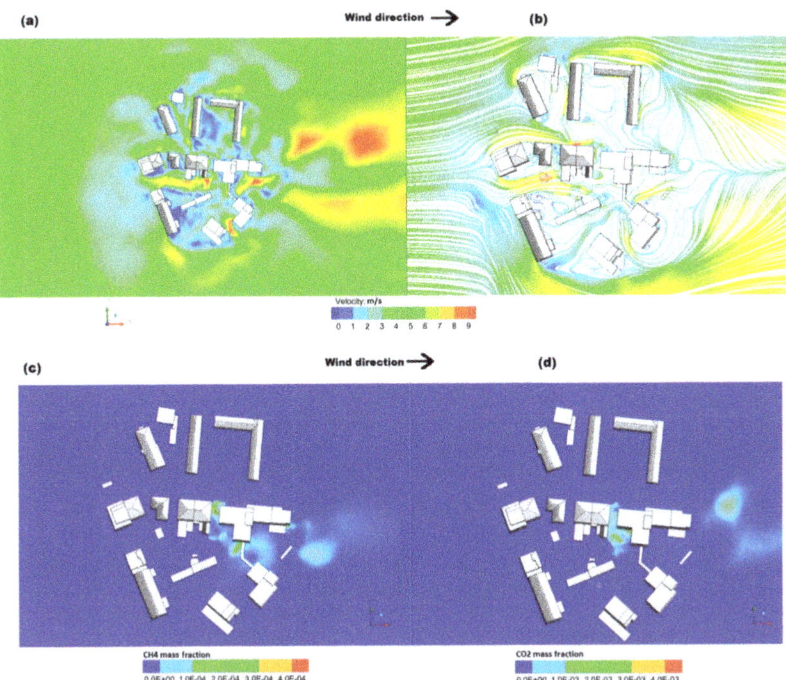


**Figure 8.** (**a**) Instantaneous velocity contour at $z = 10$ m, (**b**) streamline at $z = 10$ m, (**c**) methane ($CH_4$) concentration at the ground $z = 0$ m, (**d**) carbon dioxide ($CO_2$) concentration at the ground.

Figure 9 depicts the $CO_2$ and $CH_4$ mass fraction variations along vertical lines defined by $(x, y)$ coordinates. Six different locations are chosen to investigate the effect of the density on the dispersion of pollutants. They are located behind the building source sited at ($x = 0$ m, $y = 0$ m). All profiles of $CO_2$ mass fraction show larger concentration than $CH_4$. This is caused by the buoyancy force. The heavier gas is dragged downward (to the ground) by the effect of gravitational force. Carbon dioxide $CO_2$ is heavier than ambient air. As a result, its higher concentration is close to the ground.

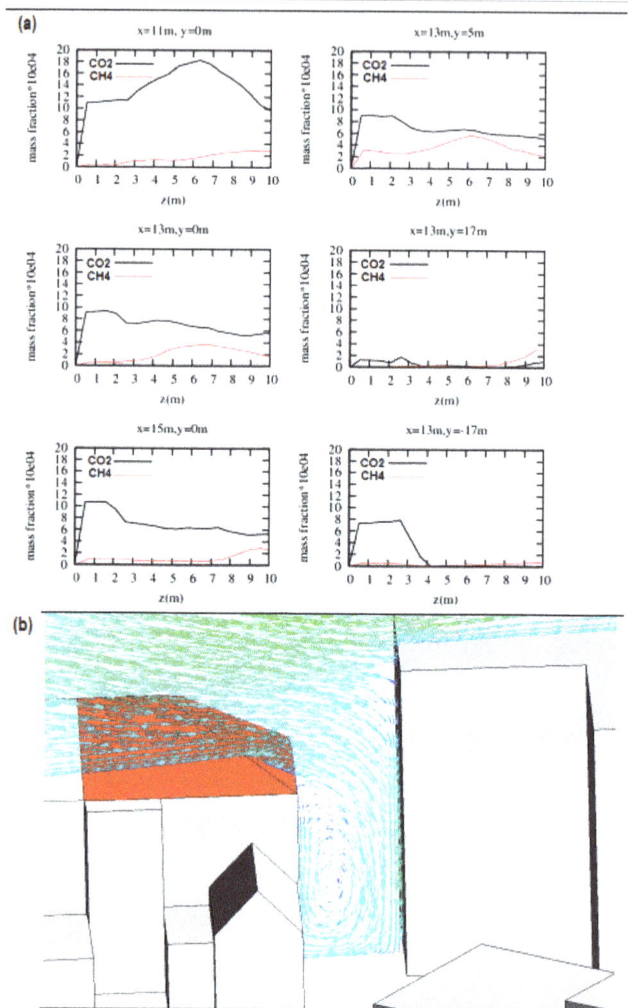

**Figure 9.** (a) $CO_2$ and $CH_4$ mass fraction distribution at different vertical lines, (b) vector showing vortex located behind building source (red roof).

Pollutant transport is also influenced by the vortex located behind the building source (see Figure 9c). This zone is created by the separation of fluid flow from the edges of the buildings. The appearance of this zone strongly affects the transport and diffusion of pollutants. Due to the recirculation of the carrier phase (air), methane is dragged downward, and $CH_4$ is accumulated at the ground.

A maximum concentration of $CO_2$ mass fraction is recorded at ($x = 11$ m, $y = 0$), which is the closest location to the wall boundary. In this region, the effect of the non-slip condition reduces the velocity magnitudes, resulting in the formation of stagnating zone in which pollutants are collected. Faraway from the recirculation zone ($x = 13$ m, $y = -17$ m) and ($x = 13$ m, $y = 17$ m), the concentration of pollutants, $CO_2$ and $CH_4$, is very low and mostly negligible.

Figure 10 shows the effect of the density ratio on the dispersion of $CH_4$ and $CO_2$. The density of pollutant-air mixture increases ($CH_4$)/decreases ($CO_2$) rapidly and reaches the density of ambient air at a height of z = 5 m, far from the injection source (Figure 10a). The density ratio controls the shape and the volume of the pollutant plume. Figure 10b,c show iso-surfaces of $CH_4$ and $CO_2$ concentration. In order to compare the dispersion of pollutant gases, an iso-value of 4% mass fraction is represented. The plume of $CH_4$ is first transported vertically by the inlet flows and then carried by the wind horizontally to cover the roof of nearby buildings. In the case of $CO_2$, the dispersion of the plume is located immediately at the top of the building that include the pollutant source. Compared to $CH_4$, the dispersion is remarkably reduced.

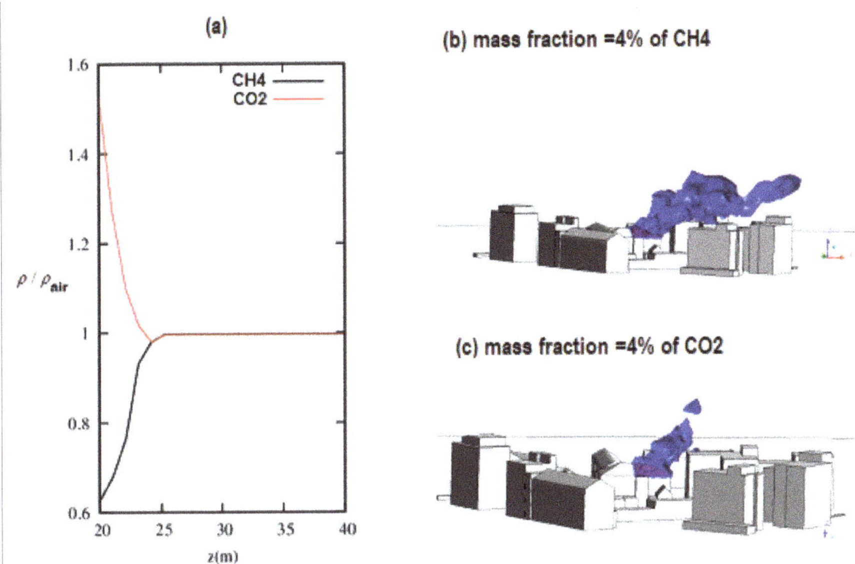


**Figure 10.** (**a**) Density ratio $\rho_{CO2}/\rho_{air}$, $\rho_{CH4}/\rho_{air}$ distribution at the vertical line (x = 0 m, y = 0 m) as a function of the height z, (**b**) iso-surface, representing the plume of $CH_4$, (**c**) iso-surface representing the plume of $CO_2$.

## 5. Conclusions

High-resolution CFD simulations of near- and far-field pollutant dispersion in a building group in downtown Hanover and Frankfurt are predicted using LES turbulence model along with the dynamic Smagorinsky SGS. The simulations focus on the concentration of pollutants in the vicinity of the release sources and the spatial distribution of the transported species following different building structures and boundary conditions. The following remarks can be drawn:

(1) Smagorinsky LES shows reasonable results of the velocities and turbulence quantities compared to experimental data of generic building configuration. The results agreement is used as validation of the flow model to investigate a larger full-scale 1:1 city geometry.

(2) Methane dispersion is controlled by the air-$CH_4$ density ratio in cases of pollutant injection close to buildings. The buoyancy forces are responsible for the stratification of gases. $CH_4$ species are first transported in the vertical direction then dragged by the carrier phase toward the outlet. The velocity direction of pollutant injection and magnitude, as well as temperature difference between $CH_4$ and ambient air, do not play important roles.

(3) If $CH_4$ is injected far from buildings, i.e., in open area in which carrier phase velocity is important, then the transport of pollutant showed different behavior. $CH_4$ was dragged horizontally much faster than in the vertical direction. It followed the street and kept almost a constant altitude until reaching the outlet.

(4) The main parameter that controls the dispersion in Frankfurt city configuration is the density ratio of $CO_2$ and $CH_4$ to ambient air. Despite the injection of both gases in the vertical direction, only $CO_2$ shows a significant concentration at the ground.

Future works should focus on the existence of vegetation, as well as the implementation of new sources that account for additional species creation, e.g., CO, $NO_x$, $SO_x$, which describe the transport means in the cities.

**Author Contributions:** Conceptualization, M.S.I. and M.C.; Methodology, M.S.I. and M.C.; Software, N.B.S.; Validation, M.S.I., M.C. and N.B.S.; Formal Analysis, M.S.I.; Investigation, M.S.I.; Resources, N.B.S.; Data Curation, N.B.S.; Writing—Original Draft Preparation, M.S.I.; Writing—Review & Editing, M.S.I. and M.C.; Visualization, M.S.I.; Supervision, M.C.; Project Administration, N.B.S.; Funding Acquisition, N.B.S.

**Funding:** This research received no external funding.

**Acknowledgments:** The numerical simulations reported in this paper were supported by the laboratory of the Research Unit Mechanical Modelling, Energies and Materials National School of engineer of Gabes and the CADFEM-AN company.

**Conflicts of Interest:** The authors declare no conflict of interest.

## References

1. Hagler, G.S.; Tang, W. Simulation of rail yard emissions transport to the near-source environment. *Atmos. Pollut. Res.* **2016**, *7*, 469–476. [CrossRef]

2. Hidouri, A.; Gazzah, M.H.; Sassi, M. Numerical investigation of piloted turbulent reacting methane/air jet. *Comput. Therm. Sci. Int. J.* **2016**, 1–12. [CrossRef]

3. Hidouri, A.; Chrigui, M.; Boushaki, T.; Sadiki, A.; Janicka, J. Large Eddy Simulation of Two Isothermal and Reacting Turbulent Separated Oxy-Fuel Jets. *Fuel* **2017**, *192*, 108–120. [CrossRef]

4. Mouldi, C.; Amsini, S.; Johannes, J. Evaporation and dispersion of N-Heptane droplets within premixed flame. *Heat Mass Transf.* **2010**, *46*, 869–880. [CrossRef]

5. Derome, D.; Kubilay, A.; Defraeye, T.; Blocken, B.; Carmeliet, J. Ten questions concerning modeling of wind-driven rain in the built environment. *Build. Environ.* **2017**, *114*, 495–506. [CrossRef]

6. Vasile, F.S.; Lucian, N.; Bianca, S.; Carol, P. Case studies of methane dispersion patterns and odor strength in vicinity of municipal solid waste landfill of Cluj–Napoca, Romania, using numerical modeling. *Atmos. Pollut. Res.* **2015**, *6*, 312–321. [CrossRef]

7. Tominaga, Y.; Blocken, B. Wind tunnel experiments on cross-ventilation flow of a generic building with contaminant dispersion in unsheltered and sheltered conditions. *Build. Environ.* **2015**, *92*, 452–461. [CrossRef]

8. Jae-Jin, K.; Jong-Jin, B. Effects of inflow turbulence intensity on flow and pollutant dispersion in an urban street canyon. *J. Wind Eng. Ind. Aerod.* **2003**, *91*, 309–329. [CrossRef]

9. Blocken, B. Computational Fluid Dynamics for Urban Physics: Importance, scales, possibilities, limitations and ten tips and tricks towards accurate and reliable simulations. *Build. Environ.* **2015**, *91*, 219–245. [CrossRef]

10. Leitl, B.M.; Kastner-Klein, P.; Rau, M.; Meroney, R.N. Concentration and flow distributions in the vicinity of U-shaped buildings: Wind-tunnel and computational data. *J. Wind. Eng. Ind. Aerod.* **1997**, *67*, 745–755. [CrossRef]

11. Meroney, R.N.; Leitl, B.M.; Rafailidis, S.; Schatzmann, M. Wind-tunnel and numerical modeling of flow and dispersion about several building shapes. *J. Wind. Eng. Ind. Aerod.* **1999**, *81*, 333–345. [CrossRef]

12. Britter, R.E.; Hanna, S.R. Flow and dispersion in urban areas. *Ann. Rev. Fluid Mech.* **2003**, *35*, 469–496. [CrossRef]

13. Blocken, B.; Stathopoulos, T.; Saathoff, P.; Wang, X. Numerical evaluation of pollutant dispersion in the built environment: Comparisons between models and experiments. *J. Wind. Eng. Ind. Aerod.* **2008**, *96*, 1817–1831. [CrossRef]

14. Tominaga, Y.; Stathopoulos, T. Numerical simulation of dispersion around an isolated cubic building: Comparison of various types of k-e models. *Atmos. Environ.* **2009**, *43*, 3200–3210. [CrossRef]

15. Chun-Ho, L.; Mary, C.B. Large-Eddy simulation of flow and scalar transport in a modeled street canyon. *Appl. Metro.* **2002**, *41*, 660–673. [CrossRef]

16. Baker, J.; Walker, H.L.; Cai, X.M. A study of the dispersion and transport of reactive pollutants in and above street canyons a large eddy simulation. *Atmos. Environ.* **2004**, *38*, 6883–6892. [CrossRef]

17. Xie, Z.T.; Castro, I.P. LES and RANS for turbulent flow over arrays of wall mounted cubes. *Flow Turbul. Combust.* **2006**, *76*, 291–312. [CrossRef]

18. Sada, K.; Sato, A. Numerical calculation of flow and stack-gas concentration fluctuation around a cubical building. *Atmos. Environ.* **2002**, *36*, 5527–5534. [CrossRef]

19. Gu, Z.L.; Jiao, J.Y.; Su, J.W. Large-eddy simulation of the wind field and plume dispersion within different obstacle arrays using a dynamic mixing length subgrid-scale model. *Bound. Lay. Meteorol.* **2011**, *139*, 439–455. [CrossRef]

20. Cheng, W.C.; Chun-Ho, L. Large-Eddy Simulation of Flow and Pollutant Transports in and Above Two-Dimensional Idealized Street Canyons. *Bound. Lay. Meteorol.* **2011**, *139*, 411–437. [CrossRef]

21. Gousseau, P.; Blocken, B.; Van Heijst, G.J.F. Large-eddy simulation of pollutant dispersion around a cubical building: Analysis of the turbulent mass transport mechanism by unsteady concentration and velocity statistics. *Environ. Pollut.* **2012**, *167*, 47–57. [CrossRef] [PubMed]

22. Yoshie, R.; Jiang, G.; Shirasawa, T.; Chung, J. CFD simulations of gas dispersion around high-rise building in non-isothermal boundary layer. *J. Wind. Eng. Ind. Aerod.* **2011**, *99*, 279–288. [CrossRef]

23. Xie, Z.T.; Hayden, P.; Wood, C.R. Large-eddy simulation of approaching-flow stratification on dispersion over arrays of buildings. *Atmos. Environ.* **2013**, *71*, 64–74. [CrossRef]

24. Xue, Y.; Zhai, Z.J. Inverse identification of multiple outdoor pollutant sources with a mobile sensor. *Build. Simul.* **2017**, *10*, 255–263. [CrossRef]

25. Liu, Y.S.; Cui, G.X.; Wang, Z.S.; Zhang, Z.S. Large eddy simulation of wind field and pollutant dispersion in downtown Macao. *Atmos. Environ.* **2011**, *45*, 2849–2859. [CrossRef]

26. Gousseau, P.; Blocken, B.; Stathopoulos, T.; van Heijst, G.J.F. Near-field pollutant dispersion in an actual urban area: Analysis of the mass transport mechanism by high resolution Large Eddy Simulations. *Comput. Fluids* **2015**, *114*, 151–162. [CrossRef]

27. Kubilay, A.; Derome, D.; Blocken, B.; Carmeliet, J. Numerical modeling of turbulent dispersion for wind-driven rain on building facades. *Environ. Fluid Mech.* **2015**, *15*, 109–133. [CrossRef]

28. Van Hooff, T.; Blocken, B. CFD evaluation of natural ventilation of indoor environments by the concentration decay method: $CO_2$ gas dispersion from a semi-enclosed stadium. *Build. Environ.* **2013**, *61*, 1–17. [CrossRef]

29. Amorim, J.H.; Rodrigues, V.; Tavares, R.; Valente, J.; Borrego, C. CFD modelling of the aerodynamic effect of trees on urban air pollution dispersion. *Sci. Total Environ.* **2013**, *461*, 541–551. [CrossRef]

30. Leitl, B.; Schatzmann, M. Cedval at Hamburg University. Available online: http://www.mi.unihamburg.de/cedval (accessed on 17 July 2019).

31. Longo, R.; Ferrarotti, M.; Schanchez, C.G.; Deurdi, M.; Parente, A. Advanced turbulence models, boundary conditions and dispersion study for flows around different configurations of ground-mounted buildings. *J. Wind. Eng. Ind. Aerod.* **2017**, *167*, 160–182. [CrossRef]

32. Sadiki, A.; Maltsev, A.; Wegner, B.; Flemming, F.; Kempf, A.; Janicka, J. Unsteady methods (urans and les) for simulation of combustion systems. *Int. J. Therm. Sci.* **2006**, *45*, 760–773. [CrossRef]

33. Tennekes, H. The Logarithmic Wind Profile. *J. Atmos. Sci.* **1973**, *30*, 234–238. [CrossRef]

*Article*

# Synchronized Multiple Drop Impacts into a Deep Pool

**Manfredo Guilizzoni** [1,*], **Maurizio Santini** [2] **and Stephanie Fest-Santini** [3]

[1] Department of Energy, Politecnico di Milano, via Lambruschini 4, 20156 Milan, Italy
[2] Department of Engineering and Applied Sciences, University of Bergamo, viale Marconi 5, 24044 Dalmine, Italy
[3] Department of Management, Information and Production Engineering, University of Bergamo, viale Marconi 5, 24044 Dalmine, Italy
* Correspondence: manfredo.guilizzoni@polimi.it; Tel.: +39-02-2399-3888

Received: 22 June 2019; Accepted: 25 July 2019; Published: 27 July 2019

**Abstract:** Drop impacts (onto dry or wet surfaces or into deep pools) are important in a wide range of applications, and, consequently, many studies, both experimental and numerical, are available in the literature. However, such works are focused either on statistical analyses of drop populations or on single drops. The literature is heavily lacking in information about the mutual interactions between a few drops during the impact. This work describes a computational fluid dynamics (CFD) study on the impact of two, three, and four synchronized drops into a deep pool. The two-phase finite-volume solver interFoam of the open source CFD package OpenFOAM® was used. After validation with respect to high speed videos, to confirm the performance of the solver in this field, impact conditions and aspects that would have been difficult to obtain and to study in experiments were investigated: namely, the energy conversion during the crater evolution, the effect of varying drop interspace and surface tension, and multiple drop impacts. The results show the very significant effect of these aspects. This implies that an extension of the results of single-drop, distilled-water laboratory experiments to real applications may not be reliable.

**Keywords:** multiple drop impact; computational fluid dynamics (CFD) simulation; volume-of-fluid; OpenFOAM; crater dimensions; vorticity

---

## 1. Introduction

A large number of phenomena, both of natural and technological interest, involve the interaction between one or—much more commonly—many liquid drops and a solid surface or a gas-liquid interface. This interaction may be mechanical, thermal, or chemical, and it usually starts with an impact of the drop onto the surface or interface. The impact velocity may be low or high, normal or oblique, and in most cases, a series of drops impact the surface at the same time or nearly simultaneously. For example, this happens in internal combustion engines, firefighting systems, surface cooling, or painting by sprays, pesticide deposition in agriculture, pollution and microbial transport in air and water, forensic analysis of blood stains, and rain effects on planes, wind turbines, and buildings. This, therefore, demonstrates the importance of a deep understanding of this phenomenon, which is still not fully understood despite its apparent simplicity and more than a century of investigation. In particular, many studies can be found in the literature about:

- The impact of a single drop onto dry solid surfaces, onto surfaces wetted by a liquid film, and onto gas-liquid interfaces alone in deep pools;
- The behavior of sprays, where the multitude of drops in the spray are considered in statistical terms, mainly with empirical models.

Detailed introductory information and reviews can be found in [1–5]. On impacts into deep pools, the work by Cole [6] is a specifically valuable reference. On the contrary, studies are quite scarce about the intermediate step, i.e., the outcomes of the combined impact of more than one drop in parallel (with simultaneous or delayed impacts) [7–10] or in a series [11–13].

As it may involve solid, one, or more liquids, and a surrounding gas, drop impact is a multiphase fluid mechanics problem, where inertial, viscous, and capillary forces merge their effects. It may also become a multi-physics problem in the case of heat or mass transfer, or chemical reactions may play a significant role (e.g., for hot or cold drops or surfaces or for drops impacting onto a chemically different liquid, for reactive wetting). Therefore, a rigorous model of the phenomenon can theoretically be created, but its analytical solution is not viable. Consequently, very simplified models and experiments have been the primary method of investigation for a long time. In recent years, the increase in the power of calculus has also made computational fluid dynamics (CFD) simulations feasible. Many remarkable studies were developed (e.g., [14–18]) using different approaches, ranging from the most used Eulerian models in finite volume or finite element frameworks, to Lattice Boltzmann and molecular dynamic simulations. Commercial, open-source, and in-house software packages were used. Specifically, the Volume-of-Fluid [19] approach is one of the most used algorithms, even though other techniques were also developed, e.g., Level-Set [20], markers [21], combined models [22–26], and (as already noted) Lattice Boltzmann [27,28] and molecular dynamics simulations (for nanodrops) [29].

In this study, CFD was used to investigate the outcomes of double and multiple synchronized drop impacts into a deep pool with the same temperature as the drops. The numerical campaign consisted of two steps: the first for validation, with simulations carried out to replicate experimental conditions that had been previously investigated by means of high-speed video acquisitions, and the second, to simulate conditions that would have been difficult to obtain in experiments. Energy conversion during the crater evolution and the effects of varying drop interspace and surface tension and multiple drop impacts were investigated. For the first phase, the agreement between simulations and experiments was satisfactory, particularly for some impact conditions. The results of the second phase evidenced the very significant effect of the investigated aspects. This outcome implies that an extension of the results from single-drop, distilled-water laboratory experiments to real applications may not be reliable.

## 2. Materials and Methods

### 2.1. Experimental Setup

The experimental set-up and procedures, including uncertainty analysis of the main parameters, were described in detail in a previous paper [10], so they will be only briefly summarized here. A thorough analysis of the experimental data against which the simulation results will be validated is available in the same paper. Drop impact experiments were performed by high-speed visualization with continuous back-light illumination. Millimeter-size deionized water droplets were produced with an on-demand droplet generator [7], which allows the synchronous detaching of two drops of equal size. Falling drops are accelerated by gravity, and during their free fall, they are detected by a light sensor, triggering the high-speed (1000 fps) charge-coupled device (CCD) camera acquisitions. Then, the drops impact the pool, which is contained in a cubic box with a side length of 40 mm. The pool depth is one order of magnitude larger than the drop diameter, to avoid any influence on the studied relevant part of the crater evolution. The drops are pigmented by a solution with well-known rheological behavior [30] and with a concentration of 1% in weight, so that the drop water can be distinguished from the pool water in the high-speed video frames, without significantly affecting the water's density and viscosity. For each recorded sequence, the images before drop impact are post-processed by an algorithm that comprises background subtraction, contrast enhancement, image cleaning, edge detection using the Laplacian of Gaussian (LoG) method, and feature detection to determine the drop section and center of each drop [31]. Using consecutive images coupled with the camera acquisition rate and the pixel-to-meter conversion factor, the diameter and velocity of the impinging droplets can be evaluated with an accuracy of ±0.03 mm and ±0.02 m/s, respectively. The drop sphericity S is also calculated for each falling drop, defined here as the aspect ratio of the vertical to horizontal axis (thus, S > 1 represents a prolate-shaped and S < 1 represents an oblate-shaped drop) before the instant of impact, and assuming a rotational symmetry around the vertical axis. The sphericity of the falling drops before impact is crucial for obtaining accurate and repeatable experimental results, as sphericity heavily affects crater dimensions and impact-generated vorticity [32–35]. Another critical aspect is the oscillation mode of the drop at the instant of impact [36,37]. The experiments were performed at ambient pressure, with the temperature of the pool equal to 30 ± 1 °C and the temperature of the drops equal to 27 ± 1 °C. The investigated experimental conditions in terms of drop diameter D, inter-drop distance $L_0$, and impact velocity w are summarized in Table 1, along with the values of the dimensionless groups typically used for drop impact characterization.

**Table 1.** Investigated conditions.

| Case | D (mm) | w (m/s) | $L_0/D$ | S | We | Fr |
|------|--------|---------|---------|------|-----|-----|
| A | 2.27 | 1.0 | 2.0 | 0.97 | 32 | 45 |
| B | 2.29 | 1.4 | 2.0 | 0.97 | 63 | 87 |
| C | 2.32 | 2.0 | 2.0 | 0.98 | 130 | 176 |

D: drop diameter; w: impact velocity, $L_0$: inter-drop distance; S: drop sphericity; We: Weber number $\rho D w^2/\sigma$; Fr: Froude number $w^2/(gD)$, where $\rho$ density, $\sigma$ surface tension, g gravity.

### 2.2. Numerical Simulations

Concerning the numerical setup, simulations were performed using the interFoam solver of the OpenFOAM® open source CFD toolbox [38]. interFoam is an isothermal finite volume (FV) solver based on the volume-of-fluid (VOF) method and implementing a model to include surface tension at the interface [39]. OpenFOAM® was selected due to its free and open source nature and because of its many favorable reviews [40–43] and successful cases of use, which are described in the literature [42–48]. The model implemented in interFoam includes the continuity and momentum equations for a Newtonian and incompressible fluid, whose density and viscosity are calculated on the basis of an indicator function named volume fraction $\gamma$. The latter assumes a value of 0 for one phase, 1 for the other, and between 0 and 1 in the interfacial regions, and it is transported by the fluid velocity field w. Volume tracking and interface reconstruction (typically as the isosurface at $\gamma = 0.5$), with no explicit interface tracking, is, therefore, performed. The advantage of the VOF method with respect to the two-fluid models is that a single set of equations must be solved. The expressions of all the used equations, a discussion of the meaning of the included terms, and further details about the interFoam solver and models can be found in [47,49,50]. Very promising modified versions of interFoam were also presented in the literature [48,51], but the source code was not made publicly available, so the version included in the official OpenFOAM® distribution was used.

As the investigation is at a constant temperature and the impact velocities are low, both phases were assumed to be incompressible and in laminar motion, in agreement with all the previously cited papers in this field. For each phase, the values of all the relevant thermophysical properties were taken at the respective temperatures measured during the experiments. The initial conditions of the two phases and their positions within the domain were fixed using the setFields OpenFOAM® utility. The implicit Euler scheme (first-order accurate) was used for the time derivative, as it proved to offer better results in comparison with the second-order Crank-Nicholson discretization schemes that were tested during preliminary simulations, selecting different blending factors between 0.5 and 1. The conventional advection term was discretized using Gauss schemes: limited Van Leer for the volume fraction and limited linear for the velocity. For the latter, variations of the limiter parameter were tested, but the best results were obtained by keeping it equal to 1. Finally, the OpenFOAM® specific "interface compression" scheme [52] proved to be the best choice for the discretization of the compression term. The maximum allowed Courant–Friedrichs–Lewy (CFL) number was 0.3. In fact, the solver is allowed to adapt the time step to keep the CFL number under the desired limit, and the requirement for the interface compression scheme used in interFoam for 3D cases is to limit it under 0.3 for 3D cases [40,52]. Some volume fraction sub-cycles are also performed to further improve the accuracy. Three-dimensional simulations were performed within a domain that is a rectangular cuboid having dimensions 0.015 m × 0.015 m × 0.020 m. This domain represents 1/4 of the real pool system, as the computational effort was reduced by exploiting two symmetry planes: the longitudinal plane passing through the two impact points and the transversal plane orthogonal to the previous one, passing through the midpoint between the impact points [10]. To reduce the domain height, drops' detachment from the drop generator and their fall towards the pool surface were not simulated. The drops were directly initialized as spheres near the free surface of the pool, with an initial velocity corresponding to the experimental one. This approach has a weakness in the fact that drop oscillations after the detachment from the needle are not considered. As already mentioned, drop shape and oscillatory behavior have an important influence on the impact outcomes (as it was confirmed also by the preliminary simulations described in the following section), but in the experiments, the drops' shapes were also nearly spherical, so this approximation seemed to be acceptable for these cases. The used mesh was a purely structured one, with hexahedral cells and grading along the axis connecting the drop centers (with a size ratio of 10 between the last and the first cell) to better capture the thin neck that separates the craters created by neighboring impact drops. In the other directions, the aspect ratio was fixed to 1 for all the cells. For the simulations involving three drops with one centered in the origin of the axes, no grading was used, and the mesh resolution was doubled along both the horizontal axes (x,z).

Domain characteristics, together with the imposed boundary conditions, the initial position of the interfaces, and the mesh details are summarized in Figure 1.

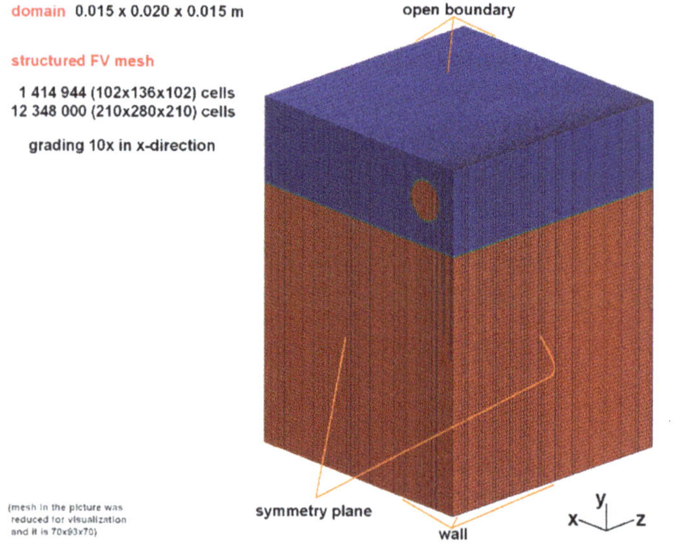


**Figure 1.** Domain dimensions, boundary conditions, and mesh details.

Single drop impacts were also performed for comparison, leaving everything equal to the double impact cases apart from the initial drop position, which was centered on the y-axis.

## 3. Results and Discussion

### 3.1. Validation against Experimental Data

To validate the numerical simulations, crater evolution and drop-formed vortex structures were compared with those observed by means of high-speed video acquisitions. Details about the latter can be found in [10]. All these simulations were performed using the properties of distilled water for the liquid phase. Simulation results were sliced along the longitudinal plane and saved as images with the same pixel-to-meter ratio of the experimental images, then, the time steps corresponding to the experimental frames were selected and superposed to the latter. Figures 2 and 3 show the results, up to time $\tau = 8$ ms for all three investigated cases and for the entire simulated time for cases A ($w = 1.0$ m/s) and B ($w = 1.4$ m/s). This comparison is the most severe in evaluating the agreement, as it allows a direct visualization of the correspondence between the experimental features of the craters and the numerical ones. The agreement is very good for all the investigated cases during the inertial phase of the impact (Figure 2) and for case B during the crater recession (Figure 3, column b). On the contrary, some significant discrepancies appear during this second phase for case A (Figure 3, column a). This will be commented on in the following section, when discussing the crater depth profiles. The results after $\tau = 8$ ms are not shown for case C because, with all the used meshes, the two craters merge, such that the crater depth is still reproduced well, but the crater shape is substantially altered, and the agreement is completely lost. With the VOF approach, air regions coming into contact merge instantaneously because the underlying model is not able to represent retarded coalescence (modified models would be needed, e.g., see [53]). On the other hand, real-world interfaces may resist even with direct contact—at least for a certain time—if the pressure of the contacting regions is not too different (e.g., bouncing bubbles or drops). Thus, in the experiments, the very thin neck between the two crates in case C is conserved, while in the simulations, a much finer mesh than that used for the simulations would have been needed to avoid crater contact and merging (OpenFOAM® is very slow in adaptive remeshing, so this feature could not be exploited).

Different mesh sizes were evaluated to analyze the mesh independence of the solution, which, for fine details, is always an issue when using the VOF technique. Mesh sizes were varied between 176 kcell and 12.35 Mcell, but meshes under 1 Mcell proved unsuitable because the already mentioned neck between the craters also disappeared (resulting in a coalescence of the cavities not observed in the experiments) for cases A and B. Thus, only the finest two meshes (1.4 Mcell and 12.35 Mcell) were used for the final simulations. For the latter, a good overall mesh independence was found, even if some details are still captured differently. No mesh is always the most accurate (as can be seen in Figures 2 and 3 for case B).

The vortices generated by drop impacts—peculiar and different from the single-drop impact case [10]— were visualized in the experiments by dying the drop water. In this way, the dye highlights the drop water, distinguishing it from the pool water, and allowing to follow its motion. In the simulations, this could be accomplished by numerically "dying" the drop water with a passive tracer, by using a multiphase solver (e.g., multiphaseInterFoam) or simply by tracing the flow streamlines [18]. An example of the latter approach is shown in Figure 4, which includes two pictures referring to case A at $\tau = 18$ ms. These two images were obtained by tracing streamlines with different origins, thereby evidencing how the streamline aspect heavily depends on the latter. The complex structure of the simulated flow streamlines is consistent with the experimental observation of vortex structures in the same region [10]. This behaviour is also in agreement with the experimental findings by other authors [54], who showed that after a certain time from the drop impact (for the investigated case, this time should be around 14 ms) the original azimuthal vorticity is progressively tilted and becomes a stream-wise vorticity. Despite the validity of this approach, using it to cover all the domain regions with good resolution would require a very large number of streamlines, such that they would be difficult to follow. Therefore, another visualization technique was preferred. Pictures showing the calculated vorticity (the curl of the velocity vector field) on the longitudinal plane were compared with the experimental images. Even if the vortex and vorticity are not related in a biunique way (due to irrotational vortices or vortex parts), vorticity offers a picture of the regions where vortices are most likely to occur, with a resolution equal to the number of mesh cells. Figure 5 shows the results for cases A and B. As can be seen, the agreement between simulations and experiments is also good for regions of high vorticity. The discrepancies are due to the differences in the crater evolution between the simulations and experiments. These differences influence the motion of the drop water under the crater itself. In any case, they are not significant enough to undermine the correct identification of the regions reached by the drop water and its swirling motion.

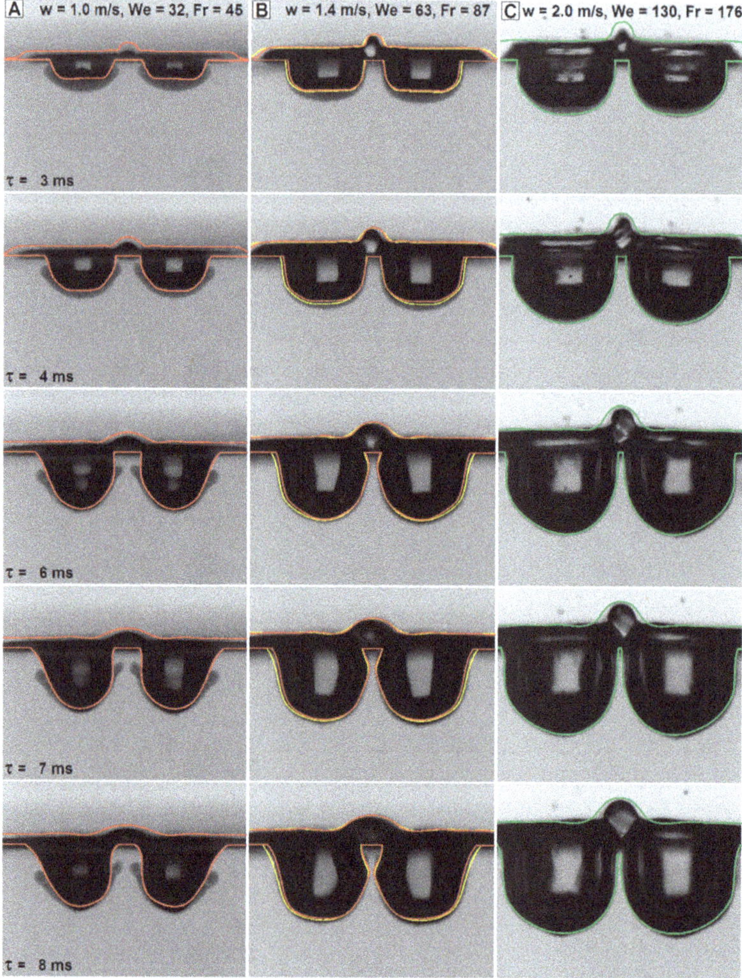



**Figure 2.** Superposition of the numerical crater contours to the experimental high-speed photos, for the three investigate cases A (w = 1.0 m/s), B (w = 1.4 m/s), C (w = 2.0 m/s), time up to 8 ms. In case B, the red line shows the results of the 12.35 Mcell simulation, while the yellow line shows the results of the 1.4 Mcell simulation.

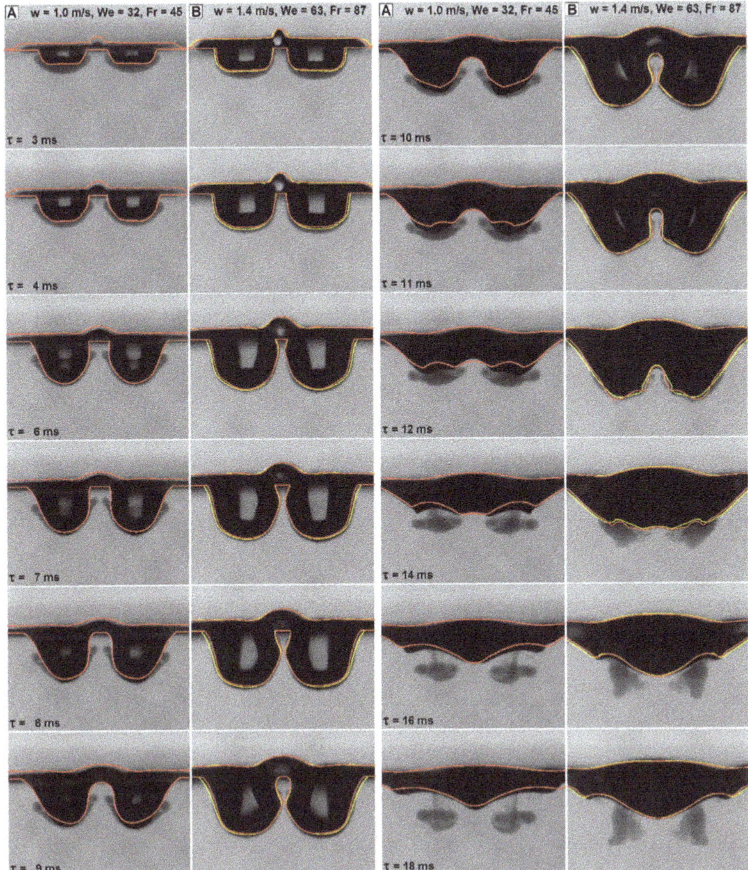



**Figure 3.** Superposition of the numerical crater contours to the experimental high-speed photos, for case A and case B (for the latter: red line, results from the 12.35 Mcell simulation; yellow line, results for the 1.4 Mcell simulation); time up to 18 ms.

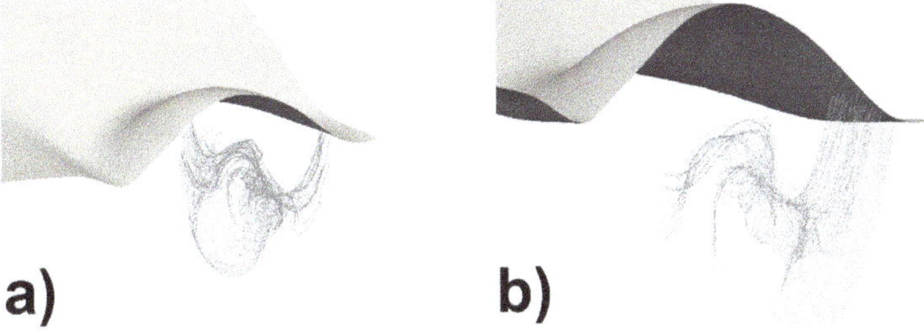



**Figure 4.** Simulated flow streamlines in a region where vortices are experimentally observed (case A, w = 1.0 m/s, τ = 18 ms).

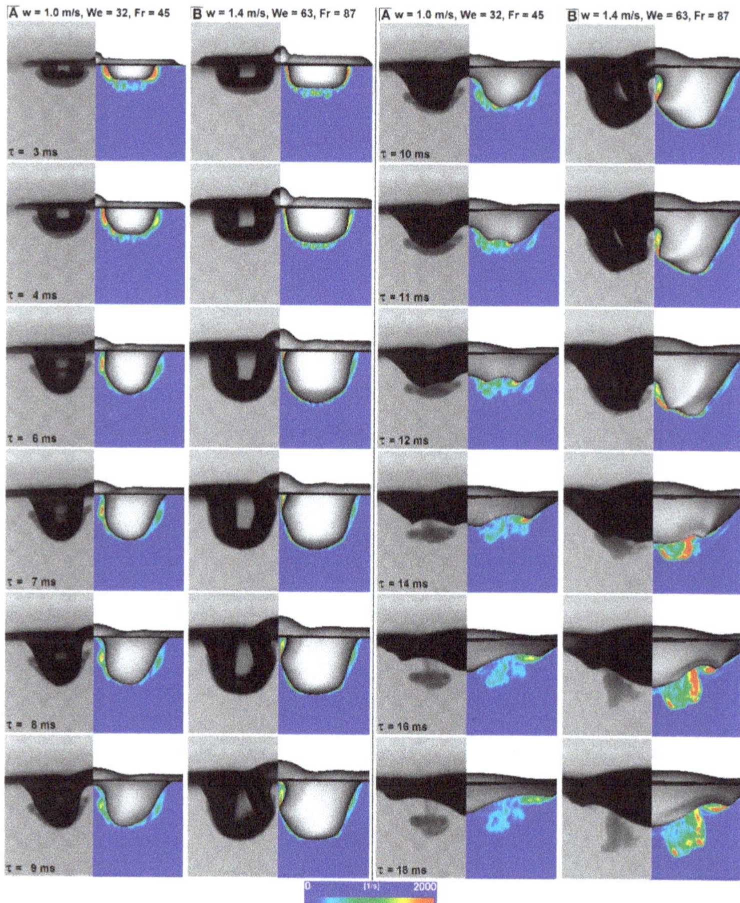


**Figure 5.** Comparison between the experimentally detected vortex structures and the vorticity calculated by numerical simulations, for cases A and B, time up to 18 ms.

For case A in Figure 5, the drop fluid is symmetrically distributed below the liquid–air interface ($\tau = 4$ ms). Later ($\tau = 6$ ms), the drop fluid begins to roll up. The continuous deformation of the liquid–air interface generates an outer vortex structure in the form of a vortex filament, which is also visible in the numerically calculated vorticity in the same region. The roll-up is continuous while the expanding crater causes downward convection of the dyed drop fluid. Asymmetries of the outer vortex rings, having their origins in different convection velocities, are also found in the numerical results. When the crater recedes ($\tau = 9$ ms), a central vortex structures is formed. Both vortex structures, the central vortex and the outer vortex, separate from the receding crater and form a unique complex vortex blob that continuous to move downward. This complex structure could not be captured in the simulation, but the region of vorticity is represented very well.

For case B in Figure 5, the generation of the outer vortex structure is superposed, and the dyed liquid spreads around the cavity. The interaction between the two evolving craters causes a local enlargement of the spreading dyed drop liquid, showing a vortical structure in both investigations (the experiment and simulation ($\tau = 8$ ms)). When the cavities recede, apparently disorganized structures separate and move downward. The numerical results clearly show these downward moving structures.

In terms of the characteristic dimensions of the drop craters, the crater's maximum penetration depth $z_{max}$, the maximum crater penetration depth in the transverse plane including the impact point $z_c$ (i.e., parallel to the yz-plane and at a x-coordinate equal to half of the drop interspace [10]), and the crater penetration depth along the vertical line including the impact point $z_{uip}$, were measured from both the experiments and the simulations. In this work, these quantities are always presented in a dimensionless form, by dividing them by the drop diameter before impact. Therefore, their symbols are completed by a "dl" subscript in all the following charts. For double drop impacts, these quantities do not always coincide during the crater evolution (differently from what happens for single drop impacts). This can be clearly seen, e.g., in Figures 2 and 3. $z_{max}$ and $z_c$ are accessible both to experiments and simulations, while $z_{uip}$ can be evaluated only by simulation because its measurement requires a cross-section of the craters (along the xy plane) and not just a side view. Figures 6 and 7 show the values of these quantities for case A (w = 1.0 m/s) and case B (w = 1.4 m/s) as a function of time. As for all the charts in this paper, the initial time $\tau$ = 0 was fixed at the instant of impact. In the experimental results, the two craters are not perfectly equal, but their difference is very slight (as can be seen for $z_{max}$ for case B in Figure 7). Therefore, only the values for the left crater are shown in Figure 6. More specifically, Figure 6 reports the results for case A (w = 1.0 m/s), comparing simulations and experiments for the two values of the water–air surface tension set in the simulations.

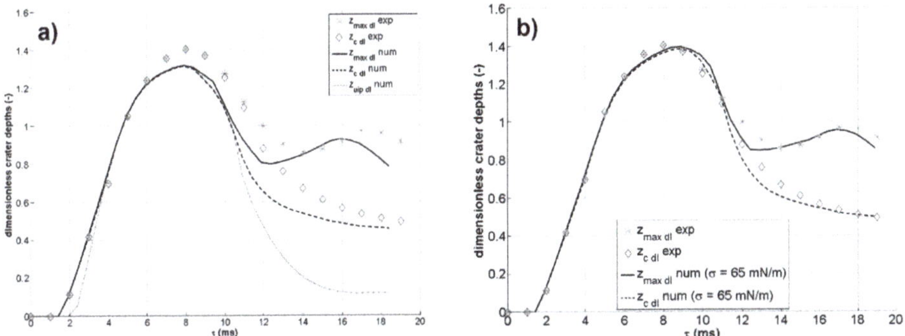

**Figure 6.** Dimensionless $z_{max}$, $z_c$ and $z_{uip}$ extracted from the experiments and from the numerical simulations for case A (w = 1.0 m/s): (**a**) the results using σ = 71 mN/m; (**b**) the results using σ = 65 mN/m.

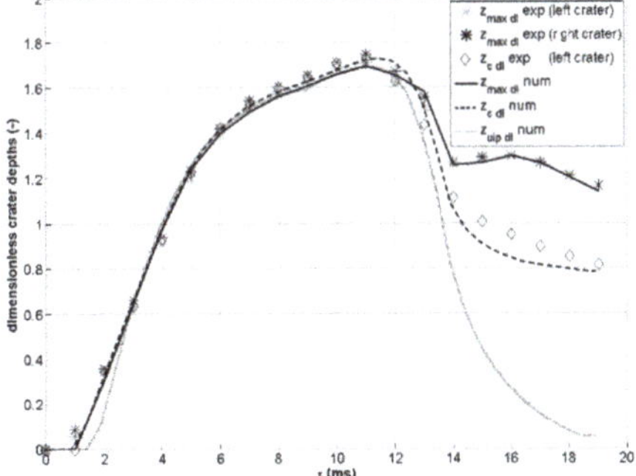

**Figure 7.** Dimensionless $z_{max}$, $z_c$ and $z_{uip}$ extracted from the experiments and from the numerical simulations for case B (w = 1.4 m/s). volume-of-fluid (VOF) results using σ = 71 mN/m.

As can be seen in Figure 6, the experimental values are better reproduced when σ = 65 mN/m (Figure 6a) instead of σ = 71 mN/m (distilled water value, Figure 6b). This seems reasonable as the dye used for the droplet in the experiments [10] contains surfactants, which may cause—despite the very dilute solution—a reduction in the drop surface tension. For case B, reported in Figure 7, the agreement is already good for simulations with $\sigma = 71$ mN/m. This can be explained by the fact that in case B, the impact velocity is higher, so the importance of the capillary actions and surface tension is reduced with respect to case A.

In any case, the overall maximum depth reached by the crater during its evolution—which is one of the most important quantities because it limits the water layer directly reached by the crater—is always reproduced very well by the simulations. The deviation when using $\sigma = 71$ mN/m is −6.1% for case A, −1.7% for case B, and −0.14% for case C. The deviation for case A is reduced to −0.68% when using $\sigma = 65$ mN/m. In all cases, the numerical simulation underestimates the experimental value, with a discrepancy that decreases with an increasing impact velocity. This is again consistent with the fact that the set surface tension is probably larger than the correct experimental value (thus, the free surface more strongly opposes pulling).

*3.2. Energy Conversion During Impact*

After validation, the numerical simulations were used to extract information about some aspects that would be difficult or very time-consuming to study experimentally. Post-processing of the numerical results was first carried out to extract the kinetic and surface energies within the domain for all the simulated time steps. Given the small involved heights, the changes in the potential energy (of the drop liquid and of the pool liquid displaced by the drop) are minor, so the latter will not be described. Kinetic energy was calculated for each cell as:

$$E_k = 1/2 \ w_{cell}^2 \left[ (\gamma \ \rho_{water} + (1 - \gamma) \ \rho_{air} \right] \tag{1}$$

where $w_{cell}$ is the velocity magnitude in the cell, $\rho_{water}$ and $\rho_{air}$ are the water and air densities, and $\gamma$ is the volume fraction. Summation over all cells was then performed to calculate the total $E_k$ over the whole domain, which, before impact, is obviously equal to the kinetic energy of the drop alone. Surface energy $E_s$ was calculated as the product of the air–water interfacial energy $\sigma$ (taken at its reference value for distilled water at the drop temperature, 71 N/m) by the total area of the air–water interface obtained as the isosurface of the color function $\gamma = 0.5$. This value was then decremented by the value of the total interfacial energy of the undisturbed pool surface, so that a direct comparison between the interfacial energy of the drop before the impact and the interfacial energy during crater evolution can be evaluated.

For both kinetic and surface energy, the values extracted from the numerical results at the time step immediately before impact were compared with the theoretical predictions for drops of known diameters and velocities to check their correctness and reliability. The average deviation was 0.78% for kinetic energy and 1.31% for surface energy, which was considered satisfactory. Figure 8 shows the results for both the double drop impacts and the corresponding (i.e., with same impact velocity) single drop impact, after normalization of the values with respect to the kinetic energy and the surface energy immediately before the impact. The time evolution of the two energy contributions appears consistent with the experimentally observed behavior and with each other. When kinetic energy increases, surface energy decreases, and vice versa. Surface energy decreases during the first instants after impact because the drop coalesces with the pool (so a part of the drop and the pool's external surfaces are merged, with a consequent reduction in the total interface area), but the crater is not formed yet. The profiles also show a very regular evolution for impact velocity. The most interesting outcome is that the evolutions for single and double impacts are extremely similar, even if minor differences are obviously present due to the more complex evolution of the crater shape in double drop cases (e.g., causing the mutual intersections between the curves). This suggests that the latter has only a minor influence on energy conversion during the crater's expansion and recession.

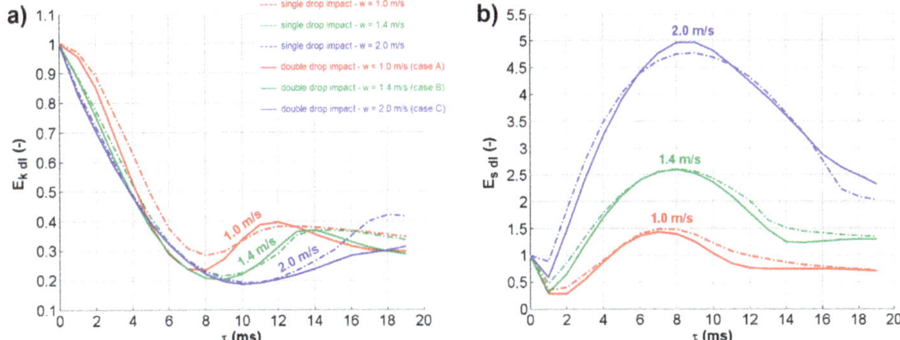

**Figure 8.** Time evolution of the normalized kinetic energy (**a**) and normalized surface energy (**b**) in the domain after double drop impact, as a function of time for single (dash-dot lines) and double (continuous lines) drop impacts.

### 3.3. Sensitivity Analysis on Surface Tension

As water is easily contaminated, particularly in real applications, its surface energy may be quite different from the distilled water case. Given the observed influence of $\sigma$ on the results, a sensitivity analysis was carried out on its value. Case B (w = 1.4 m/s) was selected as a reference case, changing only the surface tension value. It can be expected that larger surface tension values cause a more significant influence of the capillary action and thus a higher internal pressure within the original drop and a "tenser" pool interface. As can be observed from Figure 9, the trend of the inertial phase, in which the crater is expanding, is influenced only slightly by the water surface tension. The maximum expansion of the crater is reduced when the surface tension is larger (meaning that the tensing effect dominates with respect to the larger capillary pressure in the drop). The receding phase is heavily changed, and for larger surface tension values, the crater closes much faster. A few sudden changes in the path can be observed. These changes correspond to instants when the capillary waves caused very fast modification in the crater evolution. In fact, once the crater reaches its maximum depth and width, its starts receding, and the surface tension reduces the interface area. This causes the closure of the crater and perturbations of the interface that propagate as waves along the crater interface, resulting in peculiar shapes (e.g., see the time instants between $\tau = 10$ ms and $\tau = 16$ ms in Figure 3) and seemingly irregular trends in the charts. In particular, the changes in the maximum crater depth evidence the instants when different crater regions become deepest.

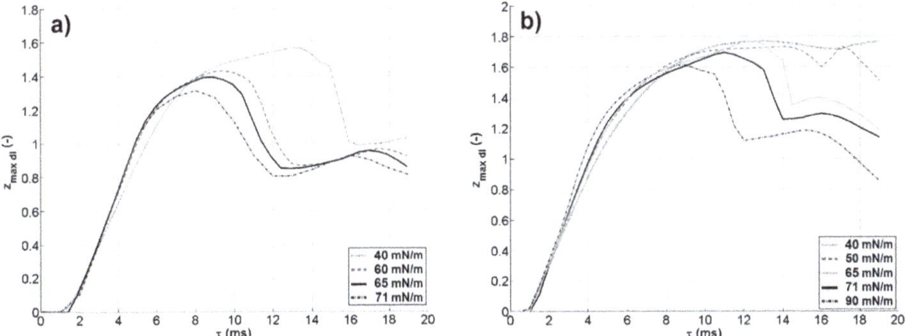

**Figure 9.** Time evolution of $z_{\text{max dl}}$ as a function of the water surface tension for case A (w = 1.0 m/s, (**a**)) and for case B (w = 1.4 m/s, (**b**)).

### 3.4. Sensitivity Analysis on Drop Spacing

The second parameter that was varied is the distance between the drop centers. This distance was changed in the range of 2.58 to 6.58 mm by modifying the reference drop distance ($L_0$ = 4.58 mm) of ±1 and ±2 mm. This resulted in a series of dimensionless drop distances equal to $L/D$ = 1.127 ÷ 2.873, $L/L_0$ =0.345 ÷ 1.437. Figure 10 shows the results in terms of $z_{max}$ (a) and $z_{uip}$ (b), including also single drop impacts for comparison. Here, too, the sudden changes in the paths are evident. Paths are consistent between the different cases, with a progressive shift of the instants at which the crater's main parameters change their trends due to the combined actions of inertia and capillarity. As expected, when the drop distance is large, the crater tends to behave as in a single drop impact. On the other hand, reduction of the interspace causes the deformation of the craters, which interact with each other first indirectly and then directly merge, affecting more and more the crater dimensions. A peculiar result is that for drops nearly touching each other, the time at which the maximum value of $z_{max}$ is reached tends to the value for a single drop with double the volume and mass (as the density is not changed), but the maximum value of $z_{max\ dl}$ is much larger than that in the corresponding single drop case. Thus, it can be concluded that direct interaction (merging) between the craters enhances crater penetration even more than a simple summation of inertial contributions would predict.

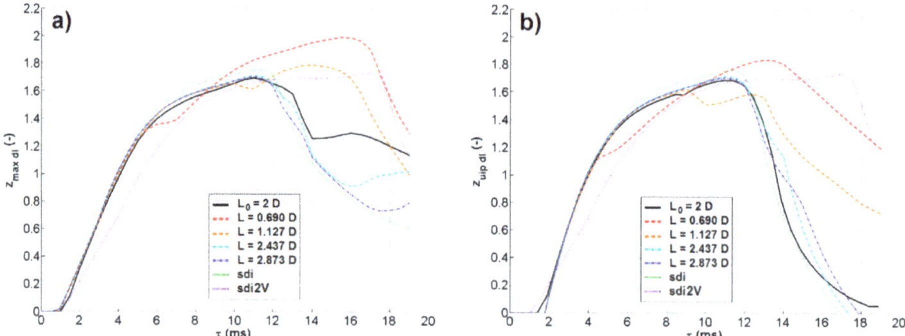

**Figure 10.** Time evolution of $z_{max\ dl}$ (**a**) and $z_{uip\ dl}$ (**b**) as a function of the distance L between the drops. Reference case ($L_0/D$ = 2) is case B; the other cases are with the drop interspace decreased and increased by 1 mm and 2 mm. Single drop impact (sdi) includes the same drop diameter and impact velocity; the single drop impact for a drop with the same impact velocity but a volume and mass double of the original one (sdi2V) is also added.

### 3.5. Three-Dimensional Visualization

Unlike the experiments, where a single side or top view of the drop impacts can be observed, from the numerical simulations, a fully 3D description of the phenomenon can be obtained. First, the air–water interface can be extracted and visualized. Figure 11 shows a rendering of the crater evolution at different time instants—for case B (w = 1.4 m/s), simulated with 12.35 Mcell. For a more intuitive visualization, the simulated domain was mirrored once to reconstruct half of the real domain. It can be noticed how not only the major features, but also the details (including the neck and the surface and capillary waves) are very well captured.

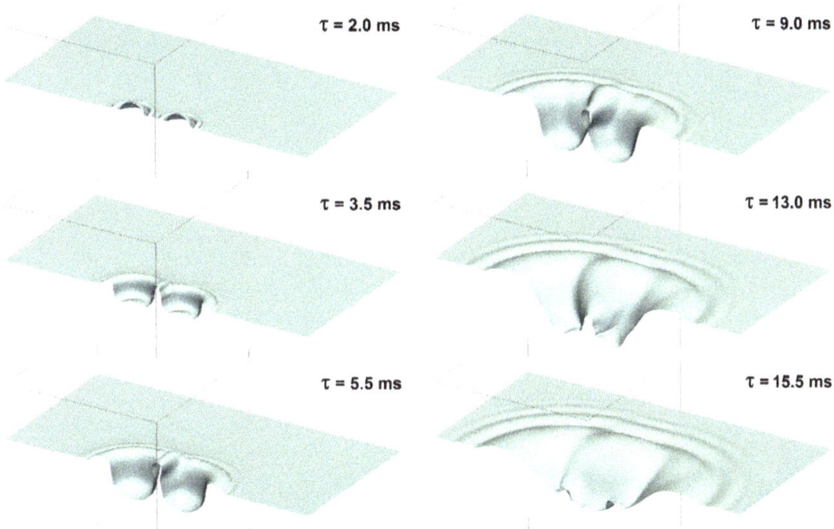

**Figure 11.** Rendering of some frames showing the crater evolution from the simulation of case B (w = 1.4 m/s), simulated with 12.35 Mcells.

### 3.6. Impact of Three and Four Synchronized Drops.

Finally, simulations of multiple drop impacts were tested in all cases for impact velocity w = 1.4 m/s. For these cases, no corresponding experimental tests were performed. Three drops in a single row, aligned along the x-axis, and four drops, two aligned along x and two along z, were tested. Four cases were simulated—one with a drop interspace L equal to $L_0$ for both three- and four-drop cases, a three-drop case with a drop interspace equal to 0.75 $L_0$, and a four-drop case with a drop interspace of 1.5 $L_0$. Figure 12 shows some rendered frames of the crater's evolution for three of these simulated cases. It can be observed how for the three-drop case with L = $L_0$ the craters do not merge, while for the other cases they do, resulting in a different evolution and a maximum crater depth that becomes much larger, as reported in Figure 13. In these cases, too, the interaction between the craters appears to have a significant effect on the maximum depth. Further studies will be needed to verify how multiple drop interactions would affect the crater depth when there is no "free space" left for expansion. On the other hand, each crater finds its way blocked by other interfaces in all directions. In this case, the crater depths might be reduced. Further studies will also be needed to investigate the effect of non-synchronous impacts.

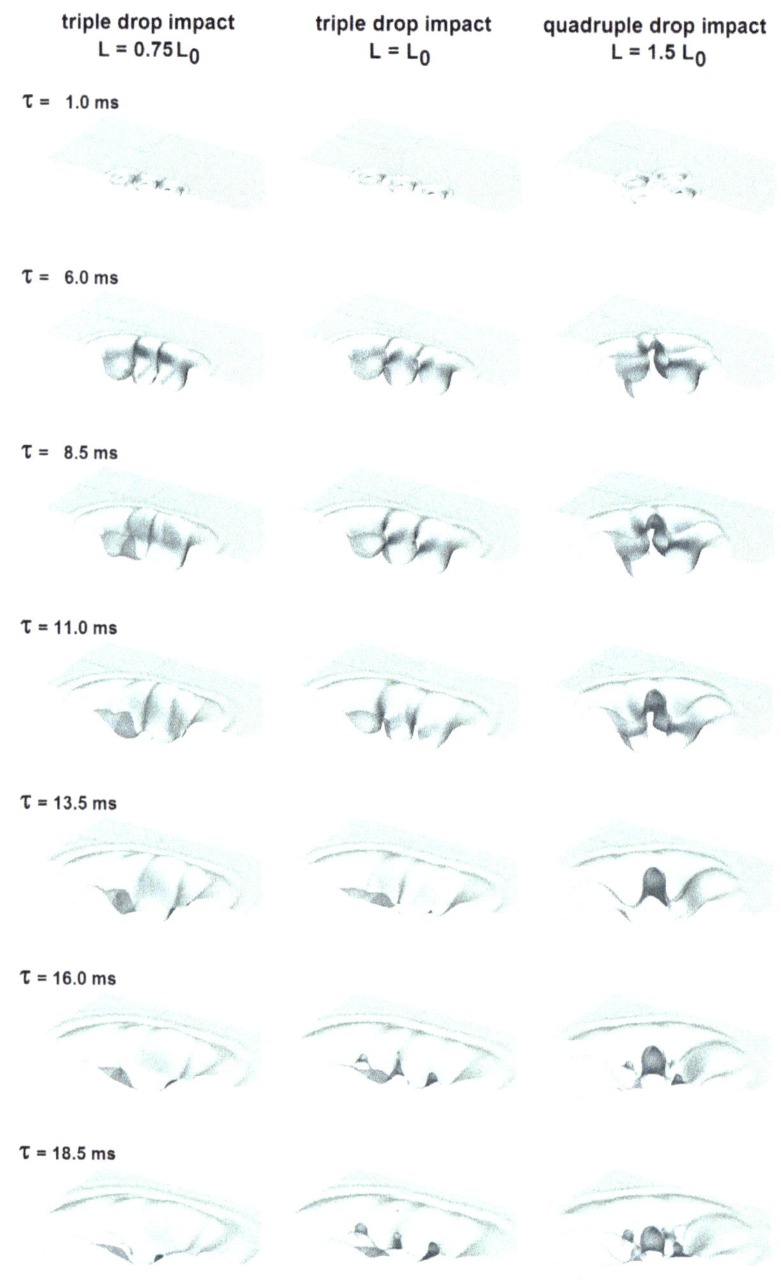



**Figure 12.** Rendering of some frames showing the crater evolution from the simulations of three-drop and four-drop cases (w = 1.4 m/s).

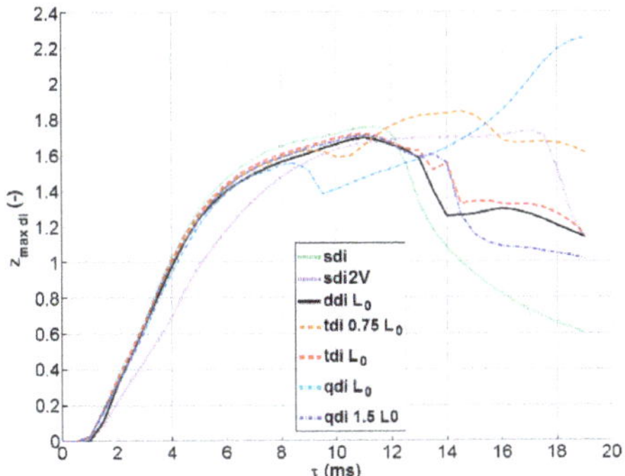

**Figure 13.** Dimensionless $z_{max}$ as a function of time from the numerical simulations for multiple drop impact cases (three drops, tdi, and four-drops, qdi) having $w = 1.4$ m/s and different inter-drop spacing. Results for single drop impacts for drops having the same volume (sdi) and two times the volume (sdi2V) are also added for comparison.

## 4. Conclusions

Numerical simulations of double and multiple synchronised drop impacts were performed using the interFoam solver of the OpenFOAM® open-source CFD package. Results for double drop impacts were validated against experimental data acquired by high-speed imaging. Crater shape and depth and vortex generation were compared between the simulation and experiments for double drop impacts, and a good agreement was found, particularly in the inertial phase (leading to the maximum crater penetration depth). Simulations were then performed to investigate energy conversion during the crater evolution, the effect of varying surface tension, and drop interspace and multiple drop impacts. The main findings are:

- Energy conversion between the kinetic and surface energies during double drop impacts shows a consistent behaviour very similar to that of single drop impacts, thus evidencing that it is not significantly influenced by the crater shape;
- Both the surface tension and the interspace have a very significant influence on the crater shape and depth. Concerning the surface tension, in real applications water is easily contaminated. This aspect must be kept in mind when using laboratory results obtained with distilled water. Concerning the drop interspace, obviously the more the drops are spaced, the more similar the crater evolution is to a single-drop impact case. More interestingly, when the drop interspace is large enough to avoid crater merging, the maximum depth reached by the crater is only slightly affected by the presence of other drops (even if the crater shape is heavily affected by the indirect (mediated by the liquid) interaction. Contrarily, when the craters merge, the maximum depth is heavily affected, with an increase that may reach more than the value corresponding to the impact of a single drop, with double the volume and mass.

In summary, direct extension of results from the study of single drop impacts to the case of multiple impacts with many unevenly spaced drops and non-simultaneous contact times (much more common in applications), appears not to be straightforward; on the contrary, neglecting mutual interactions can be quite misleading. It seems, therefore, mandatory to perform further studies including both more drops and different drop interactions (delayed impacts, different drop interspaces, etc.) to cast further light on these aspects. This would lead to the possibility of developing new models and correlations for multiple drop impacts, which are more reliable for use in all the applications fields where the phenomenon is of importance.

For such studies, improvements in the VOF approach may also be needed:

- Retarded coalescence or other related models should be added, to be able to manage situations in which interfaces are very near without need of extremely fine meshes. Otherwise, a much faster self-adaptive remeshing should be implemented;
- Hybrid methods, e.g., a combined VOF-Level Set, could be used to better manage the interface curvature and, consequently, reduce spurious velocities.

In addition, comparison with other methods (e.g., diffuse interface methods) could be interesting.

**Author Contributions:** Conceptualization, M.G., M.S., and S.F.-S.; Formal analysis, M.G., M.S., and S.F.-S.; Funding acquisition, S.F.-S.; Investigation, M.G., M.S., and S.F.-S.; Methodology, M.G.; Software, M.G.; Validation, M.G., M.S., and S.F.-S.; Visualization, M.G.; Writing—original draft, M.G., M.S., and S.F.-S.; Writing—review and editing, M.G., M.S., and S.F.-S.

**Funding:** This research received no external funding. The Research Grant of Stephanie Fest-Santini was financed within "Progetto ITALY® —Azione Giovani in Ricerca" of the University of Bergamo (Italy).

**Acknowledgments:** The contribution of Viola Knisel in performing the numerical simulations is gratefully acknowledged.

**Conflicts of Interest:** The authors declare no conflict of interest. The funders had no role in the design of the study; in the collection, analyses, or interpretation of data; in the writing of the manuscript, or in the decision to publish the results.

## References


1. Worthington, A.M. On impact with a liquid surface. *Proc. R. Soc. Lond.* **1882**, *34*, 217–230. [CrossRef]
2. Worthington, A.M. *A Study of Splashes*; Longmans Green and Co.: London, UK, 1908.
3. Rein, M. Phenomena of liquid drop impact on solid and liquid surfaces. *Fluid Dyn. Res.* **1993**, *12*, 61–93. [CrossRef]
4. Yarin, A.L. Drop impact dynamics: Splashing, spreading, receding, bouncing. *Annu. Rev. Fluid Mech.* **2006**, *38*, 159–192. [CrossRef]
5. Ashgriz, N. (Ed.) *Handbook of Atomization and Sprays: Theory and Applications*; Springer Science & Business Media: Berlin, Germany, 2011; ISBN 9781441972644. [CrossRef]
6. Cole, D. The Splashing Morphology of Liquid-Liquid Impacts. Ph.D. Thesis, James Cook University, Townsville, Australia, 2007.
7. Santini, M.; Cossali, G.E.; Marengo, M. Splashing characteristics of multiple and single drop impacts onto a thin liquid film. In Proceedings of the International Conference on Multiphase Flows, Leipzig, Germany, 9–13 July 2007.
8. Bisighini, A.; Cossali, G.E. High-speed visualization of interface phenomena: Single and double drop impacts onto a deep liquid. *J. Vis.* **2011**, *14*, 103–110. [CrossRef]
9. Guo, Y.; Chen, G.; Shen, S.; Zhang, J. Double droplets simultaneous impact on liquid film. *IOP Conf. Ser. Mater. Sci. Eng.* **2015**, *88*, 012016. [CrossRef]
10. Santini, M.; Fest-Santini, S.; Cossali, G.E. Experimental study of vortices and cavities from single and double drop impacts onto deep pools. *Eur. J. Mech. B/Fluids* **2017**, *62*, 21–31. [CrossRef]
11. Roisman, I.V.; Prunet-Foch, B.; Tropea, C.; Vignes-Adler, M. Multiple Drop Impact onto a Dry Solid Substrate. *J. Colloid Interface Sci.* **2002**, *256*, 396–410. [CrossRef]
12. Ray, B.; Biswas, G.; Sharma, A.; Welch, S.W.J. CLSVOF method to study consecutive drop impact on liquid pool. *Int. J. Numer. Methods Heat Fluid Flow* **2013**, *23*, 143–158. [CrossRef]
13. Bouwhuis, W.; Huang, X.; Chan, C.U.; Frommhold, P.E.; Ohl, C.-D.; Lohse, D.; Snoeijer, J.H.; van der Meer, D. Impact of a high-speed train of microdrops on a liquid pool. *J. Fluid Mech.* **2016**, *792*, 850–868. [CrossRef]
14. Oguz, H.N.; Prosperetti, A. Bubble entrainment by impact of a liquid drop on liquid surfaces. *J. Fluid. Mech.* **1990**, *219*, 143–179. [CrossRef]
15. Rieber, M.; Frohn, A. A numerical study on the mechanism of splashing. *Int. J. Heat Fluid Flow* **1999**, *20*, 455–461. [CrossRef]
16. Bussmann, M.; Mostaghimi, J.; Chandra, S. On a three-dimensional volume tracking model of droplet impact. *Phys. Fluids* **1999**, *11*, 1406. [CrossRef]

17. Morton, D.; Rudman, M.; Liow, J.M. An investigation of the flow regimes resulting from splashing drops. *Phys Fluids* **2000**, *12*, 747. [CrossRef]

18. Thoraval, M.-J.; Takehara, K.; Etoh, T.G.; Popinet, S.; Ray, P.; Josserand, C.; Zaleski, S.; Thoroddsen, S.T. von Karman Vortex Street within an Impacting Drop. *Phys. Rev. Lett.* **2012**, *108*, 264506. [CrossRef]

19. Hirt, C.W.; Nichols, B.D. Volume of Fluid (VOF) Method for the Dynamics of Free Boundaries. *J. Comp. Phys.* **1981**, *39*, 201–225. [CrossRef]

20. Sussman, M.; Smereka, P.; Osher, S.J. A level set approach for computing solutions to incompressible two-phase flow. *J. Comp. Phys.* **1994**, *114*, 146–159. [CrossRef]

21. Popinet, S.; Zaleski, S. A front tracking algorithm for accurate representation of surface tension. *Int. J. Numer. Meth. Fluid.* **1999**, *30*, 775–793. [CrossRef]

22. Bonometti, T.; Magnaudet, J. An interface-capturing method for incompressible two-phase flows. Validation and application to bubble dynamics. *Int. J. Multiph. Flow* **2007**, *33*, 109–133. [CrossRef]

23. Fuster, D.; Agbaglah, G.; Josserand, C.; Popinet, S.; Zaleski, S. Numerical simulation of droplets, bubbles and waves: State of the art. *Fluid Dyn. Res.* **2009**, *41*, 065001. [CrossRef]

24. Albadawi, A.; Donoghue, D.B.; Robinson, A.J.; Murray, D.B.; Delauré, Y.M.C. Influence of surface tension implementation in Volume of Fluid and coupled Volume of Fluid with Level Set methods for bubble growth and detachment. *Int. J. Multiph. Flow* **2013**, *53*, 11–28. [CrossRef]

25. Menard, T.; Tanguy, S.; Berlemont, A. Coupling level set/VOF/ghost fluid methods: Validation and application to 3D simulation of the primary break-up of a liquid jet. *Int. J. Multiph. Flow* **2007**, *33*, 510–524. [CrossRef]

26. Tryggvason, G.; Bunner, B.; Esmaeeli, A.; Juric, D.; Al-Rawahi, N.; Tauber, W.; Han, J.; Nas, S.; Jan, Y.-J. A Front-Tracking Method for the Computations of Multiphase Flow. *J. Comp. Phys.* **2001**, *169*, 708–759. [CrossRef]

27. Zheng, H.W.; Shu, C.; Chew, Y.T.; Sun, J.H. Three-dimensional lattice Boltzmann interface capturing method for incompressible flows. *Int. J. Numer. Meth. Fluids* **2008**, *56*, 1653–1671. [CrossRef]

28. Huang, J.J.; Shu, C.; Chew, Y.T. Lattice Boltzmann study of bubble entrapment during droplet impact. *Int. J. Numer. Meth. Fluids* **2011**, *65*, 655–682. [CrossRef]

29. Gentner, F.; Rioboo, R.; Baland, J.P.; De Coninck, J. Low Inertia Impact Dynamics for Nanodrops. *Langmuir* **2004**, *20*, 4748–4755. [CrossRef] [PubMed]

30. Telis-Romero, J.; Cabral, R.; Gabas, A.L.; Telis, V. Rheological properties and fluid dynamics of coffee extract. *J. Food Process Eng.* **2001**, *24*, 217–230. [CrossRef]

31. Nixon, M.S.; Aguado, A.S. *Feature Extraction and Image Processing*, 2nd ed.; Academic Press: Cambridge, MA, USA, 2008; ISBN 9780080556727.

32. Rodriguez, F.; Mesler, R. The penetration of drop-formed vortex rings into pools of liquid. *J. Colloid Interface Sci.* **1988**, *121*, 121–129. [CrossRef]

33. Chapman, D.; Critchlow, P. Formation of vortex rings from falling drops. *J. Fluid Mech.* **1967**, *29*, 177–185. [CrossRef]

34. Thoraval, M.-J.; Li, Y.; Thoroddsen, S. Vortex-ring induced large bubble entrainment during drop impact. *Phys. Rev. E* **2016**, *93*, 033128. [CrossRef]

35. Saylor, J.; Grizzard, N. The optimal drop shape for vortices generated by drop impacts: The effect of surfactants on the drop surface. *Exp. Fluids* **2004**, *36*, 783–790. [CrossRef]

36. Zou, J.; Ji, C.; Yuan, B.G.; Ren, Y.L.; Ruan, X.D.; Fua, X. Large bubble entrapment during drop impacts on a restricted liquid surface. *Phys. Fluids* **2012**, *24*, 057101. [CrossRef]

37. Wang, A.-B.; Kuan, C.-C.; Tsai, P.-H. Do we understand the bubble formation by a single drop impacting upon liquid surface? *Phys. Fluids* **2013**, *25*, 101702. [CrossRef]

38. OpenFOAM®. The Open Source CFD Toolbox. Available online: http://www.openfoam.org (accessed on 1 February 2019).

39. Brackbill, J.U.; Khote, D.B.; Zemach, C. A Continuum Method for Modeling Surface Tension. *J. Comp. Phys.* **1992**, *100*, 335–354. [CrossRef]

40. Gopala, V.R.; van Wachem, B.G.M. Volume of fluid methods for immiscible-fluid and free-surface flows. *Chem. Eng. J.* **1998**, *141*, 204–211. [CrossRef]

41. Deshpande, S.S.; Anumolu, L.; Trujillo, M.F. Evaluating the performance of the two-phase flow solver interFoam. *Comp. Sci. Discov.* **2012**, *5*, 014016. [CrossRef]

42. Trujillo, M.F.; Lewis, S.R. Thermal boundary layer analysis corresponding to droplet train impingement. *Phys. Fluids* **2012**, *24*, 112102. [CrossRef]

43. Deshpande, S.S.; Trujillo, M.F. Distinguishing features of shallow angle plunging jets. *Phys. Fluids* **2013**, *25*, 082103. [CrossRef]

44. Costa, A.B.; Graham Cooks, R. Simulated splashes: Elucidating the mechanism of desorption electrospray ionization mass spectrometry. *Chem. Phys. Lett.* **2008**, *464*, 1–8. [CrossRef]

45. Berberovic, E.; van Hinsberg, N.P.; Jakirlic, S.; Roisman, I.V.; Tropea, C. Drop impact onto a liquid layer of finite thickness: Dynamics of the cavity evolution. *Phys. Rev. E* **2009**, *79*, 036306. [CrossRef]

46. Roisman, I.V.; Weickgenannt, C.M.; Lembach, A.N.; Tropea, C. Drop impact close to a pore: Experimental and numerical investigations. In Proceedings of the ILASS—Europe 2010, 23rd Annual Conference on Liquid Atomization and Spray Systems, Brno, Czech, 6–9 September 2010.

47. Trujillo, M.F.; Alvarado, J.; Gehring, E.; Soriano, G.S. Numerical simulations and experimental characterization of heat transfer from a periodic impingement of droplets. *J. Heat Transf.* **2011**, *133*, 122201. [CrossRef]

48. Brambilla, P.; Guardone, A. Automatic tracking of corona propagation in three-dimensional simulations of non-normal drop impact on a liquid film. *Computing* **2013**, *95*, 415–424. [CrossRef]

49. Deshpande, S.S.; Trujillo, M.F.; Wu, X.; Chahine, G. Computational and experimental characterization of a liquid jet plunging into a quiescent pool at shallow inclination. *Int. J. Heat Fluid Flow* **2012**, *34*, 1–14. [CrossRef]

50. Márquez Damián, S. Description and Utilization of InterFoam Multiphase Solver. Universidad Nacional del Litoral, Argentina. Available online: http://infofich.unl.edu.ar/upload/5e6dfd7ff282e2deabe7447979a16d49b0b1b675.pdf (accessed on 1 February 2019).

51. Raeini, A.Q.; Blunt, M.J.; Bijeljic, B. Modelling two-phase flow in porous media at the pore scale using the volume-of-fluid method. *J. Comp. Phys.* **2012**, *231*, 5653–5668. [CrossRef]

52. Jasak, H.; Weller, H.G. *Interface-Tracking Capabilities of the InterGamma Differencing Scheme*; Technical Report; Imperial College of Science, Technology and Medicine, University of London: London, UK, 1995.

53. Krause, F.; Li, X.; Fritsching, U. Simulation of droplet-formation and -interaction in emulsification processes. *Eng. Appl. Comput. Fluid Mech.* **2011**, *5*, 406–415. [CrossRef]

54. Peck, B.; Sigurdson, L. The vortex ring velocity resulting from an impacting water drop. *Exp. Fluids* **1995**, *18*, 351–357. [CrossRef]

*Article*

# An Explicit Meshless Point Collocation Solver for Incompressible Navier-Stokes Equations

**George C. Bourantas [1,*], Benjamin F. Zwick [1], Grand R. Joldes [1], Vassilios C. Loukopoulos [2], Angus C. R. Tavner [1], Adam Wittek [1] and Karol Miller [1]**

[1]  Intelligent Systems for Medicine Laboratory, The University of Western Australia, 35 Stirling Highway, Perth, WA 6009, Australia; benjamin.zwick@research.uwa.edu.au (B.F.Z.); grand.joldes@uwa.edu.au (G.R.J.); angus.tavner@uwa.edu.au (A.C.R.T.); adam.wittek@uwa.edu.au (A.W.); karol.miller@uwa.edu.au (K.M.)
[2]  Department of Physics, University of Patras, Patras, 26500 Rion, Greece; vxloukop@physics.upatras.gr
*   Correspondence: george.bourantas@uwa.edu.au

Received: 8 July 2019; Accepted: 28 August 2019; Published: 3 September 2019

**Abstract:** We present a strong form, meshless point collocation explicit solver for the numerical solution of the transient, incompressible, viscous Navier-Stokes (N-S) equations in two dimensions. We numerically solve the governing flow equations in their stream function-vorticity formulation. We use a uniform Cartesian embedded grid to represent the flow domain. We discretize the governing equations using the Meshless Point Collocation (MPC) method. We compute the spatial derivatives that appear in the governing flow equations, using a novel interpolation meshless scheme, the Discretization Corrected Particle Strength Exchange (DC PSE). We verify the accuracy of the numerical scheme for commonly used benchmark problems including lid-driven cavity flow, flow over a backward-facing step and unbounded flow past a cylinder. We have examined the applicability of the proposed scheme by considering flow cases with complex geometries, such as flow in a duct with cylindrical obstacles, flow in a bifurcated geometry, and flow past complex-shaped obstacles. Our method offers high accuracy and excellent computational efficiency as demonstrated by the verification examples, while maintaining a stable time step comparable to that used in unconditionally stable implicit methods. We estimate the stable time step using the Gershgorin circle theorem. The stable time step can be increased through the increase of the support domain of the weight function used in the DC PSE method.

**Keywords:** transient incompressible Navier-Stokes; meshless point collocation method; stream function-vorticity formulation; strong form; explicit time integration

## 1. Introduction

In this paper, we describe the development of a strong form meshless point collocation method for numerically solving the two-dimensional, non-stationary, incompressible Navier-Stokes (N-S) equations in their stream function-vorticity formulation. Our method relies on the ability of meshless methods to accurately compute spatial derivatives, and on their flexibility to solve partial differential equations (PDEs) in complex geometries. The proposed numerical method can efficiently handle uniform Cartesian point clouds embedded in a complex geometry, and/or irregular sets of nodes that directly discretize the flow domain. To compute the spatial derivatives that appear in the governing flow equations, we apply an interpolation meshless scheme, which computes spatial derivatives on both Cartesian embedded grids and irregular nodal distributions.

The mathematical formulation of the N-S equations, depending on the choice of dependent variables, can be classified as (i) primitive variables, (ii) stream function-vorticity, and (iii) velocity-vorticity. The majority of numerical methods are expressed in the primitive variables (velocity-pressure) formulation [1–3]. Although the primitive variables formulation is widely employed, difficulties occur when dealing with pressure boundary conditions, and coupling of the velocity and pressure fields is challenging. The vorticity-vector-potential [4–6] and velocity-vorticity [7] formulations have a clear advantage over the primitive variables formulation, because pressure does not appear in the field equations, and the difficulty of determining the pressure boundary conditions in incompressible flows is avoided [8]. The extension to 3D flow is also straightforward, although the number of field variables increases from three to six. For 2D cases, stream function-vorticity is the preferable method for the numerical solution of the incompressible N-S equations, since the continuity equation (conservation of mass) is inherently fulfilled [4].

The non-stationary N-S equations can be numerically solved using explicit, explicit-implicit or implicit time integration schemes in the framework of finite element (FE), finite volume (FV), finite difference (FD) or spectral methods. Fletcher [9] provides an overview of finite element and finite difference techniques for incompressible fluid flow, while Gresho and Sani [10] review the field, focusing on finite elements. Standard texts on computational fluid dynamics (CFD) using finite difference and finite volume methods can be found in [11–13], and finite elements in [14–16].

Substantial effort has been invested in reducing or eliminating problems related to mesh generation. Attempts to eliminate the human and computational effort related to mesh generation have resulted in the development of meshless methods (MMs). Various MMs, developed for computational fluid dynamics (CFD), have been used to solve the incompressible N-S equations. Of particular interest is the Meshless Local Petrov-Galerkin (MLPG) method, which has been used to solve the N-S equations in their primitive variables [17], velocity-vorticity [18,19] and stream function-vorticity [20] formulations. The moving least squares (MLS) approximation method has been used to compute shape functions and derivatives. Additionally, radial basis functions (RBFs) are able to interpolate shape function and its derivatives [21]. The Local Boundary Integral Method (LBIE) has been successfully applied to solve incompressible, laminar flow cases [22], while the multi-region Local Boundary Integral Equation (LBIE) has been combined with the radial basis functions (RBF) interpolation method [23]. The mesh-free Local Radial Basis Function Collocation Method (LRBFCM) for transient heat transfer and fluid dynamics problems was presented in [24], while others [25,26] reported the use of an indirect/integrated radial-basis-function network (IRBFN) method to solve transient partial differential equations (PDEs) governing fluid flow problems. Velocity-vorticity flow equations were solved in [27–29] using a point collocation method, along with a velocity-correction method, while a meshless point collocation with both velocity and pressure correction methods [30] was applied in several fluid mechanics problems. Eulerian based meshless methods, in contrast to Lagrangian based meshless methods, such as Smoothed Particle Hydrodynamics (SPH), have been mainly applied to regular geometries (square, circles, tubes), where generating a computational grid is relatively straightforward.

In this work, we focus on the Eulerian formulation of the N-S equations. A meshless point collocation method is used for the numerical solution of the non-stationary, incompressible, laminar 2D N-S equations in their stream function-vorticity formulation in complex geometries. The stream function-vorticity formulation is an effective way to describe 2D incompressible flow, since the incompressibility constraint for the velocity is automatically satisfied and the pressure variable is eliminated. The method is generalizable to three dimensions via vector potential-vorticity formulation [4–6]. The vorticity boundary conditions are imposed in a local form. We treat the transient term using a Euler explicit time integration scheme. Since the explicit integration scheme is conditionally stable, we compute the critical time step that ensures stable numerical results using the Gershgorin circle theorem [31,32]. This circumvents the need to calculate eigenvalues, which would be time consuming. The requirement for a stable time-step is counterbalanced by the low computational cost of each step, since only one Poisson-type equation is solved for the stream function equation and the vorticity field is updated using the explicit solver (in three dimensions three Poisson-type equations need to be solved [4–6]). The proposed scheme applies to both uniform embedded and irregular nodal distributions and can be used with ease in complex geometries.

This paper is structured as follows: in Section 2, we (i) present the governing equations (ii) present a numerical scheme for the stream function-vorticity formulation based on the meshless point collocation method (iii) briefly describe the Discretization Correction Particle Strength Exchange (DC PSE) differentiation method and (iv) discuss issues related to the accuracy and the computational cost of the proposed scheme. In Section 3, we highlight the accuracy of the scheme through benchmark problems, while in Section 4, we demonstrate the accuracy, efficiency and the ease of use of the proposed scheme using numerical examples. Finally, in Section 5 we present our conclusions and preliminary results for 3D lid-driven cavity flow using the generalization of the proposed scheme via vorticity-vector potential formulation.

## 2. Methods Governing Equations and Solution Procedure

### 2.1. Stream Function-Vorticity Formulation of Navier-Stokes Equations

The governing equations for the non-stationary, viscous, laminar flow of an incompressible fluid are based on conservation of mass and momentum. The non-dimensional form of the stream function-vorticity ($\psi$-$\omega$) formulation is:

$$\frac{\partial \omega}{\partial t} + \frac{\partial \psi}{\partial y}\frac{\partial \omega}{\partial x} - \frac{\partial \psi}{\partial x}\frac{\partial \omega}{\partial y} = \frac{1}{Re}\nabla^2\omega \tag{1}$$

$$\nabla^2\psi = -\omega \tag{2}$$

where $Re = \frac{UL}{v}$ is the Reynolds number, with $U$ and $L$ being a characteristic velocity and length, respectively, and $v$ the kinematic viscosity of the fluid. Taking the spatial derivative of the stream function $\psi$ yields the velocity components $u$ and $v$, as follows:

$$u = \frac{\partial\psi}{\partial y} \tag{3}$$

$$v = -\frac{\partial\psi}{\partial x} \tag{4}$$

We require a solution to Equations (1) and (2), defined in the spatial domain $\Omega$ with boundary $\partial\Omega$, given the initial conditions at time $t = 0$

$$u = u_0 \tag{5}$$

$$\omega = \omega_0 = \nabla \times u_0, \tag{6}$$

and boundary conditions

$$u = u_{\partial\Omega} \tag{7}$$

$$\omega = (\nabla \times u_{\partial\Omega})\cdot\hat{k} \tag{8}$$

with $\hat{k}$ being the unit vector in the direction of the $z$ axis of a three-dimensional Cartesian coordinate system. We solve the governing equations using an explicit Euler time integration scheme for the transient terms and a strong form meshless collocation method to discretize the flow equations.

### 2.2. Temporal Discretization Using Explicit Euler Method

We treat the viscous term explicitly [33] to obtain the following difference formula in time:

$$\frac{\omega^{n+1} - \omega^n}{\Delta t} + \frac{\partial\psi^n}{\partial y}\frac{\partial\omega^n}{\partial x} - \frac{\partial\psi^n}{\partial x}\frac{\partial\omega^n}{\partial y} = \frac{1}{Re}\nabla^2\omega^n. \tag{9}$$

$$\nabla^2\psi^{n+1} = -\omega^{n+1} \tag{10}$$

using the notation $\omega^{n+1} = \omega(t^{n+1})$ and $\psi^{n+1} = \psi(t^{n+1})$, where $t^{n+1} = t^n + \delta t$. The proposed scheme computes the stream function (along with velocity components) and vorticity field by applying a three-step marching algorithmic procedure:

Step 1 Update vorticity at the interior grid points

$$\omega^{n+1} = \omega^n - \Delta t \frac{\partial \psi^n}{\partial y}\frac{\partial \omega^n}{\partial x} + \Delta t \frac{\partial \psi^n}{\partial x}\frac{\partial \omega^n}{\partial y} + \Delta t \frac{1}{Re}\left(\frac{\partial^2 \omega^n}{\partial x^2} + \frac{\partial^2 \omega^n}{\partial y^2}\right) \tag{11}$$

Step 2 Compute stream function by solving Poisson type problem

$$\frac{\partial^2 \psi^{n+1}}{\partial x^2} + \frac{\partial^2 \psi^{n+1}}{\partial y^2} = -\omega^{n+1} \tag{12}$$

Step 3 Update velocity $u^{(n+1)}$

$$u^{n+1} = \frac{\partial \psi^{n+1}}{\partial y} \tag{13}$$

$$v^{n+1} = -\frac{\partial \psi^{n+1}}{\partial x} \tag{14}$$

### 2.3. Discretization Corrected Particle Strength Exchange (DC PSE) Method

The accuracy of the proposed scheme relies on the accurate computation of spatial derivatives for the unknown field functions in Equations (11)–(14). In the present study, we use the Discretization Corrected Particle Strength Exchange (DC PSE) method, which, to our knowledge, is one of the most accurate meshless methods for computing spatial derivatives. The DC PSE method originated as a Lagrangian particle-based numerical method [34] and is based on Particle Strength Exchange (PSE) operators. To solve PDEs using the DC PSE meshless method, the authors in [35] reformulated the Lagrangian DC PSE method to work in the Eulerian framework. For completion, the PSE operators and the DC PSE method are described below.

#### 2.3.1. Particle Strength Exchange (PSE) Operators

The Particle Strength Exchange (PSE) method utilizes kernels to approximate differential operators that conserve particle strength in particle-particle interactions. The PSE method was proposed by Degond and Mas-Gallic [36] for diffusion and convection-diffusion problems. Eldredge et al. [37] then developed a framework for approximating arbitrary derivatives.

In general, a PSE operator $Q^\beta f(x)$ for approximating the derivative $D^\beta f(x)$ has the form

$$Q^\beta f(x) = \frac{1}{\varepsilon^{|\beta|}} \int (f(y) \mp f(x)) \eta_\varepsilon^\beta (x - y) dy \tag{15}$$

with $\eta_\varepsilon^\beta = \eta^\beta(x/\varepsilon)/\varepsilon^d$ a scaled kernel of radius $\varepsilon$, and $d$ the number of dimensions. The sign in Equation (15) is negative when $|\beta|$ is even and positive when $|\beta|$ is odd, with $\beta$ a multi-index [34]. Now, let $\beta = (\beta_1, \beta_2, \ldots, \beta_n)$, where $\beta_i$, $i = 1, 2, \ldots, n$ is a non-negative integer. Then the partial differential operator $D^\beta$ can be expressed as $D^\beta = \frac{\partial^{|\beta|}}{\partial x_1^{\beta_1} \partial x_2^{\beta_2} \ldots \partial x_n^{\beta_n}}$. The challenge is to find a kernel $\eta_\varepsilon^\beta$ that leads to good approximations for $D^\beta$. To find such kernels for arbitrary derivatives we adopt the idea in Eldredge et al. [37] and start from the Taylor expansion of a function $f(y)$ about a point $x$:

$$f(y) = f(x) + \sum_{|a| = 1}^{\infty} \frac{1}{a!}(y - x)^a D^a f(x). \tag{16}$$

Next, we subtract or add $f(x)$, depending on whether $|\beta|$ is odd or even, to both sides and convolute the equation with the unknown kernel $\eta_\varepsilon^\beta$. Considering Equation (15), this leads to:

$$Q^\beta f(x) \;=\; \frac{1}{\varepsilon^{|\beta|}} \sum_{|a|\,=\,1}^{\infty} \frac{1}{a!} D^a f(x) \int (y-x)^a \eta_\varepsilon^\beta (x-y)dy. \tag{17}$$

Introducing the continuous $a$-moments

$$M_a \;=\; \int (x-y)^a \eta^\beta (x-y)dy \;=\; \int z^a \eta^\beta (z)dz \tag{18}$$

and isolating the derivatives $D^\beta$ on the right-hand side of Equation (17) we obtain:

$$Q^\beta f(x) \;=\; \frac{(-1)^{|\beta|}}{\beta!} M_\beta D^\beta f(x) + \sum_{\substack{|a|\,=\,1 \\ a\neq\beta}}^{\infty} \frac{(-1)^{|a|}}{a!} \varepsilon^{|a|-|\beta|} M_a D^a f(x). \tag{19}$$

Finally, to approximate $Q^\beta f(x)$ with order of accuracy $r$, the following set of conditions is imposed for the moments $M_a$:

$$M_a \;=\; \begin{cases} (-1)^{|\beta|}\beta! & a = \beta \\ 0, & a \neq \beta, \quad 1 \leq |a| \leq |\beta| + r - 1. \end{cases} \tag{20}$$

In addition, if we impose

$$\int |z|^{|\beta|+r} \left|\eta^\beta (z)\right| dz < \infty \tag{21}$$

the mollification error $\epsilon_\varepsilon (x) = D^\beta f(x)$ is bounded [34]. The procedure to construct a kernel that satisfies the conditions in Equation (16) has been described in [34]. Once the kernel is defined, the operator in Equation (17) can be discretized using a midpoint quadrature over the nodes as

$$Q_h^\beta f(x) \;=\; \frac{1}{\varepsilon^{|\beta|}} \sum_{p\in N(x)} \left(f(x_p) \mp f(x)\right)\eta_\varepsilon^\beta (x - x_p)V_p \,, \tag{22}$$

where $N(x)$ is the number of all nodes in a neighborhood around $x$, which can be defined by a cut-off radius $r_c$, usually chosen such that $N(x)$ coincides to a certain level of accuracy with the kernel support, and $V_p$ is the volume associated with each particle. Given such a discretization, the discretization error $\epsilon_h(x) = Q^\beta f(x) - Q_h^\beta f(x)$ is also bounded [34].

### 2.3.2. The Discretization Corrected PSE Operators

The Discretization Corrected PSE operators were introduced by Schrader et al. [34] to reduce the discretization error $\epsilon_h(x)$ in the PSE operator approximation. We can derive the DC PSE approximation from Equation (22), aiming of finding a kernel function which minimizes the difference between this discrete operator and the actual derivative. We can achieve this, by replacing each term $f(x_p)$ in Equation (22) with its Taylor expansion about $x$. This leads to the following expression for the derivative approximation:

$$Q_h^\beta f(x) \;=\; \frac{(-1)^{|\beta|}}{\beta!} Z_h^\beta D^\beta f(x) + \sum_{\substack{|a|\,=\,1 \\ a\neq\beta}}^{\infty} \frac{(-1)^{|a|}}{a!} \varepsilon^{|a|-|\beta|} Z_h^a D^a f(x) + r_0 \tag{23}$$

with

$$r_0 = \begin{cases} 0, & |\beta| \text{ even} \\ 2e^{-|\beta|}Z_h^0 f(x) & |\beta| \text{ odd} \end{cases} \tag{24}$$

and the discrete moments defined as:

$$Z_h^a = \frac{1}{\varepsilon^d} \sum_{p \in N(x)} \left(\frac{x - x_p}{\varepsilon}\right)^a \eta^\beta\left(\frac{x - x_p}{\varepsilon}\right). \tag{25}$$

Therefore, the set of moment conditions becomes:

$$Z_h^a = \begin{cases} (-1)^{|\beta|}\beta! & a = \beta \\ 0 & a \neq \beta \qquad a_{min} \leq |a| \leq |\beta| + r - 1 \\ < \infty & \text{otherwise} \end{cases} \tag{26}$$

for all $|\beta| \neq 0$, where $a_{min}$ is one when $|\beta|$ is odd and zero when $|\beta|$ is even. The kernel $\eta^\beta$ is chosen as:

$$\eta^\beta(x, z) = \left(\sum_{|\gamma| = a_{min}}^{|\beta| + r - 1} a_\gamma(x) z^\gamma\right) e^{-|z|^2} = P(x, z)W(z), \quad z = \frac{x - x_p}{\varepsilon}. \tag{27}$$

The kernel function consists of a polynomial correction function $P(x, z)$ and the weight function $W(z)$. Different types of weight functions can be applied with the possible choices described in [38].

The unknown coefficients $a_\gamma(x)$ are obtained by requiring the kernel given by Equation (28) to satisfy the conditions in Equation (27), resulting in the following linear system of equations:

$$\sum_{|\gamma| = a_{min}}^{|\beta| + r - 1} a_\gamma(x) w_a^\gamma(x) = \begin{cases} (-1)^{|\beta|}\beta! & a = \beta \\ 0 & a \neq \beta \end{cases}, \ \forall \, a_{min} \leq |a| \leq |\beta| + r - 1 \tag{28}$$

with weights

$$w_a^\gamma(x) = \frac{1}{\varepsilon^{|a+\gamma|+d}} \sum_{p \in N(x)} (x - x_p)^{a+\gamma} e^{-\left(\frac{|x-x_p|}{\varepsilon}\right)^2} \tag{29}$$

To compute the approximated derivative $Q_h^\beta f(x)$ at node $x_p$, the coefficients are found by solving the linear system of Equation (28) for $x = x_p$. Given our choice of kernel function, the DC PSE derivative approximation becomes:

$$Q_h^\beta f(x_p) = \frac{1}{\varepsilon(x_p)^\beta} \sum_{x_q \in N(x_p)} \left(f(x_q) \mp f(x_p)\right) p\left(\frac{x - x_p}{\varepsilon(x_p)}\right) a^T(x_p) e^{-\left(\frac{|x_p - x_q|}{\varepsilon(x_p)}\right)^2} \tag{30}$$

where $p(x) = [p_1(x), p_2(x), p_l(x)]$ and $a(x)$ are the vectors of terms in the monomial basis and their coefficients, respectively. By using the DC PSE method, the spatial derivatives $Q^\beta$ up to second order are given as:

$$Q^{1,0} \equiv \frac{\partial}{\partial x}, \quad Q^{0,1} \equiv \frac{\partial}{\partial y} \tag{31}$$

and

$$Q^{1,1} \equiv \frac{\partial^2}{\partial x \partial y} = \frac{\partial^2}{\partial y \partial x}, \quad Q^{2,0} \equiv \frac{\partial^2}{\partial x^2}, \quad Q^{0,2} \equiv \frac{\partial^2}{\partial y^2} \tag{32}$$

### 2.4. Vorticity Boundary Conditions

In finite difference methods, a number of formulae can be used to impose the vorticity boundary conditions [33,39]. These formulae, generally referred to as local, compute vorticity values on a given node on the boundary from the vorticity values in the interior domain, without involving other nodes on the boundary. The most widely used method is Thom's formula [39]. Unfortunately, applying these formulae in real applications had limited success. As stated in [33], vorticity boundary conditions should be global, in the sense that computing the boundary vorticity values should involve vorticity values at nodes in the interior and the boundary. Herein, to achieve global vorticity boundary conditions, we impose vorticity boundary conditions through strong form meshless differential operators, used to compute spatial derivatives. This way, imposing vorticity boundary conditions is consistent with the rest of the algorithm (described in Section 2.2) and becomes quite straightforward.

DC PSE methods are recognized as one of the most accurate and efficient numerical methods to compute derivatives on an irregularly distributed set (cloud) of nodes [34,35,38]. For the stream function-vorticity N-S Equations, given the velocity field values computed previously by the updated stream function values $\psi^{n+1}$ (Step 2, Section 2.2), we can compute the updated vorticity values $\omega^{n+1}$ for the entire spatial domain (including boundaries) using the strong form meshless operators for first order spatial derivative as

$$\omega^{n+1} = \frac{\partial v^{n+1}}{\partial x} - \frac{\partial u^{n+1}}{\partial y} \tag{33}$$

We have already computed the updated values $\omega^{n+1}$ at the interior nodes (Step 1 in the Euler explicit solver). The updated vorticity values on the boundary nodes are computed using the updated velocity values and the DC PSE operators (described in Section 2.3) for spatial derivatives (Equation (27)).

### 2.5. Poisson Solver

To compute the stream function field values (Equation (2)), we numerically solve a Poisson-type equation. Depending on the flow problem and spatial discretization, a direct or iterative solver is chosen accordingly. Iterative and direct solvers for elliptic type PDEs are well established. Direct solvers are robust and easy to use but can be computationally costly and memory intensive when the systems of equations to be solved are very large (e.g., a few million degrees of freedom). On the other hand, a decisive advantage of iterative solvers is their low memory usage, which is significantly less than a direct solver for the same sized problems. Unfortunately, iterative solvers are less robust than direct solvers and may fail to compute the numerical solution.

It is computationally more efficient to use iterative solvers to solve the Poisson equation when a large number of nodes is used. There are many iterative solvers for Poisson type equations. For example, Gauss-Seidel or successive over relaxation (SOR) algorithms have been widely used. These solvers are accurate but relatively computationally demanding because they need $O\,(N^2)$ iterations to converge ($N$ being the total number of grid points). Multi-grid algorithms have been developed to accelerate convergence and the computational effort has been significantly reduced [40]. For equally-spaced grids, fast Fourier transform (FFT) solvers are the fastest available algorithms for solving Poisson equations [41].

In the present study, we use a direct solver. The discrete Laplace operator defined in the Poisson type equation is accurately discretized using DC PSE method, which performs accurately on uniform and locally refined Cartesian grids. The grid is fixed in time so a LU factorization can be precomputed, which reduces the computational cost significantly. For up to 4 million nodes, the factorization is fast and requires low memory. For larger number of nodes, we utilize iterative solvers, such as conjugate gradient (CG) and biconjugate gradient stabilized (BiCGSTAB), since the linear systems are always positive definite, diagonally dominant and symmetric.

### 2.6. Critical Time Step

To compute the critical time step needed to obtain stable results, we rewrite Equation (1) as

$$\omega^{n+1} \;=\; \omega^n + dt\!\left(\frac{\partial\psi^n}{\partial x}\frac{\partial\omega^n}{\partial y} - \frac{\partial\psi^n}{\partial y}\frac{\partial\omega^n}{\partial x} + \frac{1}{Re}\nabla^2\omega^n\right) \tag{34}$$

which can be written in matrix form as

$$\omega^{n+1} \;=\; [\mathbf{I} + \delta t\mathbf{A}]\omega^n \tag{35}$$

where $\omega^{n+1}$ is a column vector of dimension $N \times 1$ ($N$ is the number of nodes). The $N \times N$ matrix $\mathbf{A}$ is defined as the DC PSE approximation of the convection and diffusion terms, viz.

$$\mathbf{A} \;=\; \mathbf{K}+\mathbf{L} \;=\; \frac{\partial\psi^n}{\partial x}\frac{\partial\omega^n}{\partial y} - \frac{\partial\psi^n}{\partial y}\frac{\partial\omega^n}{\partial x} + \frac{1}{Re}\nabla^2\omega^n. \tag{36}$$

where

$$\mathbf{K} \;=\; \mathrm{diag}\!\left(\mathbf{Q}^{1,0}\psi^n\right)\!\cdot\!\mathbf{Q}^{0,1} - \mathrm{diag}\!\left(\mathbf{Q}^{0,1}\psi^n\right)\!\cdot\!\mathbf{Q}^{1,0} \tag{37}$$

and

$$\mathbf{L} \;=\; \frac{1}{Re}\!\left(\mathbf{Q}^{2,0} + \mathbf{Q}^{0,2}\right) \tag{38}$$

with $\mathrm{diag}\!\left(\mathbf{Q}^{1,0}\psi^n\right)$ and $\mathrm{diag}\!\left(\mathbf{Q}^{1,0}\psi^n\right)$ being diagonal matrices of dimension $N \times N$. The explicit method defined by Equation (34) is stable if

$$|1 + \delta t\lambda_A| \leq 1 \tag{39}$$

where $\lambda_A$ are the eigenvalues of the matrix $\mathbf{A}$. The above stability condition requires the eigenvalues $\lambda_A$ to be either real or negative, leading to

$$\delta t \leq \frac{2}{|\lambda_A|} \tag{40}$$

or complex with negative real parts, leading to

$$\delta t \leq \frac{2Re(\lambda_A)}{|\lambda_A|^2} \tag{41}$$

When the problem discretization includes a large number of nodes, matrix $\mathbf{A}$ becomes very large, and calculating the eigenvalues is not practical. The terms of matrix $\mathbf{A}$ are determined by the Laplacian operator $\mathbf{L}$, which consists of second order derivatives, and by the advection operator $\mathbf{K}$, which includes only first order derivatives. The relative weight of the two operators on the structure of matrix $\mathbf{A}$ is dictated by discretisation and velocity field values (velocity is related to stream function and vector potential field values); a more refined discretization leads to a higher weight of operator $\mathbf{L}$ (diffusion) and lower influence of operator $\mathbf{K}$ (convection), while higher Reynolds number and higher velocity field values leads to a higher influence of operator $\mathbf{K}$. When the nodal discretization is not adequately refined, the higher weight of the operator $\mathbf{K}$ can lead to eigenvalues with a positive real part; in such case explicit time integration is not possible and discretization has to be refined.

For refined discretizations, the resulting matrix is diagonally dominant, having its eigenvalues distributed close to the real axis. We can use the Gershgorin circle theorem [31] to estimate the maximum time step using Equations (36) and (37) with an upper bound of the maximum eigenvalue of matrix $\mathbf{A}$. Given the composition of matrix $\mathbf{A}$, the upper bound of the maximum eigenvalue can be computed without assembling the matrix:

$$|\lambda_A| \leq \max_i \left( \sum_j \left( \left| L_{ij} \right| + \left| K_{ij} \right| \right) \right) \tag{42}$$

Equation (42) clearly shows that the eigenvalues of matrix **A** depend on the spatial derivatives of the field variables, computed using the DC PSE method. Therefore, how derivatives are computed reflects on the critical time step. We notice that for the same Reynolds number and the same nodal discretization, the critical time step increases when spatial derivatives are computed using a large number of nodes (more than 40) in the support domain of each node, compared with the critical time step computed using a small number of nodes (around 15). In some cases, this increase in the critical time step can be up to two orders of magnitude, which decreases the computational cost dramatically.

The reason for the increased critical time step is the upwind inherent nature of meshless methods. Meshless methods are locally based methods and use neighboring nodes to compute spatial derivatives. Furthermore, since matrix **A** involves derivatives of the stream function, by computing the critical time step using the Gershgorin circle theorem we ensure that the computation resembles the estimation of the critical time step based on the nodal spacing $\left( \propto \Delta x^{-1} \right)$ and the Courant–Friedrichs–Lewy (CFL) condition $\left( \propto u \mathrm{Re} \Delta x^{-1} \right)$. Therefore, by adjusting the number of support domain nodes, we can adjust the critical time step to be comparable to that used by implicit time integration schemes.

## 3. Algorithm Verification

To demonstrate the efficiency and accuracy of the proposed numerical scheme, we first apply our method to well-established benchmark problems in computational fluid dynamics, including lid-driven cavity flow, the backward-facing step (BFS), and the unbounded flow past a cylinder.

For all the examples presented here, we examine the accuracy and the efficiency of the proposed scheme for each benchmark problem. Furthermore, we examine the dependence of the critical time step on the grid resolution, Reynolds number and nodes in the support domain. Computations were conducted using an Intel i7 quad core processor with 16 GB RAM.

### 3.1. Lid-Driven Cavity

Our first example involves incompressible, non-stationary flow in a square cavity (Figure 1). Despite its geometrical simplicity, the lid-driven cavity flow problem exhibits a complex flow regime, mainly due to the vortices formed in the center and at the corners (bottom left and right, and upper left) of the square domain. We impose no-slip boundary conditions at the bottom and vertical walls (left and right wall), while the top wall slides with unit velocity ($U = 1$), generating the interior viscous flow.

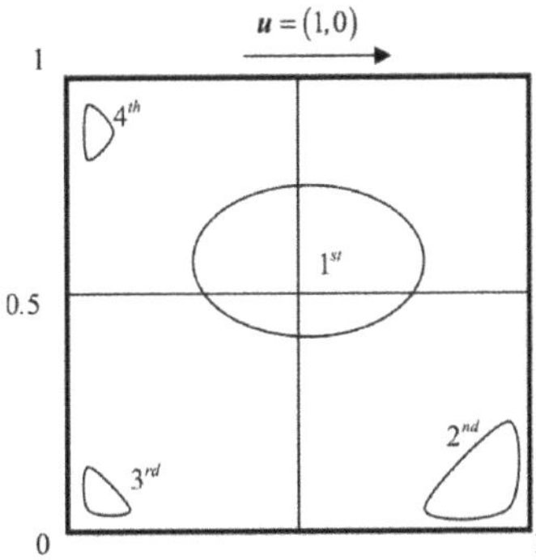


**Figure 1.** Geometry and boundary conditions for the lid-driven cavity flow problem.

In most numerical studies on the lid-driven cavity flow problem, the flow regime reaches a steady state solution for Reynolds ($Re$) numbers lower than $Re = 10{,}000$. In our study, we report on the results obtained by the proposed scheme for Reynolds numbers $Re = 7500$ and 10,000. We consider both uniform Cartesian and irregular nodal distributions. For uniform nodal distributions, our previous studies [27–29,35] show that for Reynolds numbers up to $Re = 10{,}000$, a uniform Cartesian grid with resolution of $361 \times 361$ captures the forming vortices and provides a grid-independent and accurate numerical solution. In the present study, to highlight the efficiency of the proposed scheme, we use successively finer uniform Cartesian grids, from $601 \times 601$ to $2048 \times 2408$ nodes. We notice that for $Re = 10{,}000$ the $1024 \times 1024$ grid resolution can capture all the flow scales and can be used for Direct Numerical Simulation (DNS) studies.

We use the Gershgorin circle theorem to define the critical time step for different nodal distributions (Tables 1 and 2) and for various Reynolds numbers. Table 1 lists the critical time step computed using the Gershgorin circle theorem for several grid resolutions and different Reynolds numbers (up to $Re = 10{,}000$). For the results reported in Table 1, 20 nodes were used in the support domain. We further examine the dependence of the critical time step on the number of neighboring nodes in the support domain. Table 2 lists the critical time step computed using the Gershgorin circle theorem with 40 nodes in the support domain. We can observe that the critical time step increases with the number of nodes in the support domain.

**Table 1.** Critical time step for various grid resolution and Reynolds ($Re$) numbers, using 20 nodes in the support domain.

| Grid Resolution | $Re = 5000$ | $Re = 7500$ | $Re = 10{,}000$ |
|:---:|:---:|:---:|:---:|
| $601 \times 601$ | $3.4 \times 10^{-3}$ | $5.1 \times 10^{-3}$ | $6.7 \times 10^{-3}$ |
| $1024 \times 1024$ | $1.1 \times 10^{-3}$ | $1.6 \times 10^{-3}$ | $2 \times 10^{-3}$ |
| $2048 \times 2048$ | $3.1 \times 10^{-4}$ | $4.4 \times 10^{-4}$ | $5 \times 10^{-4}$ |

**Table 2.** Critical time step for various grid resolution and $Re$ numbers, using 40 nodes in the support domain.

| Grid Resolution | $Re = 5000$ | $Re = 7500$ | $Re = 10{,}000$ |
|:---:|:---:|:---:|:---:|
| $601 \times 601$ | $7.6 \times 10^{-3}$ | $8.2 \times 10^{-3}$ | $9.2 \times 10^{-3}$ |
| $1024 \times 1024$ | $3.1 \times 10^{-3}$ | $3.5 \times 10^{-3}$ | $3.7 \times 10^{-3}$ |
| $2048 \times 2048$ | $1.2 \times 10^{-3}$ | $1.3 \times 10^{-3}$ | $9.1 \times 10^{-4}$ |

Table 3 lists the computational time (in s) for computing the spatial derivatives for various grid resolutions, and for the numerical solution of N-S equations (for each time iteration) in the case of $Re = 10{,}000$. Recall that we precompute the LU factorization for the discrete Laplacian operator defined for the Poisson type equation for the stream function. Hence, steps 1 and 2 must be performed only once at the beginning of the simulation.

**Table 3.** Computational time (in seconds) for computing (**1**) spatial derivatives, (**2**) factorization and (**3**) numerical solution for various grid resolutions.

| Grid Resolution | 1. Derivatives | 2. Factorization | 3. Solution (in s)/Iteration |
|:---:|:---:|:---:|:---:|
| $601 \times 601$ | 3.92 s | 4.97 s | 0.22 s |
| $1024 \times 1024$ | 11.10 s | 21.35 s | 0.67 s |
| $2048 \times 2048$ | 56.79 s | 274.70 s | 41.39 s |

Figure 2 shows the $u$- velocity profiles along the vertical line passing through the geometric center of the cavity, and the $v$- velocity profiles along the horizontal line passing through the geometric center of the cavity, for $Re = 7500$ and 10,000, obtained using the proposed scheme and compared to those reported by Ghia et al. [42], which are considered as benchmark solution.

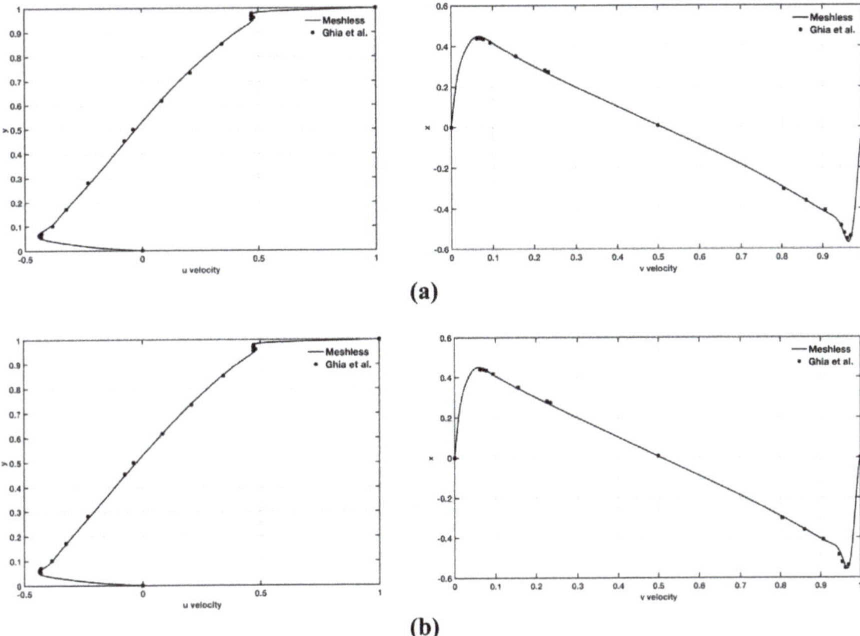

**Figure 2.** Velocity profiles along the vertical line passing through the geometric center of the cavity for *u*- velocity component (left) and horizontal line passing through the geometric center of the cavity for the *v*- velocity component (right) for (**a**) $Re = 7500$ and (**b**) $Re = 10,000$. The results we obtained in this study using our methods described in section Methods Governing Equations and Solution Procedure (indicated as "Meshless" in the velocity profiles) are compared with the results reported in Ghia et al. [42].

Table 3 demonstrates the efficiency of the proposes scheme and its ability to solve relatively large algebraic systems (Poisson equation for stream function) with low computational cost.

We set the Reynolds number to $Re = 10,000$ and the total time for the flow simulation to $T_{final} = 250$. At that time, the flow regime has all the characteristic features of steady state [42]. We use a $1024 \times 1024$ grid resolution and a time step for the simulations of $dt = 5 \times 10^{-4}$.

Figures 3–5 show stream function and vorticity contours and demonstrate the formation of secondary vortices (Figures 3 and 4) which appear as the Reynolds number increases. There are often used as qualitative results, to highlight the accuracy of the numerical method. In detail, Figure 3 shows the stream function contour plots for $Re = 7500$, while Figure 4 shows the stream function contour plots for $Re = 10,000$. Magnified views of the secondary vortices are also included. The stream function values for these contours are listed in Table 4.

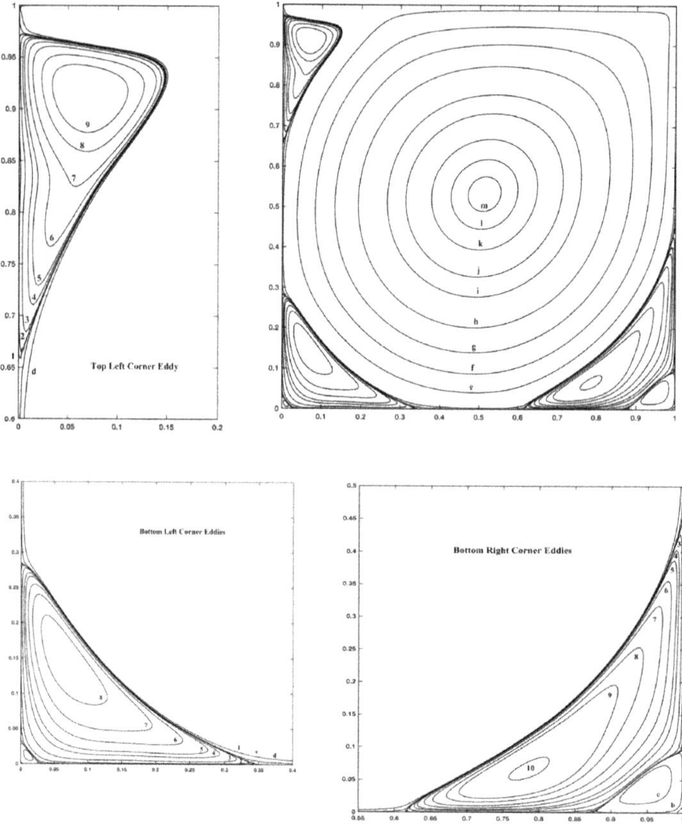



**Figure 3.** Streamline contours of primary, secondary and additional corner vortices for lid-driven cavity flow with *Re* = 7500.

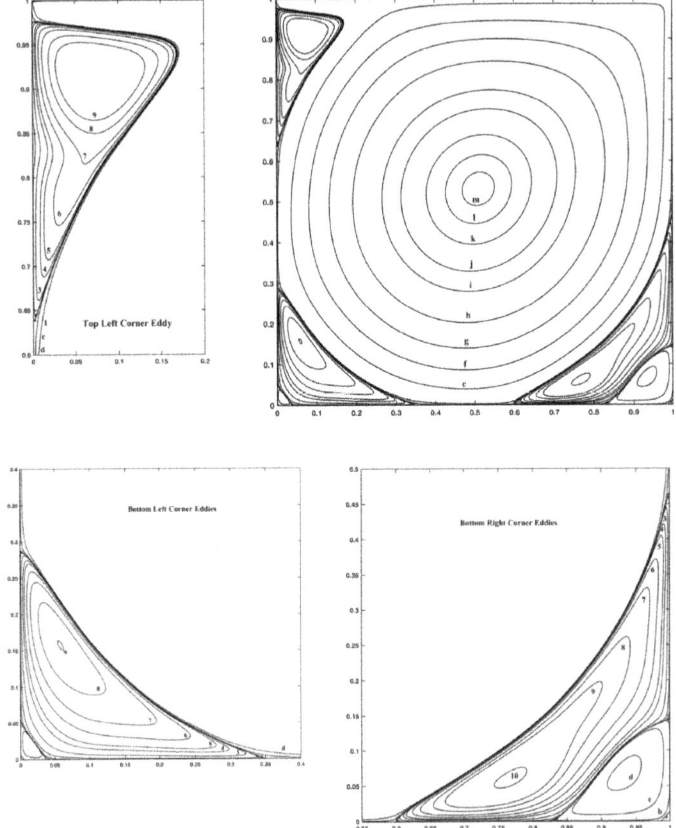

**Figure 4.** Streamline contours of primary, secondary and additional corner vortices for lid-driven cavity flow with $Re = 10{,}000$.

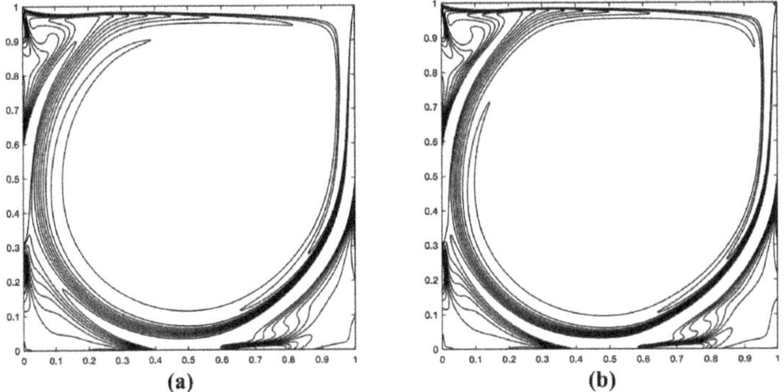

**Figure 5.** Vorticity isocontour patterns for (**a**) $Re = 7500$ and (**b**) $Re = 10{,}000$.

**Table 4.** Values for streamline and vorticity contours in Figures 3 and 4.

| Stream Function | | | Vorticity | | |
|---|---|---|---|---|---|
| Contour Level | Value of $\psi$ | Contour Number | Value of $\psi$ | Contour Number | Value of $\psi$ |
| a | $-1.0 \times 10^{-10}$ | | | | |
| b | $-1.0 \times 10^{-7}$ | 0 | $1.0 \times 10^{-8}$ | | |
| c | $-1.0 \times 10^{-5}$ | 1 | $1.0 \times 10^{-7}$ | | |
| d | $-1.0 \times 10^{-4}$ | 2 | $1.0 \times 10^{-6}$ | 0 | 0 |
| e | $-0.0100$ | 3 | $1.0 \times 10^{-5}$ | $\pm 1$ | $\pm 0.5$ |
| f | $-0.0300$ | 4 | $5.0 \times 10^{-5}$ | $\pm 2$ | $\pm 1.0$ |
| g | $-0.0500$ | 5 | $1.0 \times 10^{-4}$ | $\pm 3$ | $\pm 2.0$ |
| h | $-0.0700$ | 6 | $2.5 \times 10^{-4}$ | $\pm 4$ | $\pm 3.0$ |
| i | $-0.0900$ | 7 | $5.0 \times 10^{-4}$ | 5 | 4.0 |
| j | $-0.1000$ | 8 | $1.0 \times 10^{-3}$ | 6 | 5.0 |
| k | $-0.1100$ | 9 | $1.5 \times 10^{-3}$ | | |
| l | $-0.1150$ | 10 | $3.0 \times 10^{-3}$ | | |
| m | $-0.1175$ | | | | |

Figure 5 shows the vorticity contours for $Re$ = 7500 and 10,000. The vorticity values for the contours are listed in Table 4.

To verify the applicability of the proposed scheme in irregular nodal configurations, we compute the flow regime for $Re$ = 10,000 using randomly distributed nodes. We use 496,987 nodes to discretize the spatial domain, which provides a grid-independent numerical solution (for the uniform Cartesian grid, a grid independent solution was obtained using a 401 × 401 grid resolution with 160,801 nodes). We obtain the irregular point cloud through a 2D triangular mesh generator (MESH2D-Delaunay-based unstructured mesh-generation). To compare the results with those obtained using the uniform nodal distribution (regular Cartesian grid), we interpolate the velocity, stream function and vorticity values computed using the irregular point cloud onto the uniform nodal distribution using the moving least squares (MLS) approximation method [43] and compute the maximum normalized root mean square error (NRMSE), defined as [44],

$$NRMSE = \frac{\sqrt{\frac{1}{N} \sum_{i=1}^{N} \left( u_i^{Cartesian} - u_i^{Irregular} \right)^2}}{u_{max}^{Irregular} - u_{min}^{Irregular}} \tag{43}$$

For all field variables considered, the NRMSE was less than $10^{-4}$, which highlights the accuracy of the proposed method for irregular nodal distributions.

We further demonstrate the accuracy of the proposed scheme, providing quantitative results. We use a uniform Cartesian grid of 1024 × 1024 resolution and we set the time step to $dt = 5 \times 10^{-4}$. In Table 5, we present the strengths and positions of the primary (located in the center of the cavity), secondary (bottom right corner) and ternary vortexes (bottom and top left corners). The numerical results are in excellent agreement with the results reported in Ghia et al. [42]. Table 6 lists the $u$-velocity values along vertical line through the center of cavity, and Table 7 the $v$- velocity values along horizontal line through the center of cavity.

**Table 5.** Comparison of center of primary, secondary, and ternary vortices, and the corresponding stream function values.

| $Re$ | Reference | $\Psi_{min}$ | $(x_{1st}, y_{1st})$ | $\Psi_{max}$ | $(x_{2nd}, y_{2nd})$ | $\Psi_{min}$ | $(x_{3rd}, y_{3rd})$ | $\Psi_{max}$ | $(x_{4th}, y_{4th})$ |
|---|---|---|---|---|---|---|---|---|---|
| 7500 | Ghia [42] | $-0.1199$ | (0.5117, 0.5322) | $3.30 \times 10^{-3}$ | (0.7813, 0.0625) | $1.47 \times 10^{-3}$ | (0.0645, 0.1504) | $2.05 \times 10^{-3}$ | (0.0664, 0.9141) |
| | Present | $-0.1132$ | (0.5100,0.5333) | $3.12 \times 10^{-3}$ | (0.7783, 0.0634) | $1.36 \times 10^{-3}$ | (0.0612, 0.1554) | $1.97 \times 10^{-3}$ | (0.0614, 0.9267) |
| 10,000 | Ghia [42] | $-0.1197$ | (0.5117, 0.5333) | $3.42 \times 10^{-3}$ | (0.7656, 0.0586) | $1.52 \times 10^{-3}$ | (0.0586, 0.1641) | $2.42 \times 10^{-3}$ | (0.0703, 0.9141) |
| | Present | $-0.1183$ | (0.5109, 0.5331) | $3.36 \times 10^{-3}$ | (0.7642, 0.0582) | $1.51 \times 10^{-3}$ | (0.0583, 0.1643) | $2.41 \times 10^{-3}$ | (0.0703, 0.9142) |

**Table 6.** $u$- velocity along vertical line through the center of cavity.

| | $Re = 7500$ | | $Re = 10,000$ | |
|---|---|---|---|---|
| $y$ | Ghia [42] | Present | Ghia [42] | Present |
| 1.00000 | 1.00000 | 1.00000 | 1.00000 | 1.00000 |
| 0.9766 | 0.47244 | 0.46956 | 0.47221 | 0.47244 |
| 0.9688 | 0.47048 | 0.46818 | 0.47783 | 0.47797 |
| 0.9609 | 0.47323 | 0.47186 | 0.48070 | 0.48034 |
| 0.9531 | 0.47167 | 0.47184 | 0.47804 | 0.47814 |
| 0.8516 | 0.34228 | 0.3381 | 0.34635 | 0.34685 |
| 0.7344 | 0.20591 | 0.19527 | 0.20673 | 0.20665 |
| 0.6172 | 0.08342 | 0.08142 | 0.08344 | 0.08332 |
| 0.5000 | −0.03800 | −0.03798 | 0.03111 | 0.03154 |
| 0.4531 | −0.07503 | −0.07245 | −0.07540 | −0.07534 |
| 0.2813 | −0.23176 | −0.23905 | −0.23186 | −0.23136 |
| 0.1719 | −0.32393 | −0.31801 | −0.32709 | −0.32729 |
| 0.1016 | −0.38324 | −0.38291 | −0.38000 | −0.38012 |
| 0.0703 | −0.43025 | −0.42988 | −0.41657 | −0.41648 |
| 0.0625 | −0.4359 | −0.43050 | −0.42537 | −0.42587 |
| 0.0547 | −0.43154 | −0.42935 | −0.42735 | −0.42775 |
| 0.0000 | 0.0000 | 0.0000 | 0.0000 | 0.0000 |

**Table 7.** $v$- velocity along horizontal line through the center of cavity.

| | $Re = 7500$ | | $Re = 10,000$ | |
|---|---|---|---|---|
| $x$ | Ghia [42] | Present | Ghia [42] | Present |
| 1.00000 | 0.00000 | 0.00000 | 0.00000 | 0.00000 |
| 0.9688 | −0.53858 | −0.53889 | −0.54302 | −0.54367 |
| 0.9609 | −0.55216 | −0.55258 | −0.52487 | −0.52432 |
| 0.9531 | −0.52347 | −0.52363 | −0.49099 | −0.49056 |
| 0.9453 | −0.4859 | −0.48576 | −0.45863 | −0.45834 |
| 0.9063 | −0.41050 | −0.41082 | −0.41496 | −0.41439 |
| 0.8594 | −0.36213 | −0.36257 | −0.36737 | −0.36756 |
| 0.8047 | −0.30448 | −0.30432 | −0.30719 | −0.30739 |
| 0.5000 | 0.00824 | 0.00851 | 0.00831 | 0.00845 |
| 0.2344 | 0.27348 | 0.27378 | 0.27224 | 0.27267 |
| 0.2266 | 0.28117 | 0.28135 | 0.28003 | 0.28056 |
| 0.1563 | 0.3506 | 0.35082 | 0.35070 | 0.35078 |
| 0.0938 | 0.41824 | 0.41867 | 0.41487 | 0.41463 |
| 0.0781 | 0.43564 | 0.43591 | 0.43124 | 0.43165 |
| 0.0703 | 0.4403 | 0.44059 | 0.43733 | 0.43745 |
| 0.0625 | 0.43979 | 0.43962 | 0.43983 | 0.43953 |
| 0.0000 | 0.00000 | 0.00000 | 0.00000 | 0.00000 |

*3.2. Backward-Facing Step*

The second benchmark problem considered is the backward-facing step [45]. This problem has been studied by several researchers using different numerical methods and is considered to be a demanding flow problem to solve, mainly due to the flow separation that occurs when the fluid passes over a sharp corner and re-attaches downstream [1,45,46].

Figure 6 shows the spatial domain for the backward facing step. The coordinate system is centered at the step corner, with the $x$-coordinate being positive in the downstream direction, and the $y$-coordinate across the flow channel. The height $H$ of the channel is set to $H = 1$ (ranging from $(0, -0.5)$ to $(0, 0.5)$), while the step height and upstream inlet region are set to $H/2$. To ensure fully developed flow, the downstream channel length $L$ is set to $L = 30\,H$. The inlet velocity has a parabolic profile, with horizontal component $u(y) = 12y - 24y^2$ for $0 \le y \le 0.5$, which gives a maximum inflow velocity of

$u_{max} = 1.5$ and average velocity of $u_{avg} = 1$. At the outlet, we assume fully developed flow ($du/dx = 0$, $v = 0$). The Reynolds number is defined as $Re = u_{avg}H/v_f$, with $v_f$ being the kinematic viscosity.

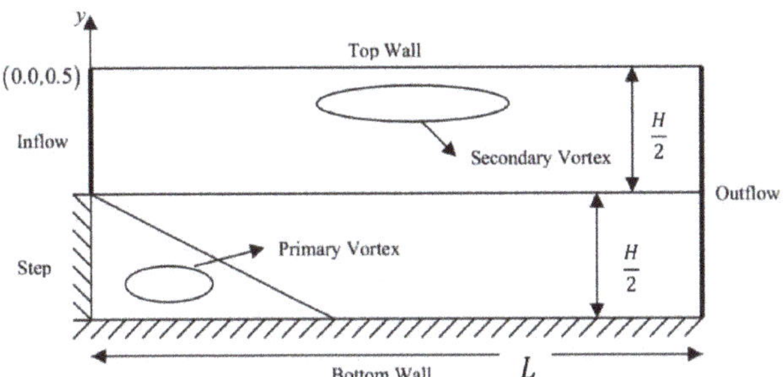


**Figure 6.** Geometry for the backward-facing step flow problem.

We first discretize the flow domain with a uniform Cartesian grid. In our previous studies [29,35], we have shown that grid with resolution $31 \times 901$ is sufficient to compute a grid-independent numerical solution for Reynolds numbers up to $Re = 800$. As in the previous section, to highlight the efficiency of our scheme, we use a denser grid with resolution $121 \times 3630$, resulting in 439,230 nodes. We set the total time $T_{total} = 250$ (to ensure that the solution will reach steady state), and the time step $dt = 10^{-4}$. The simulation terminates when the NRMSE of the time derivative of the stream function and vorticity field values in two successive time steps is less than $10^{-6}$. The time needed to create the grid was 0.023 s, and it takes 0.3 s to update the solution for each time step. We obtained a steady state solution after 100,000 time steps.

Figure 7 shows the stream function and vorticity contours for $Re = 800$. The flow separates at the step corner and vortices are formed downstream. We can observe the two vortices formed at the lower and upper wall. After reattachment of the upper wall eddy, the flow in the duct slowly recovers towards a fully developed flow. We compute the separation and reattachment points at $L_{lower} \approx 6.1$ for the lower wall separation zone, $L_{upper} \approx 5.11$ for the upper separation zone, where the separation begins at $x \approx 5.19$. Our numerical findings show good agreement with other numerical methods for 2D computations [1,46]. In [1], the authors used a finite difference method and predicted separation lengths of $L_{lower} \approx 6.0$ and $L_{upper} \approx 5.75$, while [46] using the A Fluid Dynamics Analysis Program (FIDAP) code (Fluid dynamics International Inc., Evanston, IL, USA) predicted $L_{lower} \approx 5.8$ and the upper $L_{upper} \approx 4.7$.

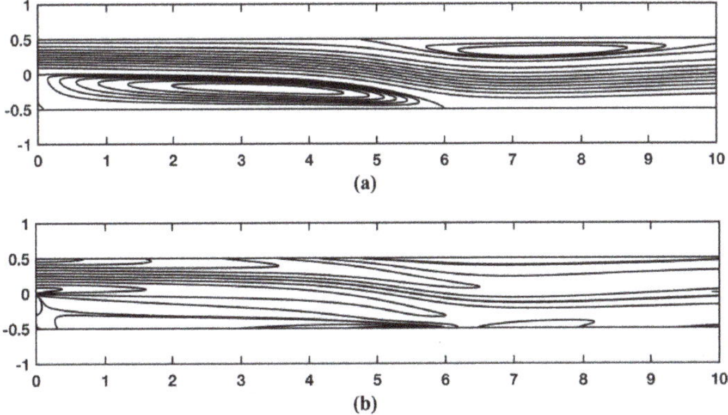

**Figure 7.** Contour plots of the (**a**) stream function and (**b**) vorticity for $Re = 800$ for the backward-facing step flow problem using the stream function and vorticity values reported in Gartling [45].

Figure 8 shows the comparisons of the *u*- and *v*- velocity components and vorticity values, between the present scheme and those obtained using the finite element method (FEM) [45], along the line $x = 7$ and $x = 15$ for $Re = 800$. Additionally, we computed the shear stress, defined as $\frac{\partial u}{\partial y}$ on the lower and upper wall, for increasingly refined nodal resolution. Figure 9a shows plots of the shear stress for the upper and lower wall, while Figure 9b displays the convergence of the upper wall shear stress when successively denser point clouds are used. We can see that shear stress converges and is accurate when compared to results of [45]. The accurate computation of spatial derivatives is a particular advantage of the strong form meshless method. In contrast, convergence of many finite element methods can be obtained only in terms of integral norms, and the point-wise convergence for the velocity gradient cannot be guaranteed even in the case of a smooth numerical solution [47].

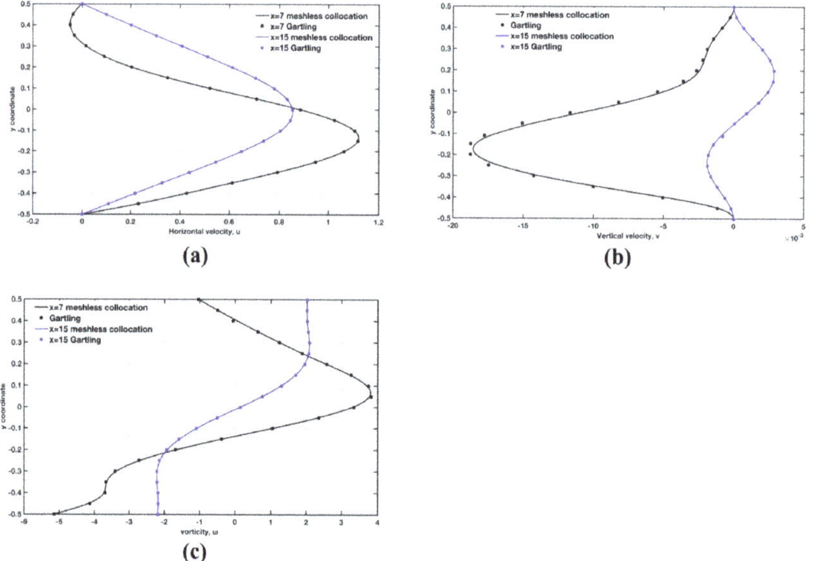

**Figure 8.** (**a**) Horizontal and (**b**) vertical velocity profiles, and (**c**) vorticity profiles at $x = 7$ and $x = 15$ for the backward-facing step flow problem with $Re = 800$.

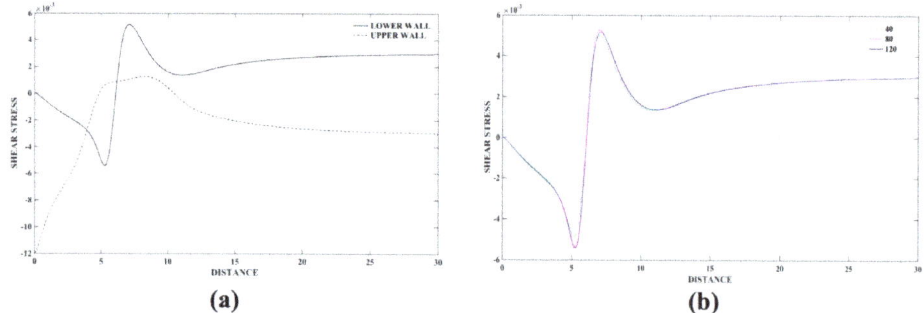

**Figure 9.** (**a**) Shear stress for the upper and lower wall for $Re = 800$ for the backward-facing step flow problem (**b**) convergence of the upper wall shear stress with successively denser point clouds (40, 80 and 120 correspond to the number of nodes in the *y*-direction).

Tables 8 and 9 list the velocity and vorticity field values at the at $x = 7$ and $x = 15$, respectively, for $Re = 800$.

**Table 8.** Cross-sectional profile at $x = 7$.

| $y$ | $u$ | | $v\left(\times 10^{-2}\right)$ | | $\omega$ | |
|---|---|---|---|---|---|---|
| | Gartling [45] | Present | Gartling [45] | Present | Gartling [45] | Present |
| 0.50 | 0.000 | 0.000 | 0.000 | 0.000 | −1.034 | −1.033 |
| 0.45 | −0.038 | −0.038 | −0.027 | −0.027 | −0.493 | −0.491 |
| 0.40 | −0.049 | −0.049 | −0.086 | −0.085 | 0.061 | 0.065 |
| 0.35 | −0.032 | −0.031 | −0.147 | −0.144 | 0.635 | 0.641 |
| 0.30 | 0.015 | 0.016 | −0.193 | −0.186 | 1.237 | 1.245 |
| 0.25 | 0.092 | 0.093 | −0.225 | −0.213 | 1.888 | 1.894 |
| 0.20 | 0.204 | 0.205 | −0.268 | 0.249 | 2.588 | 2.585 |
| 0.15 | 0.349 | 0.349 | −0.362 | −0.333 | 3.267 | 3.246 |
| 0.10 | 0.522 | 0.523 | −0.544 | −0.502 | 3.751 | 3.709 |
| 0.05 | 0.709 | 0.709 | −0.823 | −0.767 | 3.821 | 3.767 |
| 0.00 | 0.885 | 0.885 | −1.165 | −1.095 | 3.345 | 3.295 |
| −0.05 | 1.024 | 1.024 | −1.507 | −1.425 | 2.362 | 2.328 |
| −0.10 | 1.105 | 1.104 | −1.778 | −1.691 | 1.046 | 1.034 |
| −0.15 | 1.118 | 1.117 | −1.925 | −1.838 | −0.374 | −0.367 |
| −0.20 | 1.062 | 1.061 | −1.917 | −1.836 | −1.684 | −1.661 |
| −0.25 | 0.948 | 0.947 | −1.748 | −1.678 | −2.719 | −2.687 |
| −0.30 | 0.792 | 0.792 | −1.423 | −1.378 | −3.392 | −3.355 |
| −0.35 | 0.613 | 0.613 | −1.000 | −0.960 | −3.658 | −3.636 |
| −0.4 | 0.428 | 0.427 | −0.504 | −0.482 | −3.687 | −3.719 |
| −0.45 | 0.232 | 0.232 | −0.118 | −0.113 | −4.132 | −4.177 |
| −0.50 | 0.000 | 0.000 | 0.000 | 0.000 | −5.140 | −5.146 |

**Table 9.** Cross-sectional profile at $x = 15$.

| $y$ | $u$ | | $v\left(\times 10^{-2}\right)$ | | $\omega$ | |
|---|---|---|---|---|---|---|
| | Gartling [45] | Present | Gartling [45] | Present | Gartling [45] | Present |
| 0.50 | 0.000 | 0.000 | 0.000 | 0.000 | 2.027 | 2.033 |
| 0.45 | 0.101 | 0.101 | 0.021 | 0.021 | 2.013 | 2.018 |
| 0.40 | 0.202 | 0.202 | 0.072 | 0.072 | 2.023 | 2.027 |
| 0.35 | 0.304 | 0.304 | 0.140 | 0.139 | 2.058 | 2.058 |
| 0.30 | 0.408 | 0.408 | 0.207 | 0.206 | 2.090 | 2.085 |
| 0.25 | 0.512 | 0.512 | 0.260 | 0.259 | 2.075 | 2.063 |
| 0.20 | 0.613 | 0.613 | 0.288 | 0.286 | 1.959 | 1.943 |
| 0.15 | 0.704 | 0.704 | 0.283 | 0.281 | 1.703 | 1.687 |
| 0.10 | 0.779 | 0.779 | 0.245 | 0.243 | 1.298 | 1.283 |
| 0.05 | 0.831 | 0.831 | 0.180 | 0.177 | 0.761 | 0.751 |
| 0.00 | 0.853 | 0.853 | 0.095 | 0.093 | 0.141 | 0.137 |
| −0.05 | 0.844 | 0.844 | 0.003 | 0.002 | −0.500 | −0.447 |
| −0.10 | 0.804 | 0.804 | −0.081 | −0.081 | −1.096 | −1.086 |
| −0.15 | 0.737 | 0.737 | −0.147 | −0.147 | −1.588 | −1.574 |
| −0.20 | 0.649 | 0.649 | −0.185 | −0.185 | −1.939 | −1.923 |
| −0.25 | 0.547 | 0.547 | −0.191 | −0.191 | −2.139 | −2.125 |
| −0.30 | 0.438 | 0.438 | −0.166 | −0.166 | −2.213 | −2.203 |
| −0.35 | 0.328 | 0.328 | −0.119 | −0.119 | −2.210 | −2.206 |
| −0.40 | 0.218 | 0.218 | −0.065 | −0.065 | −2.184 | −2.185 |
| −0.45 | 0.109 | 0.109 | −0.019 | −0.019 | −2.174 | 2.177 |
| −0.50 | 0.000 | 0.000 | 0.000 | 0.000 | −2.185 | −2.189 |

### 3.3. Unbounded Flow Past a Cylinder

We examine the case of external flow past a circular cylinder in an unbounded domain [48]. The flow domain is large compared to the dimensions of the cylinder, as shown in Figure 10. The cylinder cross-section has a radius $R_c = 0.5$ and is located at the origin $O$ (0,0) of a square domain with dimensions $-10 \leq x \leq 30$ and $-20 \leq y \leq 20$.

The uniform flow is not perturbed near the inlet, which is far from the body. Therefore, we assume that the inflow velocity $\boldsymbol{u}_{inlet}$ behaves like the potential flow $\boldsymbol{u}_{potential}$ given as:

$$u_{potential} = \left( U_\infty \left( 1 - \frac{R_c^2}{x^2 + y^2} + \frac{2R_c^2 y^2}{(x^2 + y^2)^2} \right), U_\infty \frac{-2R_c^2 xy}{(x^2 + y^2)^2} \right) \tag{44}$$

or in terms of the stream function:

$$\psi_{potential} = U_\infty y \left( 1 - \frac{R_c^2}{x^2 + y^2} \right) \tag{45}$$

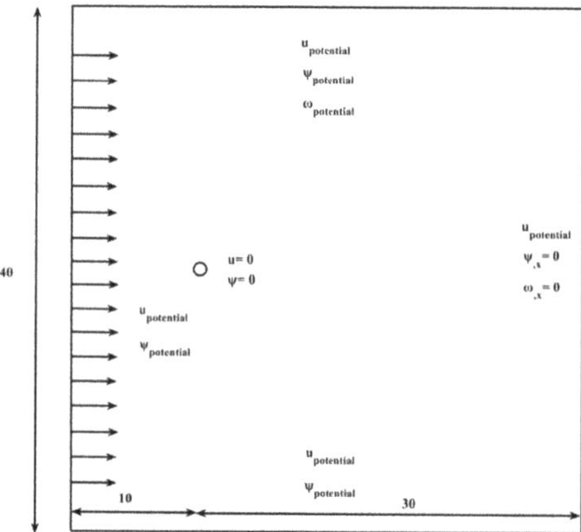


**Figure 10.** Geometry and boundary conditions for flow past a cylinder.

The boundary conditions for the stream function $\psi$ and vorticity $\omega$ on the remaining boundaries (the top and bottom wall, outlet and cylinder surface) are:

- At the inlet, top and bottom walls ($B$ = inlet, top and bottom)

$$u_B = u_{potential}\big|_B$$

$$\psi_B = \psi_{potential}\big|_B$$

- At the outlet

$$\frac{\partial \psi}{\partial x} = 0$$

$$\frac{\partial \omega}{\partial x} = 0$$

- On the cylindrical surface

$$\psi_{cylinder} = 0$$

$$u_{cylinder} = 0$$

We represent the flow domain with a uniform Cartesian embedded grid, locally refined in the vicinity of the cylinder (Figure 11a). We use 402,068 nodes (0.5 s to create the nodal distribution) and we set Reynolds number to $Re = 40$. We use a time step of $dt = 10^{-4}$ for the entire flow simulation. It takes 0.3 s to update the solution for each time step, and we obtained a steady state solution after 20,000 time steps. The spatial derivatives (up to 2nd order) are computed in 4.24 s using a C++ code. Additionally, we solve the flow problem using an irregular nodal distribution with 221,211 nodes, locally refined in the vicinity of the cylinder.

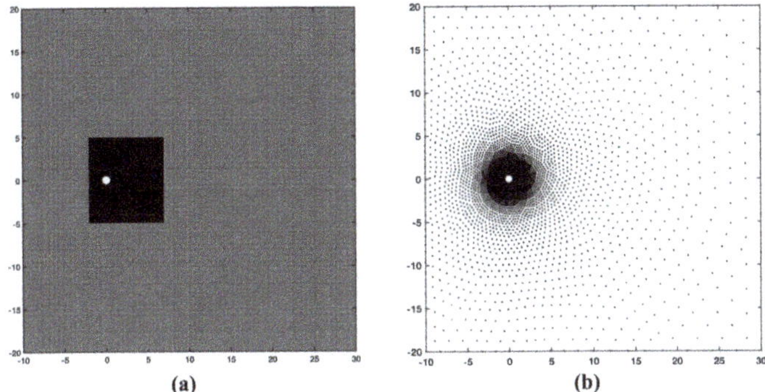

**Figure 11.** (**a**) Locally refined Cartesian and (**b**) irregular nodal configurations for the flow past cylinder flow problem.

We compute the following flow parameters: the pressure coefficient ($C_p$) on the body surface, the length ($L$) of the wake behind the body, the separation angle ($\theta_s$), and the drag coefficient ($C_D$) of the body. The drag and lift coefficient of the body are given by:

$$C_D = \frac{2F_b \cdot \hat{e}_x}{U_\infty^2 D} \tag{46}$$

$$C_L = \frac{2F_b \cdot \hat{e}_y}{U_\infty^2 D} \tag{47}$$

where $D$ is the characteristic length of the body, $\hat{e}_x$ and $\hat{e}_y$ are the unit normal vector in $x$- and $y$-direction, respectively, and $F_b$ is the total drag force acting on the body. The total drag force is given as:

$$F_b = F_p + F_f \tag{48}$$

where $F_p$, the pressure drag, can be determined from the flux of vorticity on the surface of the cylinder as:

$$F_p = -R_C \int_0^{2\pi} v_f \frac{\partial \omega}{\partial r} \hat{e}_\theta d\theta \tag{49}$$

The friction drag $F_f$ may be computed from the vorticity on the surface of the body as:

$$F_f = R_C \int_0^{2\pi} v_f \omega \hat{e}_\theta d\theta \tag{50}$$

with $\hat{e}_\theta = -\sin\theta i + \cos\theta j$. Table 5 lists the recirculation length ($L_{rec}$) of the wake behind the body, the separation angle ($\theta_s$), and the drag coefficient ($C_D$) for $Re = 40$.

The numerical findings obtained by the proposed scheme are listed in Table 10, and are in good agreement with those in the literature [48–51]. The stream function and vorticity contours around the cylinder are illustrated in Figure 12 (as in [48]).

**Table 10.** Comparison of the wake length ($L_{sep}$), the separation angle ($\theta_{sep}$), and the drag coefficient ($C_D$) for Reynolds number $Re = 40$.

| Re | Reference | $L_{sep}$ | $\theta_{sep}$ | $C_D$ |
|---|---|---|---|---|
| | Dennis and Chang [49] | 2.345 | 53.8 | 1.522 |
| | Fornberg [50] | 2.24 | N/A | 1.498 |
| 40 | Ding et al. [51] | 2.20 | 53.5 | 1.713 |
| | Kim et al. [48] | 2.187 | 55.1 | 1.640 |
| | Present | 2.30 | 53 | 1.542 |

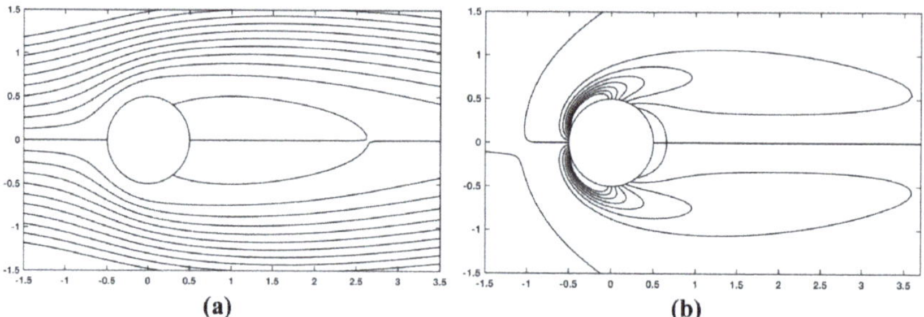

**Figure 12.** (**a**) Stream function and (**b**) vorticity contours around the body for flow behind a cylinder using a Cartesian embedded nodal distribution.

## 4. Numerical Results

In this section, we examine the applicability and reproducibility of the proposed scheme for several flow cases involving complex geometries.

### 4.1. Vortex Shedding Behind Cylinders

In the previous example (unbounded flow past a cylinder), we considered the flow past a cylinder in an unbounded domain. In this example, we study the vortex shedding behind a cylinder in a rectangular duct [52] (internal flow), as shown in Figure 13. The duct has length $L = 2.2$ m and height $H = 0.41$ m, while the cylinder is located at point $O$ (0.2, 0.2) and has diameter $D = 0.1$ m. The kinematic viscosity of the fluid is set to $v_f = 0.001$ m$^2$/s. The Reynolds number is defined as $Re = U_m D/v_f$, with the mean velocity $U_m$ given as $U_m(x,y,t) = 2U(0,H/2,t)/3$. The inflow velocity ($x$- direction component) is set to $U(0, y) = 4U_m y\left(\frac{H-y}{H^2}\right)$, and for $U_m = 1.5$ m/s yields a Reynolds number $Re = 100$. At the outlet, we consider fully developed flow ($du/dx = 0$), while at the remaining boundaries we apply no-slip boundary conditions. The total time for the simulation was set to $T_{tot} = 8$ s. The critical time step is computed and monitored for each time step (the calculation involves stream function and vorticity values) to ensure that CFL conditions are fulfilled.

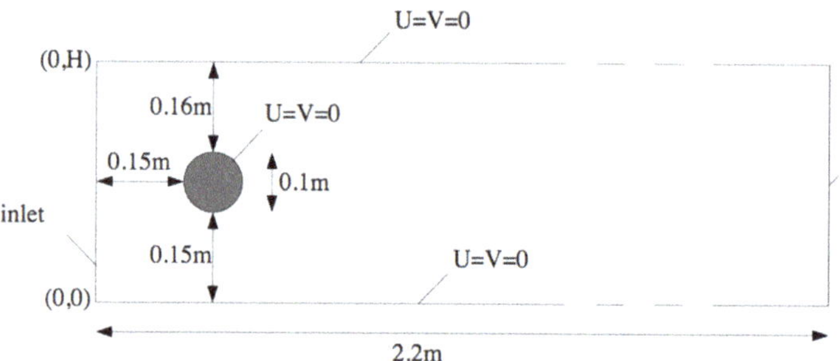


**Figure 13.** Spatial domain for flow behind a cylinder.

First, we consider a uniform Cartesian embedded grid with grid spacing $h$, defined by the average distance of the boundary nodes on the cylinder circumference. We conducted a grid independence analysis, applying successively refined Cartesian grids, starting from 34,999 (90 nodes on the circular boundary) up to 406,678 (720 nodes on the circular boundary) nodes.

The time needed to create the grid of 406,678 nodes is 0.8 s. We delete the nodes located close to the boundary (with distance less than 0.25 $h$), since they would increase the condition number of the Vandermonde matrix. A Vandermonde matrix with too large condition number would affect the accuracy of the spatial derivative computation and consequently the overall precision of the numerical simulation. A grid independent solution is obtained with 203,890 nodes (360 nodes on the circular boundary). The critical time step computed for this grid resolution is $\delta t_{critical} = 2 \times 10^{-3}$. In the simulations, we set the time step to $dt = 10^{-3}$ s. For each time step, it takes 0.38 s to update the solution.

Figure 14a shows the stream function contours at time instances $t = 2, 4$ and 8 s, while Figure 14b plots the vorticity isocontours. Figure 15 plots the drag $C_D = \frac{2F_b \cdot \hat{e}_x}{U_m^2 D}$ and lift $C_L = \frac{2F_b \cdot \hat{e}_y}{U_m^2 D}$ coefficients, computed with our meshless explicit scheme (as in the previous example). To highlight the applicability of our code for locally refined nodal distributions, we use a uniform Cartesian embedded grid, locally refined in the vicinity of the cylinder. As before, the average distance of the cylinder nodes dictates the nodal spacing $h_d$ of the Cartesian grid in the refined part of the domain. Nodes of the coarse grid have spacing $h_c = 2h$, and nodes located close to the cylinder (with distance less than $0.25\, h_d$) are deleted since they affect the condition number of the Vandermonde matrix and decrease the accuracy of the numerical results. The flow domain is represented by 321,897 nodes, which ensure a grid independent solution. For the simulations, we set the time step to $dt = 10^{-3}$ s.

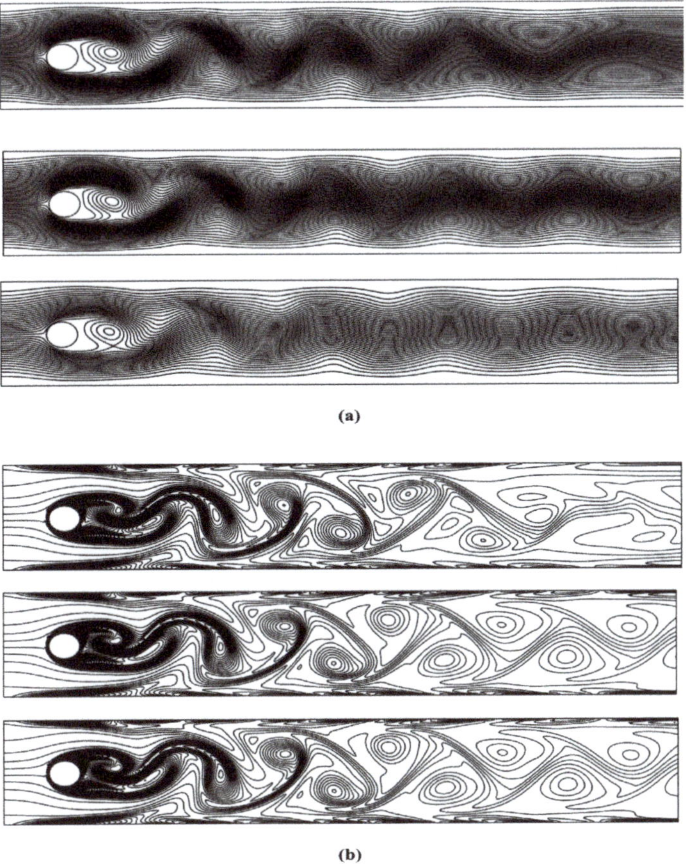

(a)

(b)

**Figure 14.** (a) Stream function and (b) vorticity contours for flow around a cylinder at time $t = 2, 4$ and 8 s.

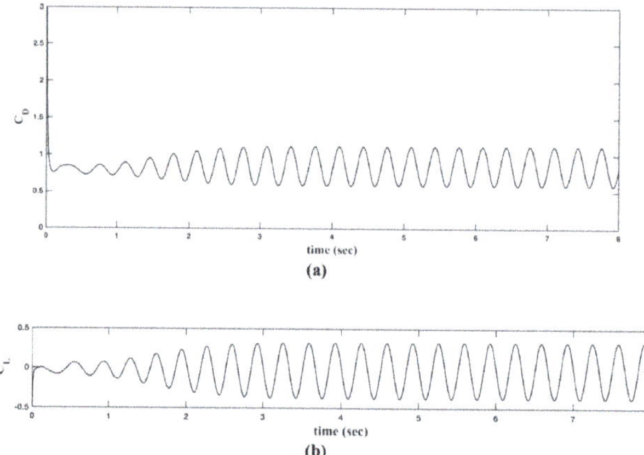

Figure 15. (a) Drag and (b) lift coefficient in time computed by the proposed scheme.

Additionally, to highlight the versatility of the proposed scheme, we examined the flow in the rectangular duct with multiple (seven in total) cylindrical obstacles. The length of the duct was increased to $L = 4.4$ m in order to ensure fully developed flow at the outlet. The boundary conditions applied are the same as before. We use Cartesian embedded and irregular nodal distributions (Figure 16) to represent the flow domain.

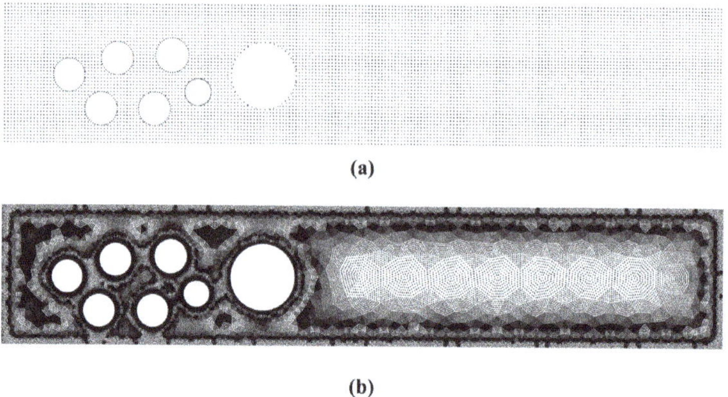

Figure 16. (a) Cartesian embedded and (b) irregular nodal distributions used in duct flow problems considering multiple cylindrical obstacles.

We conducted a comprehensive analysis in order to obtain a grid independent numerical solution. For the Cartesian grid we use 279,178 nodes (0.42 s to create the nodal distribution), while for the irregular point cloud we use 353,617 nodes (1.2 s to create the irregular nodal distribution). Both nodal distributions provided a grid independent numerical solution. The total time was set to $T_{tot} = 8$ s, the time step was set to $dt = 10^{-4}$ s. For each time step, the time needed to update the solution was 0.32 s. Figure 17 displays the stream function and vorticity contours for $Re = 100$ at time $t = 1$ and $t = 2$ s.

We further demonstrate the accuracy of the proposed scheme, by comparing our results with those computed using the finite element method (FEM). Flow equations were solved using the Incremental Pressure Correction Scheme (IPCS), which is an operator splitting method [53]. For the meshless solver, we use 353,617 nodes, generated using a triangular mesh generator, and we set the time step to $dt = 10^{-3}$ s. For the finite element model, we solve the flow equations using the generated mesh and

FEniCS (https://fenicsproject.org). We compute the Normalized Root Mean Square Error (NRMSE), as defined in Equation (43), for both velocity components ($u_x$, $u_y$), considering as gold standard the velocity field values calculated using FEniCS. Figure 18 shows the NRMSE for the velocity components for the time interval between $t = 0$ and $t = 1$ s.

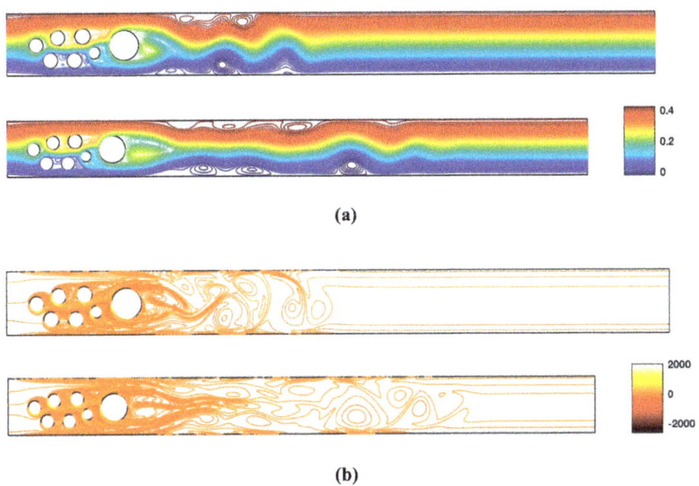

**(a)**

**(b)**

**Figure 17.** (**a**) Stream function and (**b**) vorticity contours for flow behind multiple cylinders at time $t = 1$ and $t = 2$ s.

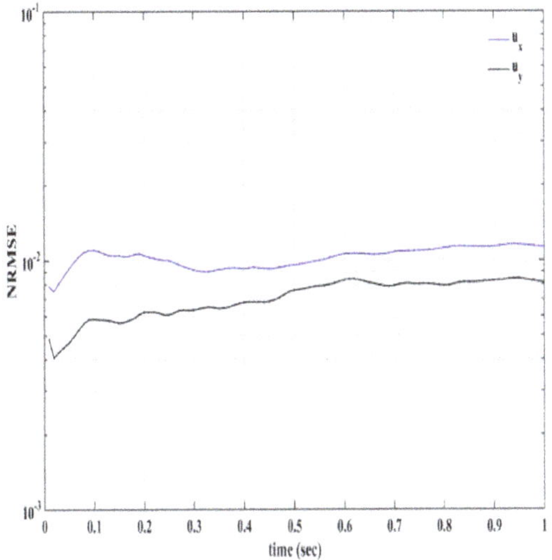

**Figure 18.** Normalized Root Mean Square Error (NRMSE) for the velocity components between the proposed scheme and the finite element solution.

### 4.2. Flow in a Duct with Irregular Geometry

Finally, we consider two flow cases in the flow domains shown in Figure 19. The first one has an irregularly shaped obstacle downstream, while in the second the flow domain is split into two branches, forming a bifurcation.

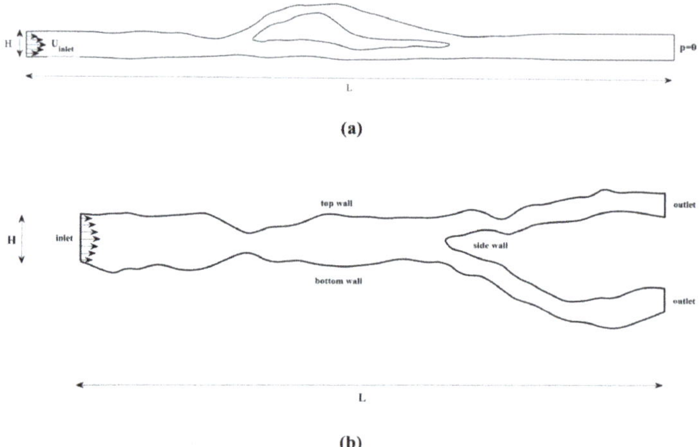

**Figure 19.** Geometry for the flow in the (**a**) bypass and (**b**) bifurcation domain.

The length $L$ is set to $L = 0.12$ and $0.041$ $m$ for the first (Figure 19a) and second (Figure 19b) flow domain, respectively. In both test cases, the height of the duct is set to $H = 0.0041$ m.

The dynamic viscosity is defined as $\mu = 0.00032$ kg/m s, and the fluid density is $\rho = 1050$ kg/m$^3$. The inflow condition $U_{inlet} = 4U_m y\left(\frac{H-y}{H^2}\right)$, with $U_m = 0.035$ m/s. The Reynolds number $Re = \rho U_m H/\mu$, with $D$ being a characteristic length defined separately for each case. At the outlet, the flow is fully developed ($du/dx = 0$). For the remaining boundaries we applied no-slip conditions. In both cases, we represent the flow domain with uniform Cartesian embedded grid, and irregular nodal distribution (Figure 20).

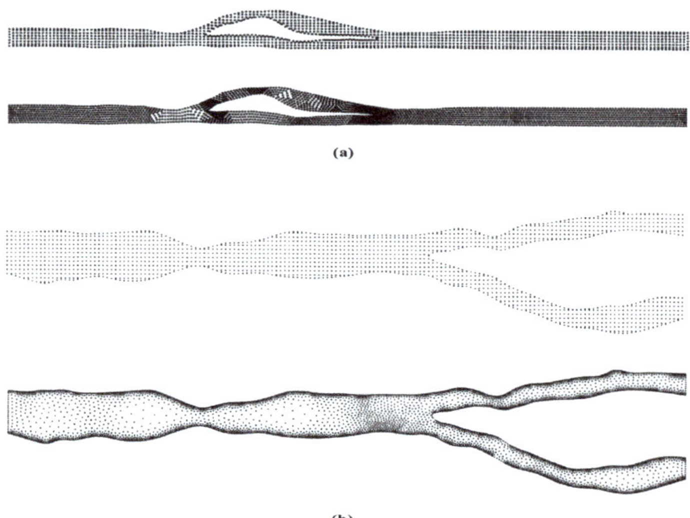

**Figure 20.** (**a**) Cartesian embedded and (**b**) irregular nodal distribution for the complex flow geometries.

First, we represent the flow domain with a uniform Cartesian embedded grid. For the irregularly shaped obstacle (bypass), we consider successively denser nodal distributions to obtain a grid independent numerical solution. We start with a grid spacing of $h = 1/8200$, $h = 1/12{,}300$ and

$h = 1/16{,}400$ for the coarse, moderate and dense Cartesian embedded grids, resulting in 51,468, 114,387 and 201,468 nodes, respectively (for the denser grid (201,468 nodes) it takes 0.25 s to create the nodal distribution). The total time for the simulation was set to $T_{tot} = 30$ s and the time step used was $dt = 10^{-4}$ s (for each time step the solution is computed in 0.35 s). Figure 21 shows the iso-contours for the stream function and vorticity field functions, at different time instances: $t = 0.5$ and $t = 1$ s, and for steady state (steady state is reached when, for two successive time instances $t^{n+1}$ and $t^n$ for both vorticity and stream function field values, $NRMSE/dt < 10^{-8}$).

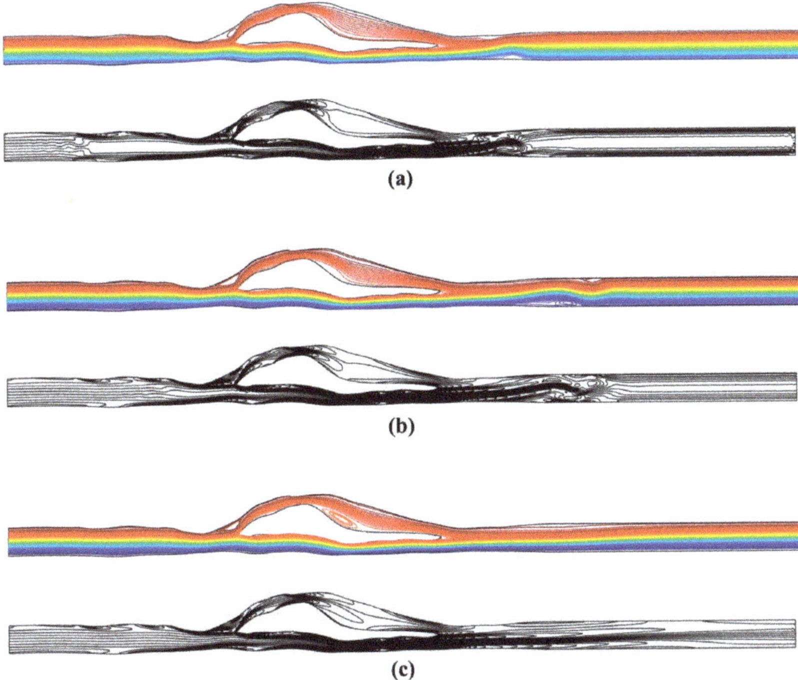

**Figure 21.** Stream function and vorticity contours for the bypass test case at (**a**) $t = 0.5$ s and (**b**) $t = 1$ s and (**c**) steady state.

We observe that two vortices are formed; the first is located at the inlet of the bypass, the second at the upper domain of the bypass which is moving towards the outlet of the flow domain. Figure 22a plots the $u$- velocity profile, computed at $x = 0.07$ m and $x = 0.08$ m at time $t = 3$ s for the coarse, moderate and dense nodal distributions, while Figure 22b plots the $u$- velocity profile, computed at $x = 0.07$ m and $x = 0.08$ m at time $t = 3$ s for the denser grid used.

Additionally, to highlight the applicability of the method to irregular nodal distributions, we solved the flow equations by representing the flow domain with nodes generated by a 2D triangular mesh generator (we use only the coordinates and not their connectivity). We used 392,122 nodes, which is higher than the number of nodes used in the Cartesian embedded grid. The numerical results obtained were compared against those obtained by the Cartesian grid and they were in excellent agreement.

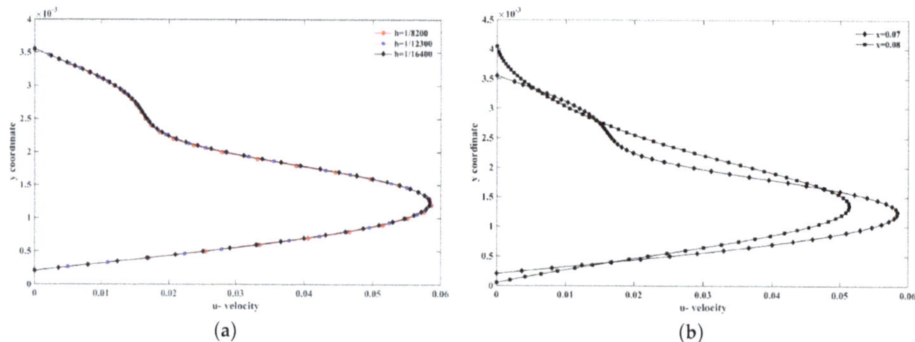

**Figure 22.** (**a**) $u$-velocity profiles at $x = 0.07$ at time $t = 3$ s for the coarse, moderate and dense nodal distributions for $Re = 474$ ($U_m = 0.035$ m/s) and (**b**) $u$-velocity profiles at $x = 0.07$ and $0.08$ for flow in the bypass geometry.

For the bifurcated geometry, we use both uniform Cartesian and irregular nodal distributions. In the case of Cartesian embedded grid, we obtain a grid independent numerical solution by using successively denser nodal distributions, with the denser one consisting of 678,967 nodes (for the fine grid it takes 0.68 s to create the nodal distribution). The total time for the simulation was set to $T_{tot} = 30$ s and the time step used is $dt = 10^{-4}$ s, which is smaller than the critical time that ensures stability for the explicit solver (the solution in each time step is computed in 0.52 s). The inflow velocity $U_{inlet} = 4U_m y\left(\frac{H-y}{H^2}\right)$, with $U_m = 0.015$ m/s results in Reynolds number $Re = 212$. Figure 23 shows the iso-contours for the stream function and vorticity field functions, at time $t = 0.5$ and $t = 1$ s and for steady state (steady state is reached when, for two successive time instances $t^{n+1}$ and $t^n$ for both vorticity and stream function field values, $NRMSE/dt < 10^{-8}$).

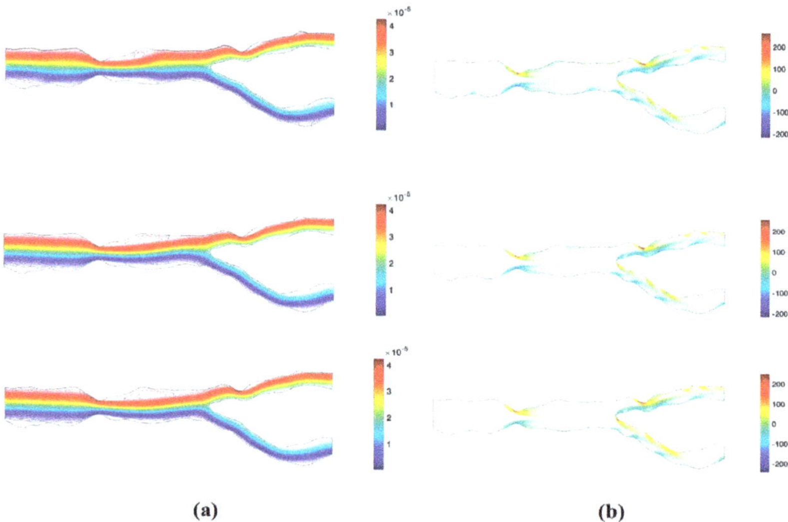

**Figure 23.** (**a**) Stream function and (**b**) vorticity contours for the bifurcation test case.

To further demonstrate the accuracy of the proposed scheme, we compare our numerical findings for both flow cases (bypass and bifurcation geometry) with those obtained using the finite element method. The flow domain is discretized using 392,122 and 678,967 nodes, respectively, and Taylor-Hood elements (as in the vortex shedding behind cylinders example). We use the IPCS flow solver to numerically solve the flow equations. We compare both numerical solutions, computing the NRMSE for the two velocity components ($u_x$ and $u_y$) on the vertices of the triangular elements (the same nodes are used in the meshless point collocation flow solver). Figure 24 shows the NRMSE for the velocity components (bypass and bifurcation flow examples) for the time interval between $t = 0$ and $t = 1$ s.

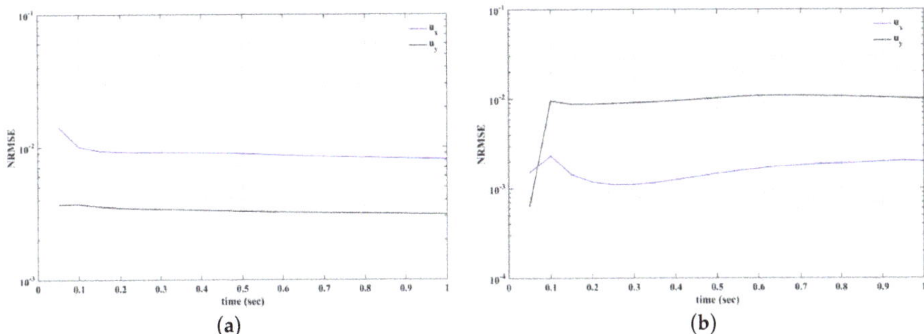

(a)        (b)

**Figure 24.** Normalized Root Mean Square Error (NRMSE) for the velocity components between the proposed scheme and the finite element solution for (**a**) bypass and (**b**) bifurcation flow cases.

## 5. Conclusions

In this contribution, we described a meshless point collocation algorithm for solving the stream function-vorticity formulation of N-S equations. We demonstrated our algorithm to work well in complex geometries like the ones shown in Section 4.1 (Vortex Shedding Behind Cylinders) and 4.2 (Flow in a Duct with Irregular Geometry). We demonstrate the accuracy of the proposed scheme by comparing our numerical results with those computed using the finite element method. An important advantage of our method is the ease and speed with which one can construct computational grids for flow domains with irregular shapes.

The meshless scheme based on DC PSE methods to compute spatial derivatives can be used in the case of Cartesian and Cartesian-embedded grids. The method works both efficiently and accurately for uniform Cartesian (embedded) grids and for irregular point clouds. We verified the accuracy of the proposed scheme through the following three benchmark problems: lid-driven cavity flow, backward-facing step and unbounded flow past a cylinder. The results were in good agreement with other published numerical studies.

Our proposed scheme offers several advantages over other commonly used methods:

- Rapid and easy generation of computational grids, as demonstrated by examples of the flow in vortex shedding behind cylinders and flow in a duct with irregular geometry in Sections 4.1 and 4.2, respectively.
- High accuracy, as demonstrated in verification examples discussed in Section 3.
- Computational efficiency, as demonstrated by time per iteration (Table 3) and total simulation time given for all examples in this paper. Our approach makes Direct Numerical Simulations (DNS) [54] possible (see lid driven cavity example with 1024 × 1024 grid resolution), even on personal computers.
- Critical time step is easily computed with low computational cost.
- Long critical time steps, comparable to those used in implicit schemes. The critical time step is related to the number of nodes located in the support domain and increases as the number of nodes increases (see Tables 1 and 2).
- The imposition of Dirichlet boundary conditions is straightforward.
- Easy and straightforward way to impose vorticity boundary conditions using spatial derivatives computed using the DC PSE method (Equation (33))

- Fast and accurate solution of the Poisson solver for the stream function equation (Equation (2)). The solution time/iteration (Table 3) is shorter compared to the solution time of other well-established numerical methods (see Figure 1 in [54]).
- Simplicity: we use a MATLAB code of ca. 150 lines to solve flow equations, and a C++ code of ca. 140 lines to compute spatial derivatives.

**Author Contributions:** Conceptualization, G.C.B.; Investigation, G.C.B.; Software, G.C.B.; Writing, G.C.B, B.F.Z. and K.M.; Writing—review and editing, G.C.B., B.F.Z., G.R.J., V.C.L., A.C.R.T, A.W. and K.M.

**Funding:** This research was supported in part by the Australian Government through the Australian Research Council's Discovery Projects funding scheme (project DP160100714).

**Acknowledgments:** This research was supported in part by the Australian Government through the Australian Research Council's Discovery Projects funding scheme (project DP160100714). The views expressed herein are those of the authors and are not necessarily those of the Australian Government or Australian Research Council.

**Conflicts of Interest:** The authors declare no conflict of interest.

## References


1. Chorin, A.J. Numerical Solution of the Navier-Stokes Equations. *Math. Comput.* **1968**, *22*, 745–762. [CrossRef]
2. Harlow, F.H.; Welch, J.E. Numerical calculation of time-dependent viscous incompressible flow of fluid with free surface. *Phys. Fluids* **1965**, *8*, 2182–2189. [CrossRef]
3. Kim, J.; Moin, P. Application of a fractional-step method to incompressible Navier-Stokes equations. *J. Comput. Phys.* **1985**, *59*, 308–323. [CrossRef]
4. Weinan, E.; Liu, J.-G. Finite Difference Methods for 3D Viscous Incompressible Flows in the Vorticity–Vector Potential Formulation on Nonstaggered Grids. *J. Comput. Phys.* **1997**, *138*, 57–82. [CrossRef]
5. Hirasaki, G.J.; Hellums, J.D. A general formulation of the boundary conditions on the vector potential in three-dimensional hydrodynamics. *Q. Appl. Math.* **1968**, *26*, 331–342. [CrossRef]
6. Wong, A.K.; Reizes, J.A. An effective vorticity-vector potential formulation for the numerical solution of three-dimensional duct flow problems. *J. Comput. Phys.* **1984**, *55*, 98–114. [CrossRef]
7. Guj, G.; Stella, F. A vorticity-velocity method for the numerical solution of 3D incompressible flows. *J. Comput. Phys.* **1993**, *106*, 286–298. [CrossRef]
8. Quartapelle, L.; Napolitano, M. Integral conditions for the pressure in the computation of incompressible viscous flows. *J. Comput. Phys.* **1986**, *62*, 340–348. [CrossRef]
9. Fletcher, C. *Computional Techniques for Fluid Dynanic*; Springer: Berlin/Heidelberg, Germany, 1988; Volumes I and II.
10. Gresho, P.M.; Sani, R.L. *Incompresible Flow and the Finite Element Method*; Wiley: Hoboken, NJ, USA, 1998.
11. Anderson, J.D. *Computational Fluid Dynamics-The Basics with Applications*; McGraw-Hill: New York, NY, USA, 1995.
12. Ferziger, J.H.; Peric, M. *Computational Method for Fluid Dynamics*; Springer: Berlin/Heidelberg, Germany, 1996.
13. Patankar, S.V. *Numerical Heat Transfer and Fluid Flow*; McGraw-Hill: New York, NY, USA, 1980.
14. Nithiarasu, P. An efficient artificial compressibility (AC) scheme based on the characteristic based split (CBS) method for incompressible flows. *Int. J. Numer. Methods Eng.* **2003**, *56*, 1815–1845. [CrossRef]
15. Quarteroni, A.; Valli, A. *Numerical Approximation of Partial Differential Equations*; Springer: Berlin/Heidelberg, Germany, 1994.
16. Turek, S. *Efficient Solvers for Incompressible Flow Problems*; Springer: Berlin/Heidelberg, Germany, 1999.
17. Lin, H.; Atluri, S.N. The Meshless Local Petrov-Galerkin (MLPG) method for solving incompressible Navier-Stokes equations. *CMES Comput. Model. Eng. Sci.* **2001**, *2*, 117–142.
18. Loukopoulos, V.C.; Bourantas, G.C. MLPG6 for the solution of incompressible flow equations. *CMES Comput. Model. Eng. Sci.* **2012**, *88*, 531–558.
19. Wu, Y.L.; Liu, G.R.; Gu, Y.T. Application of Meshless Local Petrov-Galerkin (MLPG) Approach to Simulation of Incompressible Flow. *Numer. Heat Transf. Part B Fundam.* **2005**, *48*, 459–475. [CrossRef]
20. Arefmanesh, A.; Najafi, M.; Abdi, H. Meshless local Petrov-Galerkin method with unity test function for non-isothermal fluid flow. *CMES Comput. Model. Eng. Sci.* **2008**, *25*, 9–22.

21. Šarler, B. A radial basis function collocation approach in computational fluid dynamics. *CMES Comput. Model. Eng. Sci.* **2005**, *7*, 185–193.
22. Sellountos, E.J.; Sequeira, A. A hybrid multi-region BEM/LBIE-RBF velocity-vorticity scheme for the two-dimensional Navier-Stokes equations. *CMES Comput. Model. Eng. Sci.* **2008**, *23*, 127–147.
23. Sellountos, E.J.; Sequeira, A. An advanced meshless LBIE/RBF method for solving two-dimensional incompressible fluid flows. *Comput. Mech.* **2008**, *41*, 617–631. [CrossRef]
24. Siraj ul, I.; Vertnik, R.; Šarler, B. Local radial basis function collocation method along with explicit time stepping for hyperbolic partial differential equations. *Appl. Numer. Math.* **2013**, *67*, 136–151. [CrossRef]
25. Mai-Duy, N.; Mai-Cao, L.; Tran-Cong, T. Computation of transient viscous flows using indirect radial basis function networks. *CMES Comput. Model. Eng. Sci.* **2007**, *18*, 59–77.
26. Mai-Duy, N.; Tran-Cong, T. Numerical solution of Navier–Stokes equations using multiquadric radial basis function networks. *Int. J. Numer. Methods Fluids* **2001**, *37*, 65–86. [CrossRef]
27. Bourantas, G.C.; Loukopoulos, V.C. A meshless scheme for incompressible fluid flow using a velocity–pressure correction method. *Comput. Fluids* **2013**, *88*, 189–199. [CrossRef]
28. Bourantas, G.C.; Petsi, A.J.; Skouras, E.D.; Burganos, V.N. Meshless point collocation for the numerical solution of Navier–Stokes flow equations inside an evaporating sessile droplet. *Eng. Anal. Bound. Elem.* **2012**, *36*, 240–247. [CrossRef]
29. Bourantas, G.C.; Skouras, E.D.; Loukopoulos, V.C.; Nikiforidis, G.C. Numerical solution of non-isothermal fluid flows using local radial basis functions (LRBF) interpolation and a velocity-correction method. *CMES Comput. Model. Eng. Sci.* **2010**, *64*, 187–212.
30. Bourantas, G.C.; Loukopoulos, V.C.; Chowdhury, H.A.; Joldes, G.R.; Miller, K.; Bordas, S.P.A. An implicit potential method along with a meshless technique for incompressible fluid flows for regular and irregular geometries in 2D and 3D. *Eng. Anal. Bound. Elem.* **2017**, *77*, 97–111. [CrossRef]
31. Gerschgorin, S. Uber die Abgrenzung der Eigenwerte einer Matrix. *Izv. Akad. Nauk Sssrserija Mat.* **1931**, *7*, 749–754.
32. Isaacson, E.; Keller, H. *Analysis of Numerical Methods*; Wiley: New York, NY, USA, 1966.
33. Weinan, E.; Liu, J.-G. Vorticity Boundary Condition and Related Issues for Finite Difference Schemes. *J. Comput. Phys.* **1996**, *124*, 368–382. [CrossRef]
34. Schrader, B.; Reboux, S.; Sbalzarini, I.F. Discretization correction of general integral PSE Operators for particle methods. *J. Comput. Phys.* **2010**, *229*, 4159–4182. [CrossRef]
35. Bourantas, G.C.; Cheeseman, B.L.; Ramaswamy, R.; Sbalzarini, I.F. Using DC PSE operator discretization in Eulerian meshless collocation methods improves their robustness in complex geometries. *Comput. Fluids* **2016**, *136*, 285–300. [CrossRef]
36. Degond, P.; Mas-Gallic, S. The Weighted Particle Method for Convection-Diffusion Equations. Part 2: The Anisotropic Case. *Math. Comput.* **1989**, *53*, 509–525. [CrossRef]
37. Eldredge, J.D.; Leonard, A.; Colonius, T. A General Deterministic Treatment of Derivatives in Particle Methods. *J. Comput. Phys.* **2002**, *180*, 686–709. [CrossRef]
38. Schrader, B. *Discretization-Corrected PSE Operators for Adaptive Multiresolution Particle Methods*; ETH Zurich: Zürich, Switzerland, 2011.
39. Thom, A.; Taylor Geoffrey, I. The flow past circular cylinders at low speeds. *Proc. R. Soc. London. Ser. Acontaining Pap. A Math. Phys. Character* **1933**, *141*, 651–669. [CrossRef]
40. Gupta, M.M.; Kouatchou, J.; Zhang, J. Comparison of Second- and Fourth-Order Discretizations for Multigrid Poisson Solvers. *J. Comput. Phys.* **1997**, *132*, 226–232. [CrossRef]
41. Moin, P. *Fundamentals of Engineering Numerical Analysis*; Cambridge University Press: New York, NY, USA, 2001.
42. Ghia, U.; Ghia, K.N.; Shin, C.T. High-Re solutions for incompressible flow using the Navier-Stokes equations and a multigrid method. *J. Comput. Phys.* **1982**, *48*, 387–411. [CrossRef]
43. Lancaster, P.; Salkauskas, K. Surfaces generated by moving least squares methods. *Math. Comput.* **1981**, *37*, 141–158. [CrossRef]
44. Zienkiewicz, O.C.; Zhu, J.Z. A simple error estimator and adaptive procedure for practical engineerng analysis. *Int. J. Numer. Methods Eng.* **1987**, *24*, 337–357. [CrossRef]
45. Gartling, D.K. A test problem for outflow boundary conditions—flow over a backward-facing step. *Int. J. Numer. Methods Fluids* **1990**, *11*, 953–967. [CrossRef]

46. Sohn, J.L. Evaluation of FIDAP on some classical laminar and turbulent benchmarks. *Int. J. Numer. Methods Fluids* **1988**, *8*, 1469–1490. [CrossRef]

47. Guermond, J.L.; Minev, P.; Shen, J. An overview of projection methods for incompressible flows. *Comput. Methods Appl. Mech. Eng.* **2006**, *195*, 6011–6045. [CrossRef]

48. Kim, Y.; Kim, D.W.; Jun, S.; Lee, J.H. Meshfree point collocation method for the stream-vorticity formulation of 2D incompressible Navier–Stokes equations. *Comput. Methods Appl. Mech. Eng.* **2007**, *196*, 3095–3109. [CrossRef]

49. Dennis, S.C.R.; Chang, G.-Z. Numerical solutions for steady flow past a circular cylinder at Reynolds numbers up to 100. *J. Fluid Mech.* **1970**, *42*, 471–489. [CrossRef]

50. Ding, H.; Shu, C.; Yeo, K.S.; Xu, D. Simulation of incompressible viscous flows past a circular cylinder by hybrid FD scheme and meshless least square-based finite difference method. *Comput. Methods Appl. Mech. Eng.* **2004**, *193*, 727–744. [CrossRef]

51. Fornberg, B. A numerical study of steady viscous flow past a circular cylinder. *J. Fluid Mech.* **1980**, *98*, 819–855. [CrossRef]

52. Schäfer, M.; Turek, S.; Durst, F.; Krause, E.; Rannacher, R. Benchmark Computations of Laminar Flow Around a Cylinder. In *Flow Simulation with High-Performance Computers II: DFG Priority Research Programme Results 1993–1995*; Hirschel, E.H., Ed.; Vieweg+Teubner Verlag: Wiesbaden, Germany, 1996; pp. 547–566. [CrossRef]

53. Goda, K. A multistep technique with implicit difference schemes for calculating two- or three-dimensional cavity flows. *J. Comput. Phys.* **1979**, *30*, 76–95. [CrossRef]

54. San, O.; Staples, A.E. A coarse-grid projection method for accelerating incompressible flow computations. *J. Comput. Phys.* **2013**, *233*, 480–508. [CrossRef]

*Article*

# A New Wall Model for Large Eddy Simulation of Separated Flows

**Ahmad Fakhari**

Institute for Polymers and Composites, Department of Polymer Engineering, Campus of Azurém, University of Minho, 4800-058 Guimarães, Portugal; ahmadfakhari@gmail.com; Tel.: +351-253-510397

Received: 14 September 2019; Accepted: 19 November 2019: date; Published: 28 November 2019

**Abstract:** The aim of this work is to propose a new wall model for separated flows which is combined with large eddy simulation (LES) of the flow field in the whole domain. The model is designed to give reasonably good results for engineering applications where the grid resolution is generally coarse. Since in practical applications a geometry can share body fitted and immersed boundaries, two different methodologies are introduced, one for body fitted grids, and one designed for immersed boundaries. The starting point of the models is the well known equilibrium stress model. The model for body fitted grid uses the dynamic evaluation of the von Kármán constant $\kappa$ of Cabot and Moin (Flow, Turbulence and Combustion, 2000, 63, pp. 269–291) in a new fashion to modify the computation of shear velocity which is needed for evaluation of the wall shear stress and the near-wall velocity gradients based on the law of the wall to obtain strain rate tensors. The wall layer model for immersed boundaries is an extension of the work of Roman et al. (Physics of Fluids, 2009, 21, p. 101701) and uses a criteria based on the sign of the pressure gradient, instead of one based on the friction velocity at the projection point, to construct the velocity under an adverse pressure gradient and where the near-wall computational node is in the log region, in order to capture flow separation. The performance of the models is tested over two well-studied geometries, the isolated two-dimensional hill and the periodic two-dimensional hill, respectively. Sensitivity analysis of the models is also performed. Overall, the models are able to predict the first and second order statistics in a reasonable way, including the position and extension of the downward separation region.

**Keywords:** wall layer model; LES; separated flow; body fitted; immersed boundary

## 1. Introduction

Wall bounded turbulent flows, especially at high Reynolds numbers, require a high resolution near the wall, because of the need to solve the thin viscous sub-layer [1–3]. In LES, this resolution is comparable with that required by direct numerical simulation (DNS). As far as Reynolds number increases, the cost of a wall-resolving LES augments with $Re^{2.5}$ (for a discussion, see Piomelli and Balaras [1]). Moreover, for applications where wall roughness is the rule rather than the exception, it is not feasible to describe the wall boundary in a deterministic sense.

The most practical way of simulating a wall-bounded, high Reynolds number flow, is to consider coarse grid, and model the wall shear stress with a wall function. This function is designed to mimic the stress induced by the wall, produced by the adherence condition and ruled by the turbulent field. Several wall layer models have been developed in the past (for a discussion, see Piomelli and Balaras [1] and Piomelli [2]). They are designed to work under certain conditions, and in particular they may suffer in the presence of massive separation. This subject is still challenging for improvement.

In equilibrium stress models, the grid is generated in such a way that the first near-wall computational node is placed in the log region of the boundary layer. So, based on the law of the wall, the instantaneous or mean horizontal velocity can be fitted to determine the wall stress. These models are valid only under the equilibrium assumption and work for both smooth and rough walls.

Deardorff [4] and Schumann [5] first used the concept of wall layer modeling in conjunction with LES. Applying LES on plane channel flow and annuli at large Reynolds numbers, they assumed the existence of an equilibrium-stress layer near the wall and used the outer flow velocity to calculate the wall stress based on the logarithmic law.

Deardorff [4] considered a second order velocity derivative in vertical direction and forced plane averaged stream-wise and spanwise velocity gradients to follow logarithmic law. Schumann [5] obtained the plane averaged wall shear stress from plane averaged velocity at the first grid point, using the iterative method so that the shear velocity is computed from the law of the wall from the averaged velocity. These methods are the least expensive wall layer models. They also provide roughness corrections easily from the logarithmic law modification, which is an important feature in environmental and oceanographic flows. The models suffer in cases of marginal separation, or strong pressure gradient [2].

In zonal approaches (two-layer models, TLMs), stress derives from the solution of a separate set of equations on a finer mesh close to the wall. This model was proposed by Balaras and Benocci [6] who applied turbulent boundary layer equations inside the boundary layer (inner layer) and LES on the outer region. In this model, a finer grid is considered between the wall and the first computational node of the coarser grid. A uniform pressure field is applied from LES for the inner layer, and velocity field from LES is considered at the edge of the Reynolds Averaged Navier-Stokes (RANS) region as its boundary condition. The coupling between the inner and outer regions of the boundary layer is weak, and these models have problems when a perturbation extends from the wall to the outer layer. Balaras and Benocci [6] used an algebraic eddy viscosity to parametrize the near-wall region. In addition to plane channel flow, this method was tested on square ducts and also rotating channel, the cases in which equilibrium stress model was not valid or even failed; this method showed a good accuracy in simulating these flows. These methods, able to simulate equilibrium as well as non-equilibrium flows were not designed to deal with separated flows.

Cabot [7] applied this approach to simulate flow over a backward-facing step with only 10% grid points fewer than those used in former wall-resolved LES. He found that stream-wise pressure gradient has an important effect in boundary layers of TLMs. In this case, mean velocity and skin friction coefficient were predicted well; however, the backward-facing step is just a simple case where separation is induced by the sharp discontinuity in the streamlines.

Cabot and Moin [8] used the wall layer model in conjunction with LES to simulate the flow over a backward-facing step. They considered a zonal approach using thin boundary layer equations (TBLEs) in the inner layer and LES (dynamic SGS model) in the outer region. This way, they were able to use eight uniform cells from the bottom of the wall up to half the height of the step, greatly reducing the total number of cells with respect to a stretched-grid case previously used for the wall-resolved LES.

Finally, in hybrid methods, a single grid is used and the turbulence models are different in inner and outer layers. Detached eddy simulation (DES) is a hybrid approach proposed by Spalart et al. [9] as a method to compute massive separation. According to DES, a stretched grid is used close to the wall to resolve the boundary layer. RANS is applied in the inner and LES in the outer region.

Since there is no zonal interface between the inner and the outer layer, velocity profile is smooth everywhere. Some weaknesses of this method are logarithmic layer mismatch between the two regions and high grid resolution dependency near the wall. In addition, time and length scales of the eddies in the inner layer (modelled by unsteady RANS equations with the Spalart-Allmaras 1-equation turbulent model) are much larger than those computed in the outer region by LES; this leads to much lower resolved stress by LES than the modelled stress (by RANS) at the RANS/LES interface and even farther Piomelli [2].

A review of DES and the two modifications of it, which are delayed detached eddy simulation (DDES) and improved delayed detached eddy simulation (IDDES), can be seen in Mockett et al. [10].

The total shear stress in plane channel flow only depends on the distance from the wall, therefore a reduction of eddy viscosity farther than the RANS/LES interface affects the velocity profile and generates a high gradient at the transition into LES region. This phenomenon is called the DES buffer layer. To eliminate this, Keating and Piomelli [11] added stochastic force in the interface region to accelerate resolved eddy generation, thus obtaining better results. Temmerman et al. [12] also computed the constant required to calculate eddy viscosity in the RANS region in such a way to equate the eddy viscosity of RANS and LES at the interface of their hybrid RANS/LES application.

While the hybrid RANS/LES methods work satisfactorily in flows with instabilities such as those with adverse pressure gradient and concave curvature, they are weak in attached flows, and, in general, in flows with a low level of instability; in these cases a false merging region may appear at the RANS/LES interface. Since the grid near the wall must be resolved, they are the most expensive ones in the wide family of wall layer models. In addition to the three main wall layer frameworks herein discussed, there are other models which are not mentioned here.

The near-wall modelisation must be implemented in the algorithms that solve the governing equations using different strategies. A body fitted (also called boundary fitted) case is the case in which a grid is generated to follow the geometry. Since the grid boundaries coincide with the solid surface of a solid body, boundary conditions are imposed on certain grid nodes or cell faces. This facilitates the computation of all the vectorial quantities as well as of their gradients.

In complex geometry, where a structured grid can hardly follow the boundary complexities, alternative approaches must be employed. Among others, the immersed boundary method (IBM) has proved to be simple, effective and accurate (Mittal and Iaccarino [13]). The domain grid can be either Cartesian or curvilinear. In this approach the cells are cut by the solid surface and the cell boundaries in the general cases do not coincide with the solid boundaries. Hence the fluxes at the solid surfaces are difficult to define. Finally, the velocities in near-wall computational nodes are reconstructed and thus a local frame of reference must be introduced at each near-wall node to identify directions normal and tangential to the immersed boundary.

A number of papers have been published discussing different implementations of immersed boundary methods (see Mittal and Iaccarino [13]). Among others, Roman et al. [14] extended IBM to the general case of curvilinear coordinates, and simulated both steady and unsteady flows in complex geometric configurations showing the flexibility of this methodology when compared to the Cartesian counterpart. However, few papers have dealt with the development of wall layer models in conjunction with immersed boundaries. This is a matter of great importance when flows at a high Reynolds number over complex geometry must be simulated.

Tessicini et al. [15] applied wall layer model LES in the presence of immersed boundaries to simulate the flow past a 25°, asymmetric trailing edge of a model hydrofoil. The Reynolds number, based on free stream velocity and the hydrofoil chord (C), was chosen equal to $Re_C = 2.15 \times 10^6$. They used turbulent boundary layer equations for modelling the wall layer. In general, their results were good, although they had a deviation at the second off-the-wall node, which was considered as the outer boundary for the wall model. This discrepancy was sensitive to the distance of the second node from the wall, and was more evident where the distance was larger.

Posa and Balaras [16] proposed a new near-wall reconstruction model to account for the lack of resolution and provided correct wall shear stress and hydrodynamic forces. They used a zonal approach (TLM), boundary layer equations with a finer grid in the near-wall region (called in this case the full boundary layer FBL) and LES in the outer region. They validated their model to simulate flow around a cylinder and a sphere.

Then, they considered a coarser grid, and set one node inside the boundary layer and neglected the advective term in boundary layer equations. This was justified by the fact that, if the first point off the wall is located inside the boundary layer, neglecting this term does not provide significant errors. They assumed a constant pressure gradient between the first and second nodes, and obtained tangential velocity with two approaches: The reduced diffusion model (RDM) and the hybrid reconstruction method (HRM). The Reynolds number in the cylinder case was $Re = 300$ based on free stream velocity and the cylinder diameter, and in the sphere was $Re = 1000$. The pressure coefficient for all cases was predicted well, while the skin friction was underestimated in all cases, but an improvement was observed using RDM and HRM in comparison with linear reconstruction. The prediction of separation was also improved a lot using these two approaches.

Chen et al. [17] also proposed a wall layer model based on turbulent boundary layer equations at high Reynolds numbers for implicit LES in the presence of immersed boundaries. First, they tested it on a turbulent channel flow in the range of Reynolds numbers (based on shear velocity) from $Re_\tau = 395$ to $Re_\tau = 100,000$. They used a minimum of 20 cells inside the boundary layer for lower Reynolds numbers and 40 cells for higher values. Inside the boundary layer (inner region), the pressure gradient and friction are dominant, thus advection is neglected in cases where there are slow changes in wall parallel direction and sharp changes in wall normal direction. The fact that the changes in wall normal direction are sharper at higher Reynolds numbers requires more resolution for capturing the flow in this direction.

They also simulated flow over a backward-facing step at a Reynolds number based on the step height $Re_h = 5000$ in a Cartesian grid. Finally they simulated the flow over periodic hills at $Re_H = 10,595$ based on hill height, using different resolutions, and they obtained comparatively good results. To summarize, different models have an acceptable accuracy under some special conditions; therefore, more general models are needed to work well in different situations characterized by attached as well as separated flow conditions.

In the present study, attempts are made to overcome some of the difficulties mentioned above, proposing wall layer model to be used in conjunction with LES, suited for both attached and detached flows. Since in engineering applications a combination of body fitted solid walls and immersed boundaries is often employed, a comprehensive model must include the two aspects. Here the models are designed to work in body fitted cases and in the presence of immersed boundaries, respectively. The solver for the simulations is the LES model (LES–COAST) developed over the years at the Laboratory of Environmental and Industrial Fluid Mechanics of the University of Trieste, Italy, and which has successfully been used in a number of environmental real-scale applications (see among others Roman et al. [18] and Galea et al. [19]). The paper is organized as follows: In Section 2 the governing equations are described and the numerical method employed; in Section 3, the wall layer model for body fitted grids, its improvement and application to a case characterized by the presence of a boundary layer together with massive separation, namely the case of the flow over an isolated hill, are shown; in Section 4 the model in the presence of the immersed boundary is presented, together with its application to the case of single hill and to the case of periodic hill. Concluding remarks are given in Section 5.

## 2. Governing Equations

The incompressible form of the filtered Navier-Stokes equations read as:

$$\frac{\partial \overline{u}_j}{\partial x_j} = 0, \tag{1}$$

$$\frac{\partial \overline{u}_i}{\partial t} + \frac{\partial (\overline{u}_i \overline{u}_j)}{\partial x_j} = -\frac{1}{\rho_0} \frac{\partial \overline{p}}{\partial x_i} - \frac{\partial \tau_{ij}}{\partial x_j} + \nu \frac{\partial^2 \overline{u}_i}{\partial x_j \partial x_j}. \tag{2}$$

The filter operation (denoted by overbar) allows reproduction of the space–time evolution of the large scales of motion, which are anisotropic and energetic, while the effect of the small sub-grid scales (SGS) is contained in the SGS stress terms ($\tau_{ij}$) in the momentum equation. It is here modeled through an SGS eddy viscosity model:

$$\tau_{ij} - \frac{\delta_{ij}}{3}\tau_{kk} = -2\nu_T \overline{S}_{ij},$$ (3)

where $\overline{S}_{ij}$ is the resolved-scale tensor, defined as:

$$\overline{S}_{ij} = \frac{1}{2}\left(\frac{\partial \overline{u}_i}{\partial x_j} + \frac{\partial \overline{u}_j}{\partial x_i}\right),$$ (4)

$\nu_T$ is the SGS turbulent viscosity, also known as eddy viscosity. The standard Smagorinsky model [20] is based on the equilibrium assumption. Considering the length scale $l \sim \overline{\Delta}$, the eddy viscosity can be written as:

$$\nu_T = (C_s \overline{\Delta})^2 |\overline{S}|,$$ (5)

where $C_s$ is the constant of the model (the Smagorinsky constant), and $|\overline{S}| = \sqrt{2\overline{S}_{ij}\overline{S}_{ij}}$ is the contraction of the strain rate tensor of the large scales, $\overline{S}_{ij}$. Finally, the SGS stresses are calculated as:

$$\tau_{ij} = -2\nu_T \overline{S}_{ij}.$$ (6)

The Smagorinsky constant is commonly considered to range between 0.065 and 0.2; Lilly [21] found a theoretical value of 0.18, but the Smagorinsky constant depends on the type of the flow, e.g., in shear flows it must be declined to 0.1 [22]. A comprehensive study of the Smagorinsky constant in plane channel flows can be found in Stocca [23]. The filter width is proportional to the grid size in all directions, and is equal to $\overline{\Delta} = (\Delta x \Delta y \Delta z)^{1/3}$. Near the walls, where the eddy viscosity is expected to vanish, the van Driest damping function is adopted.

In the dynamic Smagorinsky model proposed by Germano et al. [24], the coefficient is calculated dynamically by defining an additional test filter (denoted by a caret) with a width $\widehat{\Delta}$ larger than the grid filter width $\overline{\Delta}$; here $\widehat{\Delta} = 2\overline{\Delta}$. The dynamic Smagorinsky coefficient is computed in a least-square sense over tensor components:

$$C = -\frac{1}{2}\frac{\langle L_{ij}M_{ij}\rangle}{\langle M_{ij}M_{ij}\rangle},$$ (7)

where $<>$ denotes spatial average and $L_{ij}$ resolved turbulent stresses:

$$L_{ij} \equiv \widehat{\overline{u}_i \overline{u}_j} - \widehat{\overline{u}}_i \widehat{\overline{u}}_j, \quad M_{ij} \equiv 2(\Delta^2 \widehat{|\overline{S}|\overline{S}_{ij}} - \widehat{\Delta}^2 |\widehat{\overline{S}}|\widehat{\overline{S}}_{ij})$$ (8)

and then eddy viscosity can be calculated as below.

$$\nu_T = C\widehat{\Delta}^2 |\overline{S}|.$$ (9)

The advantage of the dynamic model over the standard one is that the eddy viscosity vanishes near the walls and in laminar flows [24]; so, the model allows to treat transitional and wall bounded flow without the need to use special treatments of the constant. In order to follow the solid boundaries, the numerical model uses the curvilinear-coordinate form of the Navier-Stokes equations frame. Spatial discretization in the computational space is carried out using second order central finite differences. Temporal integration is carried out by using the second order accurate Adams-Bashforth scheme for the convective term and the implicit Crank-Nicolson scheme for the diagonal viscous terms. A collocated grid is considered where pressure and Cartesian velocity components are defined at the cell centers, and the volume fluxes are defined at the midpoints of the cell faces. For the numerical model see Zang et al. [25], whereas for the implementation of the SGS models see Armenio and Piomelli [26]. A new parallel version of the model has been developed over the years and a version suited for environmental applications (LES-COAST) is available [27].

## 3. Wall Layer Model for Body Fitted Geometry

A basic equilibrium wall stress model is present in the solver. The wall stress is obtained from instantaneous horizontal velocity at the first off-wall centroid based on the law of the wall:

$$
u_p^+ = \begin{cases} \dfrac{1}{\kappa} ln(y_c^+(1)) + B & if \ \ y_c^+(1) > 11 \\ y_c^+(1) & if \ \ y_c^+(1) \leq 11 \end{cases} \tag{10}
$$

where

$$
u_p^+ = \frac{u_p}{u_\tau} = \frac{\sqrt{u^2 + w^2}}{u_\tau}
$$

is the modulus of instantaneous non-dimensional velocity at the first off-wall computational node $P$, which has a distance $y_c(1)$ from the wall; $\kappa = 0.41$ is the von Kármán constant, and B varies between 5 and 5.5 [2]. Here $B = 5.1$ was used for all the simulations. The friction velocity $u_\tau$ is calculated from the velocity $u_p$ at each near-wall computational node, and depends on the distance from the wall $y_c^+(1)$, either from the linear or logarithmic law. Then, wall shear stress $\tau_w$ is calculated from friction velocity; $\tau_w = \rho u_\tau^2$. More details of the procedure can be found in Fakhari [28].

In addition to this, the knowledge of the contraction of the resolved strain rate tensor is required in order to use an SGS eddy viscosity at the wall in the Smagorinsky model. Since the tangential velocity at the wall is not determined, the evaluation of the leading terms of $\overline{S}_{ij}$ becomes increasingly wrong with decreasing grid resolution.

To overcome this problem, the leading terms of the strain rate tensor $\overline{S}_{ij}$ are set analytically based on the location of $y_c(1)$. If the first point $P$ is in the logarithmic region, the leading terms of the strain rate tensor are:

$$
\overline{S}_{12} = \frac{u_\tau}{\kappa y_c(1)} \frac{\overline{u}}{u_p} + \frac{\partial \overline{v}}{\partial x} , \quad \overline{S}_{32} = \frac{u_\tau}{\kappa y_c(1)} \frac{\overline{w}}{u_p} + \frac{\partial \overline{v}}{\partial z} . \tag{11}
$$

Consequently the value of the eddy viscosity near the wall adjusts consistently with the imposed stress. The wall layer model mentioned above is for smooth surfaces. If we deal with a rough surface with the roughness height $y_0$, the velocity profile is:

$$
u_p^+ = \frac{1}{\kappa} ln(\frac{y_c(1)}{y_0}) \quad y_c(1) > y_0. \tag{12}
$$

The same procedure based on Equation (11) can be used to modify the leading terms of $\overline{S}_{ij}$.

This wall layer model is very economic and reasonably accurate in attached flows. Its main drawback is the simulation of separated flows [1,2] as will be discussed with details in Section 3.2. The logarithmic law does not capture separation; therefore, in detached flows a stretched grid near-wall is required to have more resolution there to set the near-wall computational nodes inside the viscous layer.

Cabot and Moin [8] used wall layer models with LES to simulate flow over a backward-facing step. They considered a zonal approach using the thin boundary layer equation (TBLE) in the inner layer and LES with dynamic SGS model in the outer region. They used 8 uniform cells from the bottom of the wall up to half the height of the domain, much less than in a former stretched-grid which was used for a wall-resolving LES. The authors introduced a dynamic treatment of the von Kármán constant to equate the stress predicted by TBLE in the inner layer ($\kappa y \widehat{\overline{u}}_\tau \widehat{\overline{S}}_{ij}$) in a least-square sense to the total one (resolved + SGS) computed by LES in the outer region ($-\widehat{\overline{u}_i\overline{u}_j} - \widehat{\tau}_{ij}$). The dynamic $\kappa$ was used to calculate the eddy viscosity in the inner region, $\nu_t = \kappa y u_\tau D$, $D = [1 - exp(-y^+/A^+)]^2$ in which $A^+ = 17$, but in that case since $y^+ \gg A^+$ they considered $D$ as unity. In Cabot and Moin [8] the use of dynamic $\kappa$ was justified by the fact that their wall model (TBLE) would carry shear stress both in the RANS eddy viscosity and the RANS convective terms, thus they needed a reduced RANS eddy viscosity, and by using the dynamic $\kappa$ which had the values less than the von Kármán constant, and reduced the shear stress.

### 3.1. Model Optimization for Body Fitted Geometry

In this work the context is different. It is known that the equilibrium stress wall function works well in equilibrium flows, and has a drawback in the presence of separation [1,2]. In addition, in separated regions the velocity profile does not follow the standard log law; therefore, a modification of the standard log law is required. Here, the aim of using dynamic $\kappa$ is not to reduce stress, but to have it deviate from the standard log law. The dynamic $\kappa$ is calculated here to match analytical stress from the boundary layer and the stress computed with dynamic Smagorinsky using a coefficient:

$$\kappa = -coeff \cdot \frac{< y_c(1)\widehat{\overline{u}}_\tau \widehat{\overline{S}}_{ij}(\widehat{\overline{u}_i\overline{u}_j} + \widehat{\tau}_{ij}) >}{< y_c(1)^2 \widehat{\overline{u}}_\tau^2 \widehat{\overline{S}}_{ij}\widehat{\overline{S}}_{ij} >}, \tag{13}$$

in which $<>$ denotes averaging over the directions of homogeneity. First, a simulation of a periodic open channel flow was carried out in order to obtain a cross-sectional plane of data instantaneously to be used as the inflow over the hill, as will be discussed in Section 3.2. For the simulation of periodic open channel flow, a plane averaged approach was performed to compute $coeff$ in such a way that the dynamic $\kappa$ becomes equal to the von Kármán constant. In LES–COAST, $coeff = 0.4$ was obtained, and this will be explained in Section 3.2.

### 3.2. Application of the Wall Layer Model for Body Fitted Geometry

The wall layer model is here tested on a detached flow case. The accuracy of the wall layer model is checked and then the model is improved to increase the accuracy of the wall function. Most of the wall layer models have weakness in capturing the points of flow separation and re-attachment, separately, especially when the slope of a solid wall changes gradually. Thus, this work focuses on flow simulation over a single two-dimensional hill, which, from one side, is a very challenging problem for wall layer models and, from the other side, literature data are available.

The geometry and boundary conditions are collected from ERCOFTAC classic database, environmental flows area, case 69 (http://ERCOFTAC/case69) [29]. The hill height is $H = 0.117$ m and the amplitude of the hill is $a = 3H$ (Figure 1a). The wind tunnel experiments were carried out by Khurshudyan et al. [30]. A wall-resolved LES of turbulent boundary layer over a 2D hill was also performed by Chaudhari et al. [31] and Chaudhari [32] for two different aspect ratios $a = 3H$ and $5H$. Since they faced difficulty in validating the reattachment point for the aspect ratio of $a = 3H$ with the experiment, this case was specifically chosen for this study as a challenging problem.

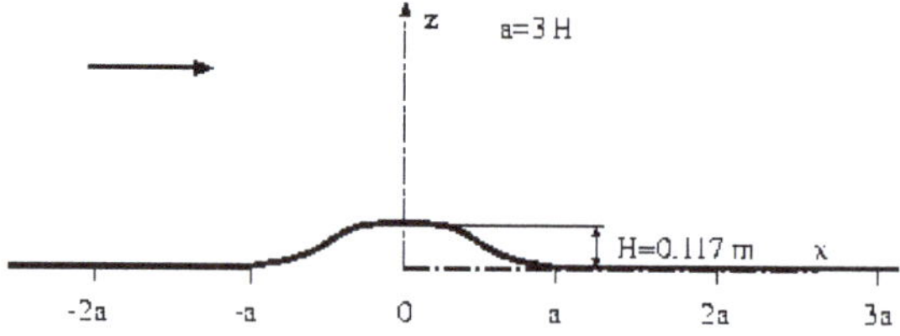



(**a**) Hill geometry in which $H$ is the hill height, $a$ is the aspect ratio, with courtesy of ERCOFTAC, Environmental Flows [29].

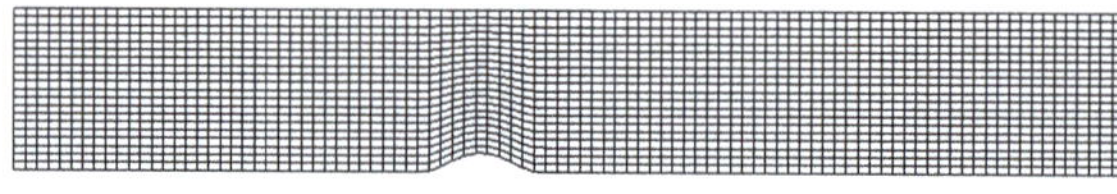

(**b**) Computational domain.

**Figure 1.** The hill geometry and the domain grid with resolution of $96 \times 20 \times 32$ cells in streamwise, wall normal and spanwise directions, respectively, for simulation of flow over hill using dynamic $\kappa$ (body fitted approach).

A domain dimension with streamwise length $L_x = 7.12\,\delta$, height $\delta$ and width (in spanwise direction) $L_z = \delta$ is considered where $\delta = 1$ m.

The boundary conditions are given as follows: At the inlet, inflow is obtained from simulation of a periodic open channel flow with the same cross-sectional grid resolution; a cross-sectional plane was considered to collect the instantaneous data at each time step after the channel flow reached a steady state and the collected data was used as inflow for the flow simulation over the hill. The roughness characteristic height of the wall for this case is $y_0 = 1.57$ mm, shear velocity $u_\tau = 0.178$ m/s and the Reynolds number based on shear velocity is $Re_\tau = u_\tau \delta / \nu = 1187$; at the outlet, radiative boundary conditions were given; at the upper boundary the free slip condition is imposed; periodic conditions are applied along the spanwise direction. The dynamic SGS model wis employed for simulations with body fitted grid, with the constant obtained from spanwise averaging; simulations were run with a fixed Courant number equal to 0.2.

The results of this simulation, which was initially performed using $96 \times 32 \times 32$ cells in streamwise, wall normal and spanwise directions, respectively, show that the wall layer model described in Section 3, with a fixed value of the von Kármán constant, is not able to capture flow separation when a uniform grid is employed (where $y_c^+(1) = 18.5$). Conversely, stretching the grid to set the first computational node off the wall inside the viscous layer allows obtaining flow separation. Figure 2 shows a comparison between the experiment by Khurshudyan et al. [30] and three simulations: The case with uniform grid, and two cases with stretched grids in which the first centroids are at $y_c^+(1) = 4$ and $y_c^+(1) = 7$, respectively. All the grids share the same number of grid cells and the resolution in streamwise and spanwise directions, while the two grids are stretched vertically using the Vinokur approach [33] to have the largest cells at the top of the domain with $\Delta y^+ \approx 64$. The free surface velocity is used as $U_{ref}$ to normalize the velocity plots.

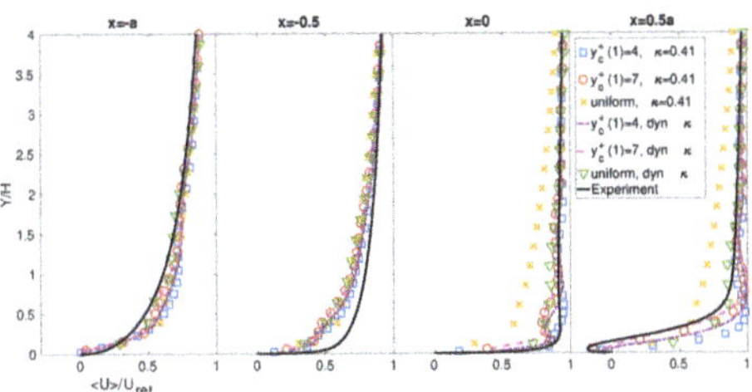

(**a**) Mean stream-wise velocity upstream, over and slightly downstream from the hill.

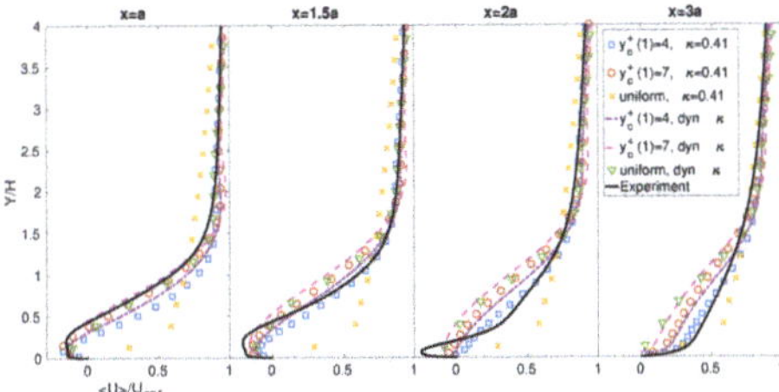

(**b**) Mean stream-wise velocity downstream from the hill.

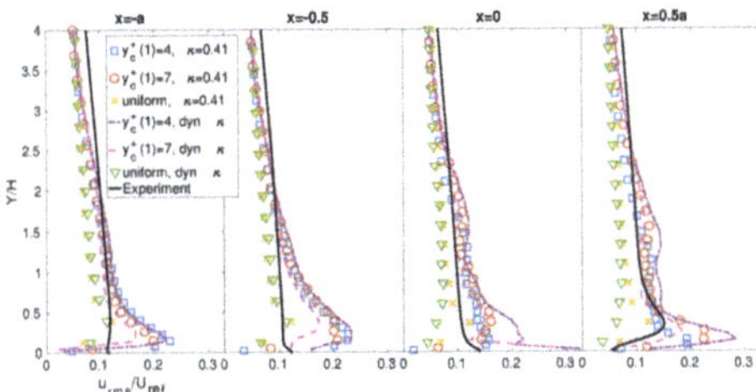

(**c**) RMS of streamwise velocity fluctuations upstream, over and slightly downstream from the hill.

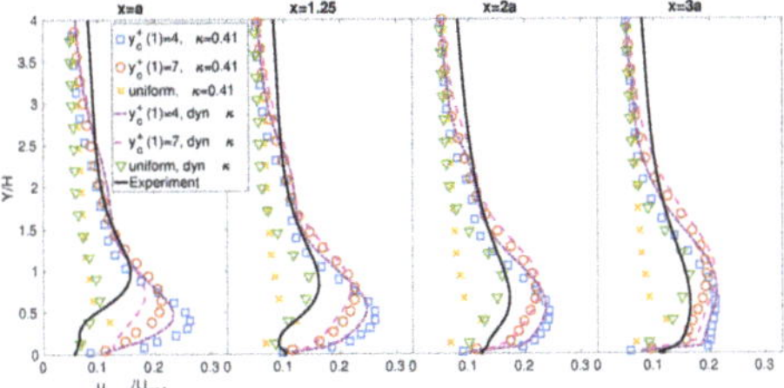

(**d**) RMS of streamwise velocity fluctuations downstream from the hill.

**Figure 2.** Mean streamwise velocity and root mean square (RMS) of its fluctuations obtained from simulation using 32 cells in the wall normal direction for the cases with $y_c^+(1) = 4$ and $y_c^+(1) = 7$ compared to the experiment by Khurshudyan et al. [30]; the hill starts from $x = -a$ and ends at $x = a$.

For the cases with fixed value of $\kappa$, the starting point of flow separation is at $x = 0.5a$ using the stretched grids; the reattachment point is predicted well in these cases. This test reveals that the resolutions in streamwise and spanwise directions are sufficient to capture boundary layer separation. Overall, separation is predicted with a grid that for the vertical resolution near the wall resembles that of a wall-resolving LES. This is opposite the expectations from a wall layer model, which is supposed to work well with a poor resolution in the wall region.

The improved model which makes use of the dynamic evaluation of the von Kármán constant $\kappa$ is tested on the uniform mesh. In this case separation is captured; at $x = 0.5a$ the first centroid is too far from the wall to capture the starting point of separation, while at the other locations, separation is predicted correctly as well as boundary layer reattachment.

Applying the dynamic evaluation of $\kappa$ in the case of the stretched grids reveals that in the case with $y_c^+(1) = 4$ the mean velocity profile at $x = 0$ demonstrates a better acceleration at the first centroid compared to the case with a fixed value of $\kappa$; moreover, at $x = 0.5a$, which is the starting point of separation, the negative velocity displays a larger magnitude at the few cells above the wall, which is closer to the experimental values, while a bit farther from the wall, the case with constant $\kappa$ is in a better agreement with the experiment. In the separation region and even after reattachment, the mean velocity at the near-wall centroid is smaller than using the fixed value of $\kappa$, and the reattachment point is predicted correctly for $y_c^+(1) = 4$. The case with $y_c^+(1) = 7$ demonstrates higher velocity at the near-wall computational node using dynamic $\kappa$ in the beginning of separation ($x = 0.5a$), while at $x = a$ and beyond, the velocity gets smaller than the fixed $\kappa$, and the reattachment point is delayed.

The root mean square (RMS) of streamwise velocity fluctuations exhibits an overestimation close to the wall for stretched grids using both fixed and dynamic $\kappa$, while the uniform grid with dynamic $\kappa$ slightly underestimates this quantity in most of the regions. This overestimation is due to the fact that the equilibrium stress model for the rough surface computes the shear velocity and wall shear stress from the logarithmic law (Equation (12)), while the near-wall computational nodes in the stretched grids are located inside or near the viscous sub-layer.

Figure 3 contains mean velocity profiles related to the present simulations, the wall-resolved LES by Chaudhari et al. [31] and the experiment by [30] at three critical locations: The hill crest, the starting point of separation and just beyond the reattachment point. For the wall-resolved LES case, the data were obtained by extraction of significant points from their plots. The mean velocity profile shows that at the top of the hill, the flow acceleration for wall-resolved LES agrees well with the experiment. Moreover, the point of separation is captured well, but the flow experiences an early reattachment at $x = 2a$.

The wall-resolved LES underestimates root mean square of streamwise velocity fluctuations except at the start of separation $x = 0.5a$ close to the wall, as depicted in Figure 3b.

Figure 3a shows that, overall, the present results exhibit a reasonable accuracy, also in view of the grid herein employed. It is to be noted that the reference wall-resolved LES by Chaudhari et al. [31] with $495 \times 70 \times 136$ numerical grids (48 times more expensive than the current wall layer model simulation) had an early prediction of reattachment point, while in this simulation there is a better agreement of the reattachment point with the experiment.

Successively, a different grid resolution was applied in the vertical direction ($96 \times 20 \times 32$ cells in streamwise, wall normal and spanwise directions, respectively) on the same geometry with the same boundary conditions. The domain grid is shown in Figure 1b, with the first computational node situated in the logarithmic region at $y_c^+(1) \approx 30$.

The simulations were carried out as follows. First, an open channel flow at $Re_\tau = 1187$ based on the friction velocity $u_\tau$ and the channel height was simulated, with a grid resolution of $40 \times 20 \times 32$ cells in streamwise, wall normal and spanwise directions, respectively. Before the simulation, a test was carried out to compute the dynamic (Smagorinsky) model constant and also $coeff$ of Equation (13) applying a plane average approach, and using $coeff = 0.4$ as a target value of the von Kármán constant $\kappa$.

Thereafter, in another attempt, spanwise averaging of the numerator and denominator of Equation (13) with $coeff = 0.4$ and also the dynamic model constant was performed. Note that $\kappa$ must be positive since it is used in the denominator to calculate friction velocity and strain rate tensor (see Section 3). Setting the minimum value of 0.06 avoided unrealistic values of velocity due to the near-zero or negative value of $\kappa$. Moreover, larger threshold values were tested to check the sensitivity to this lower bound; the analysis showed that it does not affect the results, hence the method is robust from this aspect.

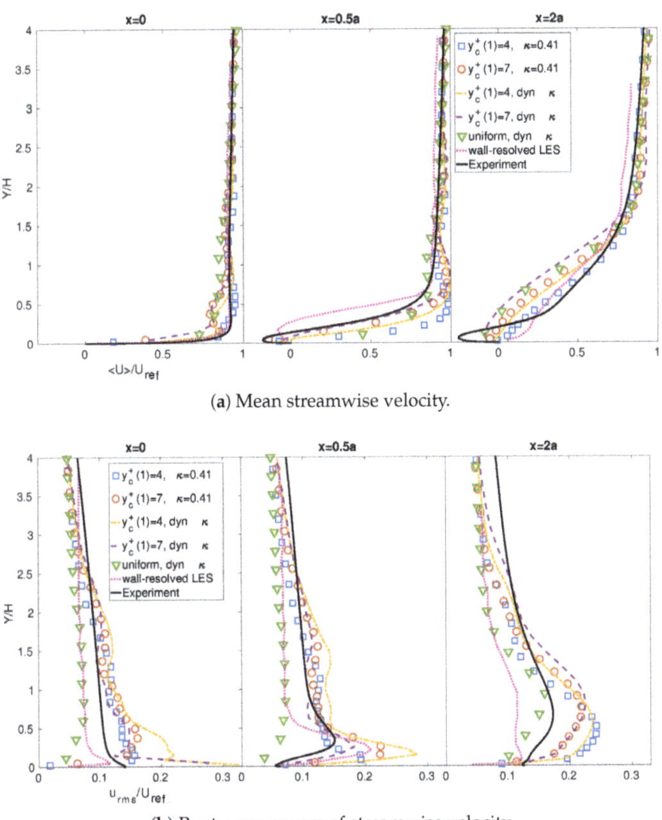

(a) Mean streamwise velocity.

(b) Root mean square of streamwise velocity.

**Figure 3.** Mean streamwise velocity and RMS of its fluctuations at three critical positions from simulation using 32 cells in the wall normal direction for the cases with $y_c^+(1) = 4$ and $y_c^+(1) = 7$ using constant and dynamic $k$, and uniform grid with dynamic $\kappa$ compared to resolved LES by Chaudhari et al. [31] and the experiment by Khurshudyan et al. [30] at the hill crest, the beginning of separation and just before the reattachment points.

Figure 4a shows time- and spanwise-averaged velocity profiles at different positions along the streamwise direction. As expected (Figure 4a), the current resolution is too coarse, since the first computational node is out of the separation region; this explains the absence of separation at $x = 0.5a$. However, separation occurs at $x = 2a$ which has a delay with respect to both the wall-resolved LES and the experiment, and reattachment is obtained at $x = 3a$ which is closer to the experiment, while the wall-resolved LES exhibits an earlier reattachment.

Figure 4b shows the root mean square of streamwise velocity fluctuations. Generally there is an underestimation of the velocity fluctuations especially right before and at the starting point of the separation region. Chaudhari et al. [31] also underestimated the RMS of the streamwise velocity fluctuations uphill, and overestimated it in the separation region.

As was discussed in the previous section, the use of dynamic $\kappa$ in this method is different than in Cabot and Moin [8]. To understand this better, instantaneous values of the dynamic $\kappa$ along $x$ for the flow over the single hill of this simulation in Figure 5. Since $\kappa$ is in the numerator to calculate the shear velocity and wall shear stress ($u_\tau = u * \kappa / ln(y/y_0)$), on top of the hill, there is an increase of $\kappa$, which gives a better acceleration rather than the standard log law ($\kappa = 0.41$). Then, at the starting point of separation, $\kappa$ reduces and therefore the wall shear stress goes close to zero. Thereafter, inside the separation region, at most of the spots $\kappa$ is large, leading to a negative value of the wall shear stress with a larger magnitude, and finally it goes close to 0.41 after reattachment. In this way, the equilibrium stress model is applicable on separated flows.

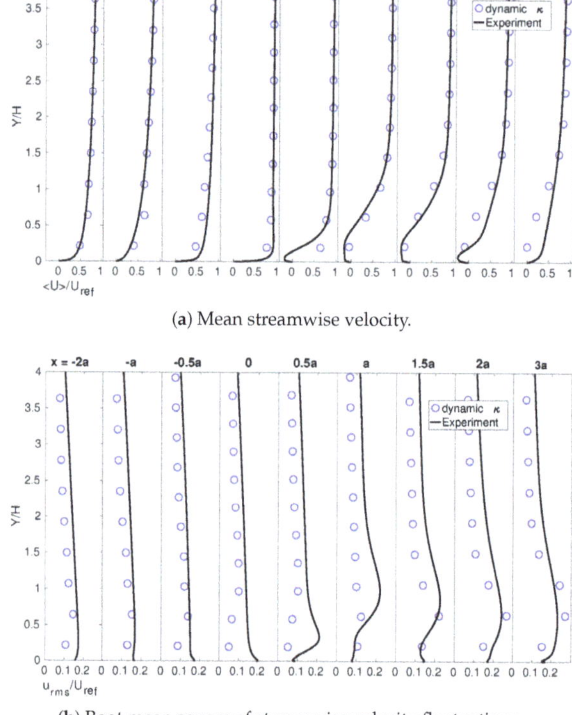

(a) Mean streamwise velocity.

(b) Root mean square of streamwise velocity fluctuations.

**Figure 4.** Mean streamwise velocity and RMS of its fluctuations obtained from simulation using 20 cells in the wall normal direction, applying dynamic $\kappa$ compared to the experiment by Khurshudyan et al. [30]; the hill starts from $x = -a$ and ends at $x = a$.

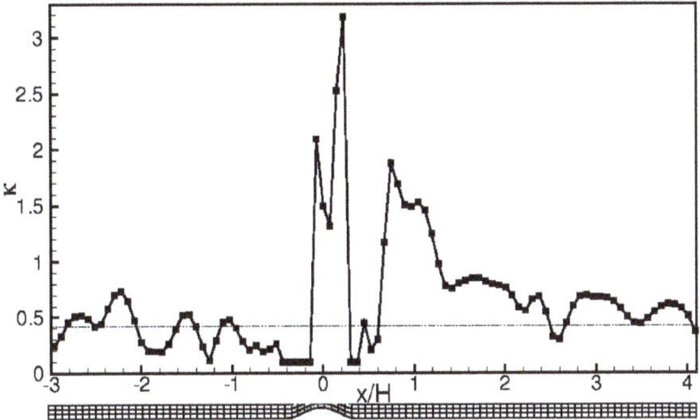

**Figure 5.** Instantaneous values of dynamic $\kappa$ on a stream line in the current simulation of flow over the single hill.

## 4. Wall Layer Model for Immersed Boundary Methodology (IBM)

This part of the work takes advantage of the model developed in Roman et al. [34]. The velocity reconstruction method is briefly recalled here. With reference to Figure 6, the velocity at the projection point (PP) is interpolated from velocities at its surrounding points (empty square points in Figure 6). Then, shear velocity at the PP point is computed. Finally, based on the distance of the first centroid off-the-wall, i.e., the IB node, from wall in wall units, the velocity at the IB node is calculated using the law of the wall:

$$\overline{u}_{IB} = \begin{cases} \overline{u}_{PP} - \dfrac{1}{\kappa}\sqrt{\dfrac{\tau_w}{\rho}}\ln(\dfrac{d_{PP}}{d_{IB}}) & if\ d_{IB}^+ > 11 \\[2ex] \dfrac{d_{IB}u\tau^2}{\nu} & if\ d_{IB}^+ \le 11 \end{cases} \tag{14}$$

Considering that the velocity is known at the PP point, a parabolic interpolation is carried out to compute the wall normal velocity at the IB node:

$$\overline{u}_{n,IB} = \overline{u}_{n,PP}\frac{d_{IB}^2}{d_{PP}^2}. \tag{15}$$

A reconstruction of shear stress at the cell face is also done using a RANS-like eddy viscosity. The eddy viscosity at the IB nodes is calculated analytically from mixing the length theory from the equation

$$\nu_t = C_w \kappa u_\tau d_{IB}, \tag{16}$$

where $\kappa$ is the von Kármán constant, and $C_w$ is an intensification coefficient to be determined such that the Reynolds shear stress is a fraction of the wall shear stress near the wall (where $\nu_T$ is set). Consequently, this coefficient can be written as:

$$C_w = \frac{\tau_F d_F}{\tau_w d_{IB}}, \tag{17}$$

where the index $F$ denotes a quantity calculated at the cell face, $d_F$ is the distance between the cell face and the immersed surface. A more detailed description can be found in Roman et al. [34] and Roman [35]. This wall layer model is very economic since the velocity is just interpolated from the projection point PP, and works well in attached flows using a very coarse grid [34]. Since the interpolation is based on the logarithmic law, and the velocity direction at the IB node always follows the velocity at the PP, this method has a drawback in separated flows even setting the IB node inside the viscous layer.

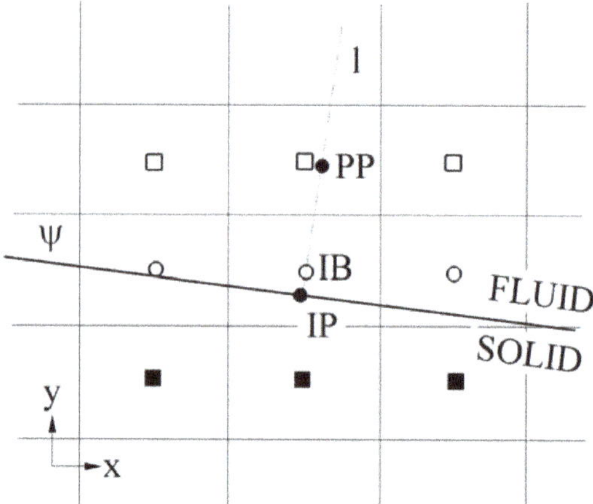


**Figure 6.** Discretization of a fluid–solid interface with the immersed boundary method by Roman et al. [34]. Solid squares, empty squares and empty circles represent solid, fluid and IB nodes respectively. Reproduced with permission from AIP Publishing [4714120594491].

Simulation of the flow over the hill, applying Cartesian and curvilinear grids, using the Roman et al. [34] model in its original formulation, does not allow to obtain acceptable results, since the flow does not separate even with increasing resolution close to the immersed boundary and with setting the of the IB node inside the viscous layer [28]. Here, first the Roman et al. [34] model on a standard open channel flow is re-analysed and successively the flow over the hill is investigated.

*4.1. Calibration of the Wall Layer Model with IBM*

In order to check the accuracy of the current IBM, an open channel is considered with a resolution of $48 \times 20 \times 32$ cells in the streamwise, wall normal and spanwise directions, respectively, and a dimension of $7.12\delta \times 1\delta \times 1\delta$ in which $\delta$ is the height of the open channels. The lower wall is reproduced with immersed boundaries and the simulation is carried out imposing a non-dimensional driving pressure gradient $(dp/dx = 1)$. The friction Reynolds number is $Re_\tau = 2000$, giving the IB node as located at $y^+ = 50$.

The velocity profile scaled with the friction velocity collapses over the theoretical line as has already been shown in Roman et al. [34]. However, considering the non-scaled averaged velocity, the method underestimates the velocity compared to the logarithmic law at the same rate as the shear velocity, thus demonstrating a loss of momentum. In particular, the wall shear stress is underestimated by approximately 16% in the simulation. This has to be attributed to the current implementation of IBs, which does not conserve momentum perfectly. This is a common issue in a number of IB methodologies, although more recent implementations appear more conservative (Kim et al. [36], Meyer et al. [37] and Rapaka and Sarkar [38]).

After this observation, several cases of open channel geometries are created sharing the same dimension and resolution; the only difference stands in the fact that the immersed boundary surface was moved slightly to check whether the position of the IB node affects the results. Five cases are considered:

Cases 1 to 4 in which the IB node was located in the logarithmic region, and case 5 in which it was in the viscous layer. The positions of the immersed boundary for the five cases are displayed in Figure 7 and details are given in Table 1. Simulations are hence carried out for all cases at $Re_\tau = 2000$, with the same boundary conditions as the original case. The results show an underestimation of friction velocity for cases 1 to 4 in which the IB node is set in the logarithmic region, and an overestimation of shear velocity for case 1 which has the IB node in the viscous layer (see Table 1), indicating that the IBM of Roman et al. [34] is sensitive to the position of the immersed surface with respect to the grid line, and it is geometry dependent. Figure 8a displays a non-scaled mean velocity profile for all cases compared to the theoretical velocity based on the law of the wall.

Apart from the shear velocity deviation, the non-dimensional velocity profile compared well with the wall law for the first four cases (Figure 8b). A calibration of the IBM is carried out to avoid momentum loss. For this purpose, the coefficient which is used to calculate eddy viscosity at the IB nodes ($C_w$) in Equation (16) is varied to reach the target value of the wall stress.

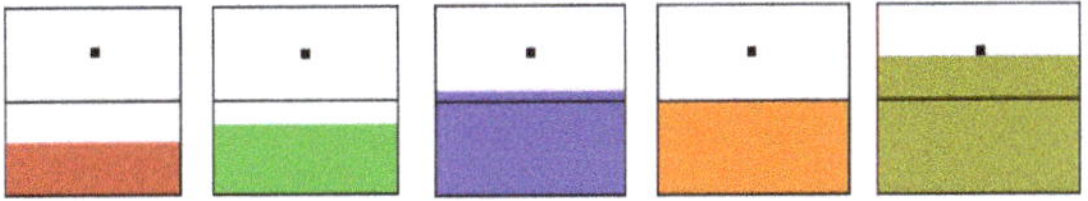

**Figure 7.** Five different channels using the immersed boundary as the lower wall in different positions, cases 1 to 5 in order from left to right. Filled squares display IB node.

**Table 1.** Different geometries description of IB cases and shear velocity obtained from simulation.

| Case | IB Surface Position | $d_{IB}^+$ | $u_\tau$ |
|---|---|---|---|
| 1 | just above centroid | 101 | 0.8883 |
| 2 | between centroid and grid line | 82 | 0.8816 |
| 3 | just above grid line | 43.5 | 0.8541 |
| 4 | overlapping grid line | 50 | 0.8393 |
| 5 | just below centroid | 5 | 1.2740 |

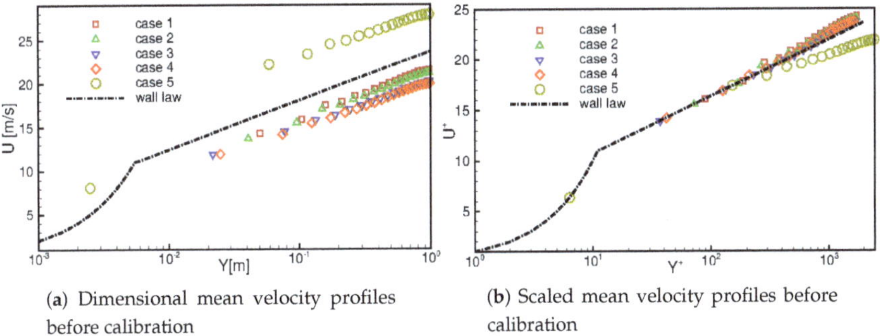

(**a**) Dimensional mean velocity profiles before calibration

(**b**) Scaled mean velocity profiles before calibration

**Figure 8.** *Cont.*

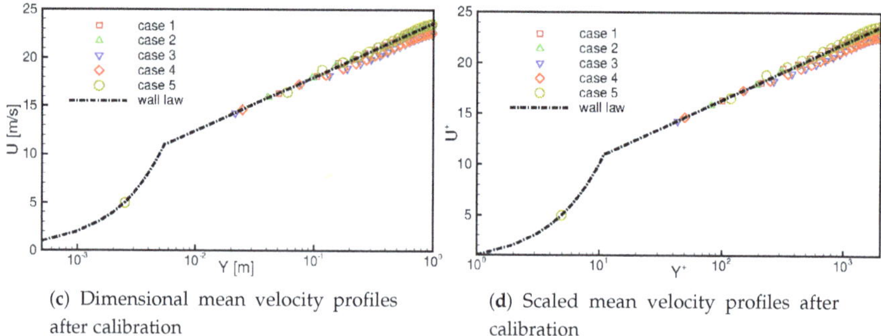

(**c**) Dimensional mean velocity profiles after calibration

(**d**) Scaled mean velocity profiles after calibration

**Figure 8.** Mean velocity for IBM cases before and after calibration versus law of the wall.

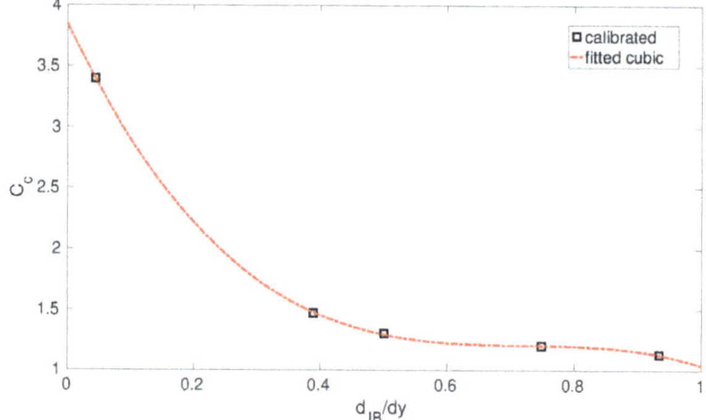

**Figure 9.** Cubic line fitted to the calibrated values computed for the 5 cases.

Table 2 shows the optimal coefficients and friction velocities obtained after running the simulations for all five cases. Further, it has been verified that changes in the coefficients by 10% produced very small variations in the shear velocity, thus showing the robustness of the method.

**Table 2.** Optimal coefficient for different positions of the IB.

| Case | $d_{IB}/dy$ | Coefficient ($C_{optimal}$) | $u_\tau$ |
|------|-------------|------------------------------|----------|
| 1 | 0.93318 | 1.13 | 0.9987 |
| 2 | 0.74818 | 1.20 | 1.0043 |
| 3 | 0.38909 | 1.47 | 0.9966 |
| 4 | 0.5 | 1.3 | 0.9974 |
| 5 | 0.04386 | 3.4 | 0.9942 |

The simulations are then repeated at $Re_\tau = 4000$ to check the accuracy of the calibration; it appeared insensitive to variations in the Reynolds number [28]. The coefficient can be written as a function of the non-dimensional distance from the wall ($d_{IB}/dy$), which is within the interval of 0–1. As it can be seen in Figure 9, a cubic law can be fitted to the numerical points is written below.

$$C_c = -6.9576 \left( \frac{d_{IB}}{dy} \right)^3 + 15.051 \left( \frac{d_{IB}}{dy} \right)^2$$
$$-10.902 \left( \frac{d_{IB}}{dy} \right) + 3.8493 \tag{18}$$

The dimensional and non-dimensional mean velocity profiles after calibration for all the cases are shown in Figure 8c,d, respectively. All cases are non-dimensionalized with shear velocity obtained from the mean velocity at the IB nodes based on the law of the wall. An improvement can be observed after calibration, especially for case 5 in which the IB node is located close to the immersed boundary. Finally the RANS-like eddy viscosity at the IB nodes can be calculated from the calibrated coefficient $C_c$, which is obtained from the cubic fitting:

$$\nu_T = C_c k u_\tau d_{IB}. \tag{19}$$

Calibration of the coefficient allows to improve the prediction of the wall shear stress, hence, the velocity profile. The results of the simulations for all cases show that the mean velocity and root mean square of velocity fluctuations are in good agreement with the law of the wall and also the DNS of plane channel flow of Hoyas and Jimenez [39] for a channel at $Re_\tau = 2000$ [28]. Only for cases 3 and 4 in which IB nodes are at the beginning of the log-region ($d_{IB}^+ \leq 50$) is there a deviation of the mean velocity profile. However, this is a case that can be hardly encountered in real-scale flows, characterized by very high values of $Re$. It is important to point out that calibration may not be necessary in the presence of IB methodologies strictly conservative for momentum.

While the wall layer model predicts the velocity profile well, it underestimates the wall normal velocity fluctuations. Thus, random fluctuations near the wall are added to improve the model. In hybrid LES/RANS and DES, stochastic forcing is needed to reduce the effect of excessive damping coming from the RANS eddy viscosity [11,12]. In the present model, stochastic forcing is more likely required because of the grid coarseness, which is, however, a standard in simulations for engineering purposes.

Taylor and Sarkar [40] used random stochastic forcing in the wall normal direction after finding that the near-wall model (NWM) LES with the dynamic eddy viscosity model (DEVM) underrated vertical velocity fluctuations in their body fitted plane-channel-flow simulation.

This forcing term was applied to the right-hand side of the momentum equation in the wall normal direction to increase velocity fluctuations in this direction and control the Reynolds stress in order to allow it to fit the logarithmic velocity profile. This term is written as:

$$f_y(x,y,z) = \pm R * A(y), \tag{20}$$

where $R$ is a random number between 0 and 1, and $A(y)$ is an amplitude function. At each time step, the latter can be obtained from summation of the amplitude at the previous iteration, and another term including the error function:

$$A(y)^{n+1} = A(y)^n + \frac{u_\tau \epsilon(y)}{\tau}, \tag{21}$$

where the error function is the difference between the resolved shear stress and the theoretical value based on the logarithmic law:

$$\epsilon(y) = \frac{ky}{u_\tau} \left( \frac{d\langle u \rangle}{dy}^2 + \frac{d\langle w \rangle}{dy}^2 \right)^{\frac{1}{2}} - 1,$$ (22)

where $\tau$ is a relaxation time. This approach is correct if the first computational node is in the logarithmic region since in Equation (22) the error is established upon $ky/u_\tau$, which is the theoretical resolved shear stress based on the logarithmic law. However, this is not a serious issue, since the model is designed to work for high $Re$ number flows. Finally, the sign of the function is computed to reduce the correlation between streamwise and vertical velocity fluctuations ($u'$ and $v'$) if the error is positive, and to enhance this correlation if the error is negative:

$$sgn\left(f_y(x,y,z)\right) = -sgn\left(\frac{d\langle u_s \rangle}{dy}\right) * sgn\left(\epsilon(y)\right)$$
$$*sgn\left(u_s'(x,y,z)\right),$$ (23)

where $u_s$ is the velocity in the direction of the mean wall shear stress.

The stochastic random force in case 4 is imposed instantaneously at every iteration of the simulation. Since the IB node is located in the logarithmic region, the horizontal velocity is interpolated based on the logarithmic law from the projection point PP; adding this force to the first two nodes does not improve the velocity profile. Considering one more point to impose the forcing term on, it makes the deviation from the log law disappear (Figure 10).

This happens because the stochastic forcing allows to obtain a more accurate prediction of the resolved Reynolds shear stress (associated to the resolved velocity fluctuations) as can be seen in Figure 11.

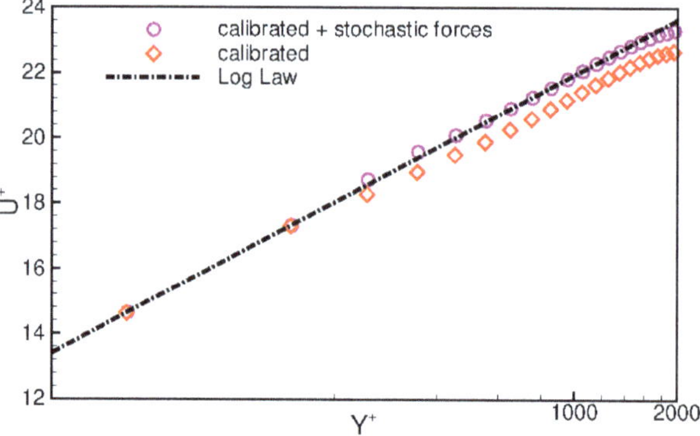


**Figure 10.** Mean velocity profile before and after adding stochastic forces for case 4 compared to the log law.

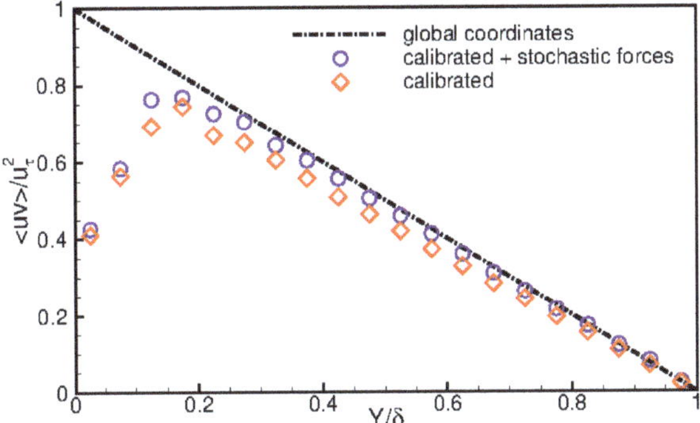



**Figure 11.** Reynolds shear stresses in global coordinates compared to the calibrated IBM for case 4.

The relaxation time plays an important role in the simulation. It has to be large enough to let the flow adapt itself to the force. On the other hand, if the relaxation time is too large the effect of force is diluted. In the present simulations, $0.0006\,u_\tau/\delta \le 1/\tau \le 0.002\,u_\tau/\delta$ gave the best results. Values larger than $0.002\,u_\tau/\delta$ increased the velocity continuously since the flow did not have time to adjust itself to the force; values smaller than $0.0006\,u_\tau/\delta$ displayed a very small improvement in velocity profile.

This method works very well when the IB node is located in the logarithmic region and in attached flows since a logarithmic law for the velocity profile is expected. However, in the present form, it is not able to predict massive separation.

Steady two-dimensional separation is known to be related to the sign of the streamwise pressure gradient. After neglecting advective terms, the boundary layer equation in the streamwise direction simplifies as:

$$\nu\left(\frac{\partial^2 u}{\partial y^2}\right)_{wall} = \frac{1}{\rho}\frac{\partial p}{\partial x}. \tag{24}$$

Considering the flow over a curved surface, a negative streamwise pressure gradient develops upward, associated to an acceleration of the flow. Thereafter, the change of curvature makes the pressure gradient become unfavourable, and from the continuity equation it contributes to an enhancement of the boundary layer thickness. Hence the second derivative of velocity is positive and therefore there is a change of sign of the velocity curvature inside the boundary layer. At a point called inflection, the second derivative of streamwise velocity becomes zero; $\partial^2 u/\partial y^2 = 0$. If the unfavourable pressure gradient is strong enough, the flow next to the wall changes direction giving rise to a recirculation region. As already noted, in this method the velocity at the IB node is calculated from the friction velocity at the projection point. This means that $u_{IB}$ is always in the same direction of $U_{PP}$, and the flow is not allowed to separate, since the above-described physical mechanism is not accounted for. For this reason, the method of calculation of the velocity at the IB point is modified as follows. The boundary layer equation close to the wall, after neglecting advective terms is:

$$\nu_T\left(\frac{\partial^2 u}{\partial y^2}\right)_{IB} \sim \left(\frac{1}{\rho}\frac{\partial p}{\partial x}\right)_{IB}. \tag{25}$$

Using the central difference scheme (CDS) in order to discretize the second derivative of velocity for a non-uniform grid (Ferziger and Peric [41]) considering PP, IB and IP nodes, it can be written as:

$$\left(\frac{\partial^2 u}{\partial y^2}\right)_{IB} \approx \frac{u_{PP}(d_{IB}) + u_{IP}(d_{PP-IB}) - u_{IB}(d_{PP})}{\frac{1}{2}(d_{PP})(d_{PP-IB})(d_{IB})}, \tag{26}$$

in which $d_{PP-IB}$ is the distance between IB and PP points. Finally the streamwise tangential velocity at the IB node can be written as a function of the tangential pressure gradient instead of the friction velocity;

$$u_{IB} = u_{PP}\left(\frac{d_{IB}}{d_{PP}}\right) - C\left(\frac{\partial p}{\partial x}\right)_{IB}\frac{(d_{PP-IB})(d_{IB})}{\nu_T}, \tag{27}$$

where the eddy viscosity is also the calibrated value obtained from Equation (19).

Using a coefficient (C) of the order of $10^{-3}$ inside the second term containing the pressure gradient at the IB avoids problems in regions characterized by the presence of large values of the pressure gradient with small values of turbulent eddy viscosity. Many simulations have been carried out to find out the best criterion for using this new scheme. Finally, applying Equation (27) under the following conditions:

$$\left(\frac{\partial p}{\partial x}\right)_{IB} > 0, \tag{28}$$

$$d_{IB}^+ < 40, \tag{29}$$

gave the best results for simulating separated flows. The two criteria physically mean that the new scheme to calculate tangential velocity at the IB is used if the tangential pressure gradient at the IB is unfavourable, and the distance of the IB node from the immersed boundary is such that the velocity point does not belong to the log region of the velocity profile.

*4.2. Flow Simulation over a Single Hill Using IBM*

This new model is applied for simulation of the flow over a two-dimensional hill. First an open channel is constructed in which the immersed boundary is set as the lower wall. After the periodic open channel flow reached a steady state, instantaneous data are collected from a cross sectional plane at different time steps. Then the bottom of the domain including the hill shape is created as the immersed boundary, and the data that are obtained from the plane channel flow are used as inflow for the flow simulation over the hill.

The simulation is performed using two different grids (Cartesian and curvilinear) at $Re_\tau = 1187$. Applying the new scheme, it is feasible to capture separation in the simulation with a grid resolution of $128 \times 40 \times 32$ cells in the streamwise, wall normal and spanwise directions, respectively. With this resolution, the IB node upward and downward from the hill is located at $d_{IB}^+ \approx 15$. The mean velocity profiles for both grids are compared with the body fitted approach and the experiment by Khurshudyan et al. [30] as shown in Figure 12a. In the simulation for both grids, the starting point of separation is predicted well at $x = 0.5a$. The reattachment points for both cases are anticipated before $x = 2a$; therefore, at $x = 2a$ the mean velocities are positive, although very small close to the immersed boundary.

In addition, for the Cartesian grid a simulation without calibration is performed to check the effect of eddy viscosity calibration on flow separation; as it can be observed in Figure 12a, the case without calibration is not able to capture the starting point of separation properly, and in general it underestimates separation.

The root mean square of streamwise velocity fluctuations of the present simulations (Figure 12b) are in a good agreement with the experimental data. While there is an overestimation of fluctuations beyond the starting point of separation, the behavior of this quantity is similar to that of the experiment. Comparing the results of the present model with the one developed for the body fitted grid, it appears that the present IBM anticipates the starting point of separation in better agreement with the wall-resolved LES and the experiment, while the prediction of the reattachment point is similar to the wall-resolved LES, earlier than in the body fitted approach and the experiment. The present IBM gives more energetic velocity fluctuations than the body fitted approach. Moreover, comparing this result with the wall-resolved LES by Chaudhari et al. [31], it can be stated that the mean velocity profile has similar behavior: Predicting the starting point of separation well, and undergoing an early reattachment. The RMS of streamwise velocity fluctuations is also similar to the wall-resolved LES; depicting marginal underestimation before and after the separation region especially far from the wall, and an overestimation inside the separation region near the wall.

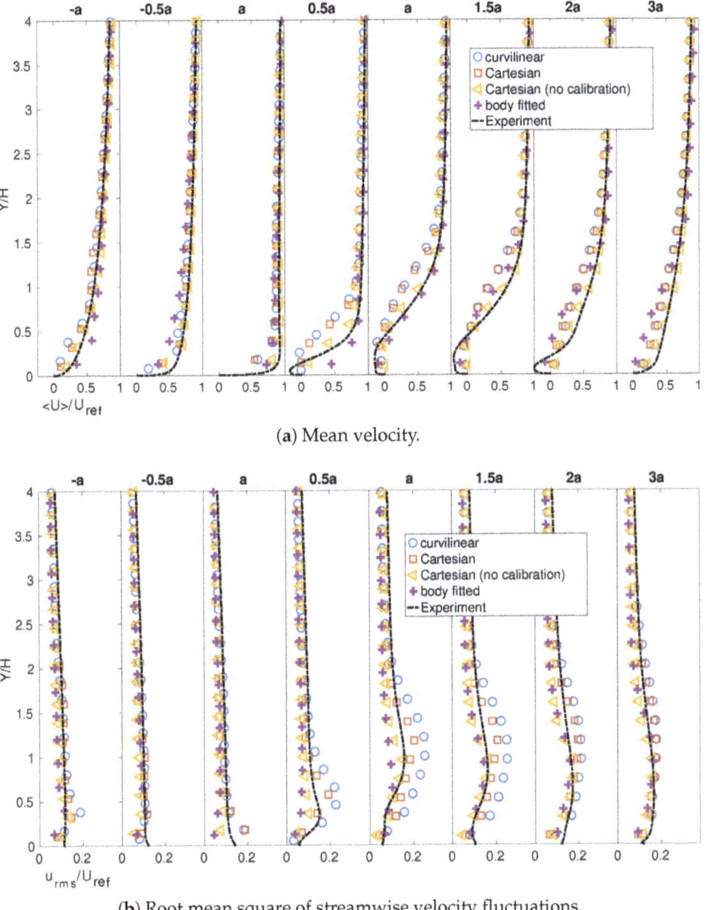

(a) Mean velocity.

(b) Root mean square of streamwise velocity fluctuations.

**Figure 12.** The results for the simulation of flow over a single hill using new the IBM scheme in Cartesian and curvilinear grids compared to the body fitted approach and experiment by Khurshudyan et al. [30]; the hill starts at $x = -a$ and ends at $x = a$.

### 4.3. Flow Simulation over 2D Periodic Hills

Finally, the new scheme is tested on flow over two-dimensional periodic hills with polynomial shape (Almeida et al. [42]). It is Case 81 of the ERCOFTAC classic database (http://ERCOFTAC/cases/case81)[43]. The shape of the domain is shown in Figure 13.

The edges of the domain are at the top of the hill. The hill height is $h = 28$ mm, and the crests of the two successive hills are separated by $L_x = 9h$. The height of the channel is equal to $L_y = 3.035h$, and the channel width is $L_z = 4.5h$. The reference for comparing the results are the mean velocity and RMS of the velocity fluctuations obtained from resolved LES by Temmerman and Leschziner [44] available in ERCOFTAC, and for separation and reattachment points, obtained from resolved LES by Fröhlich et al. [45] from the NASA Langley Research Center database (http://nasa/LES/2dhillperiodic) [46].

The computational domain is created in two different ways, using Cartesian and curvilinear grids, respectively (see Figure 14). The latter is shaped in such a way as to follow the hill geometry. In both cases, the bottom surface is created as an immersed boundary. The Reynolds number based on the hill height and bulk velocity on top of the first hill is equal to $Re = U_b h/\nu = 10595$, where $\nu$ is the molecular viscosity. The flow is periodic in the $x$ and $z$ directions, using immersed boundary at the bottom and free-slip surface at top.

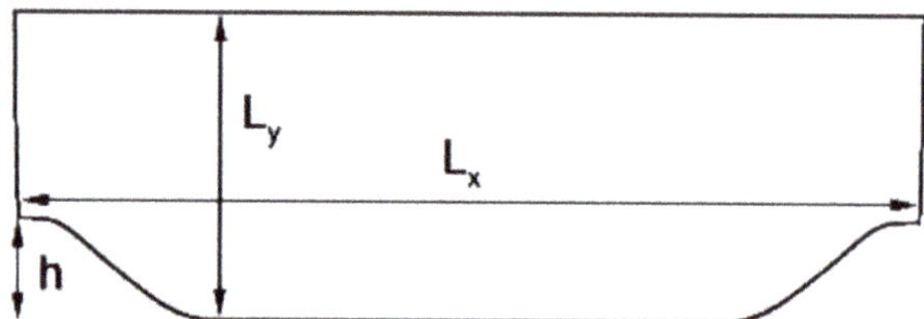


**Figure 13.** Domain characteristics for the simulation of flow over 2D periodic hills, with courtesy of ERCOFTAC [43].

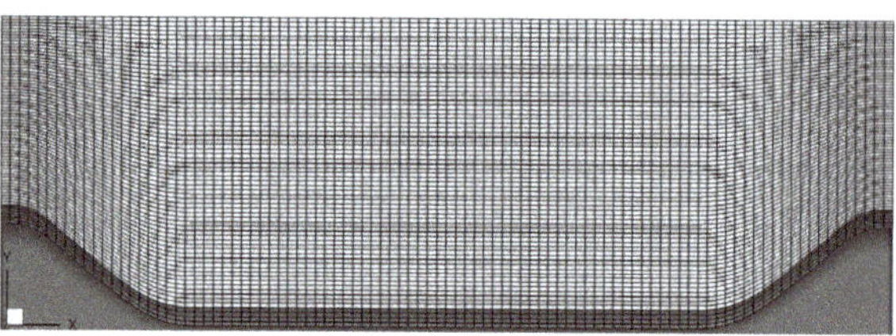

(**a**) Curvilinearn grid.

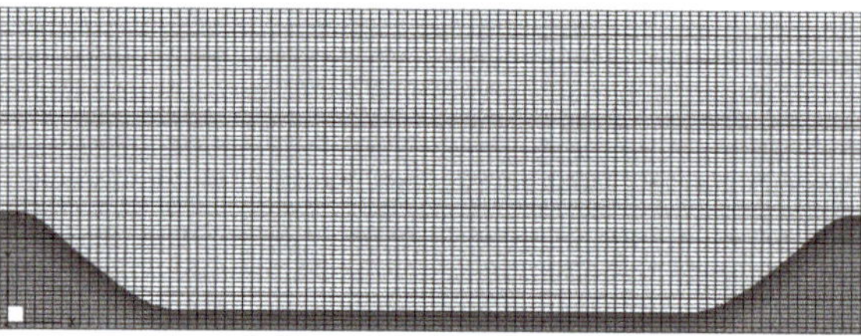

(**b**) Cartesian grid.

**Figure 14.** Two different grids for simulation of periodic hills using immersed boundaries, $96 \times 64 \times 32$ cells in $x$, $y$ and $z$ directions, respectively.

A domain with grid resolution of $96 \times 64 \times 32$ cells out of immersed boundary in streamwise, wall normal and spanwise directions, respectively, is constructed. This is equivalent to the coarsest grid of Chen et al. [17], who simulated flow over periodic hills using immersed boundary and applied a zonal approach (TLM); in particular, the authors used the turbulent boundary layer equations close to the immersed boundary and LES in the interior region.

The robustness of the method to reproduce separation is tested under a large variation of the threshold values employed in Equation (29). The results of this test for the threshold in the range of 30–50 are shown in Figure 15. The mean velocity profile in Figure 15a shows that in all cases separation is captured at $x = 1a$, but using $d_{IB}^+ < 50$ makes the reattachment point shift back slightly with respect to the other cases and the wall-resolved LES. The mean velocity profile for $d_{IB}^+ < 40$ is marginally closer to the resolved LES compared to $d_{IB}^+ < 30$ (which overestimates it narrowly). In addition to the mean velocity profile, Reynolds shear stresses for $d_{IB}^+ < 40$ demonstrates better agreement to the wall-resolved LES in the separation region and after reattachment (Figure 15b). Overall, it can be stated that the method is robust for large variations of the threshold value chosen to switch from one to the other law for the velocity at the first grid point. Large variations of this value produce marginal variations in the results.

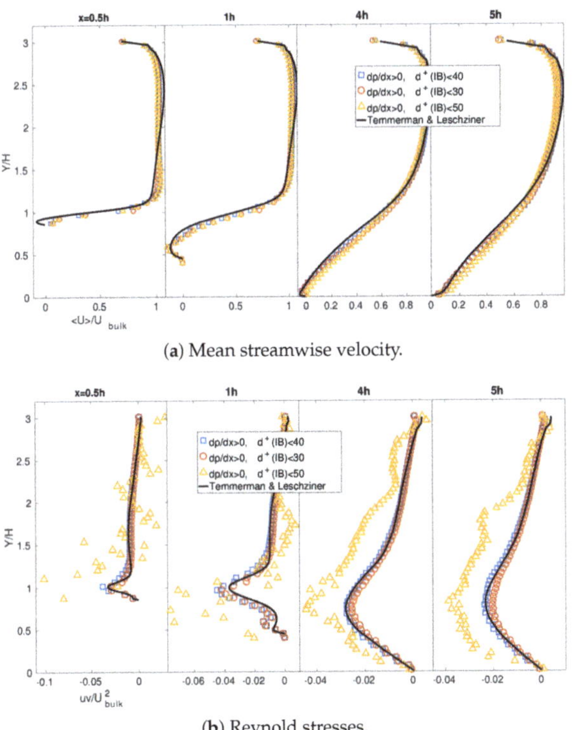

(**a**) Mean streamwise velocity.

(**b**) Reynold stresses.

**Figure 15.** Mean streamwise velocity and Reynolds shear stresses for a Cartesian grid compared to resolved LES by Temmerman and Leschziner [44], applying three different criteria to use the new scheme; $(\partial p / \partial x)_{IB} > 0$ and: $d_{IB}^+ < 40$ (square), $d_{IB}^+ < 30$ (circle), $d_{IB}^+ < 50$ (delta).

Figure 16 depicts statistics of the simulations of flow over periodic hills compared to the wall-resolved LES by Temmerman and Leschziner [44] at ten different streamwise locations. Velocities are made non-dimensional with $U_b$ which is the bulk velocity on the first hill crest; moreover, Reynolds stress and turbulent kinetic energy are non-dimensionalized with $U_b^2$.

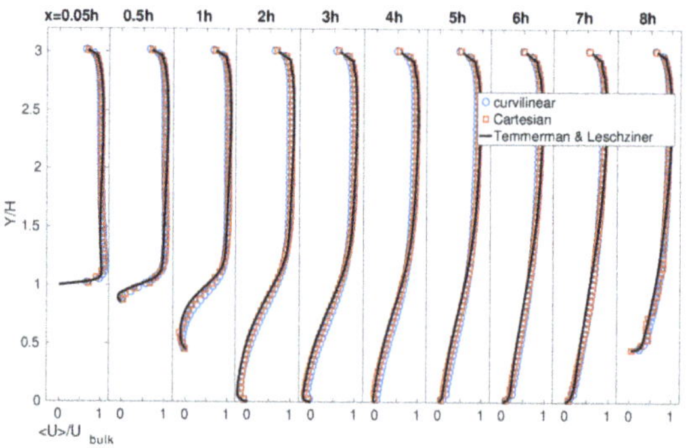

(**a**) Mean streamwise velocity profiles.

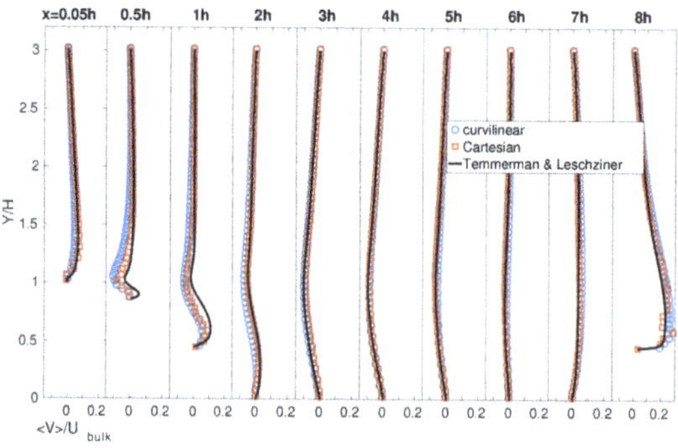

(**b**) Mean vertical velocity profiles.

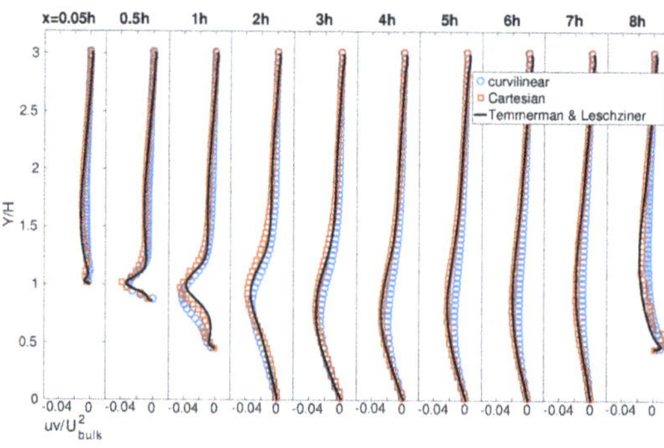

(**c**) Reynolds shear stress profiles.

**Figure 16.** *Cont.*

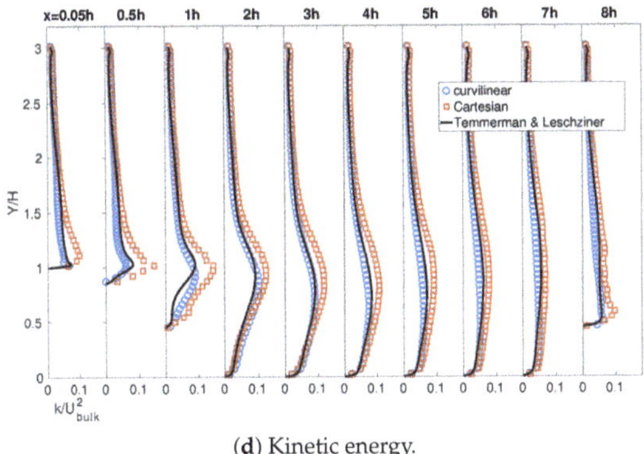



(**d**) Kinetic energy.

**Figure 16.** Results for simulation of flow over periodic hills at different positions, using Cartesian and curvilinear grids compared to the resolved LES by Temmerman and Leschziner [44].

Figure 16a shows mean streamwise velocity profiles (averaged in time and the spanwise direction) at different streamwise locations. They are overall in good agreement with the wall-resolved LES by Temmerman and Leschziner [44]. While the mean velocity for the Cartesian grid at $x = 0.5h$ matches the profile obtained from the wall-resolved LES, the velocity is not negative at the IB node, while the curvilinear grid displays a negative velocity at the IB node in this location. Separation is captured in both Cartesian and curvilinear grid cases, and the reattachment point for the Cartesian grid is a bit earlier than the resolved LES, while the flow in the curvilinear grid reattaches earliest before $x = 4h$.

The mean vertical velocity profiles in Figure 16b show a slight underprediction for both grids downhill, but in the other locations they matched the wall-resolved LES data.

Figure 16c displays the Reynolds shear stress for both cases (based on the resolved fluctuations). It behaves similarly to that of the resolved LES, while a marginal underestimation is observed before the end of downhill at $x = 1h$ near the immersed boundary, especially in the case with curvilinear grid.

As illustrated in Figure 16d, turbulent kinetic energy for the Cartesian grid is slightly overpredicted with respect to the wall-resolved LES, while the curvilinear grid demonstrates a closer value to the reference data. Overall, the behavior of this quantity is similar to that of the wall-resolved LES for both cases.

Table 3 displays the separation and reattachment points in the current simulation, the wall-resolved LES of Fröhlich et al. [45] and WMLES (using two-layer model) of Chen et al. [17]. In the present simulation, the boundary layer separation occurs at $x \approx 0.54h$ with the Cartesian grid and at $x \approx 0.47h$ with the curvilinear grid; in the wall-resolved LES it happens at $x \approx 0.22h$. Therefore a delay is observed to predict the separation point. Moreover, the intensity of the recirculation bubble predicted by the optimized IBM simulations is smaller than that of the wall-resolved LES (comparing the present simulation values with the large negative values of the reference simulation). The wall-resolved LES shows the reattachment point at $x \approx 4.69h$, while in the current simulation the boundary layer reattaches a bit earlier in both cases; the Cartesian grid predicts the reattachment point at $x \approx 4.20h$, while the curvilinear grid anticipates it at $x \approx 3.81h$.

Overall, the agreement of the simulation result is acceptable, considering the simplicity of the model, which does not require the solution of an additional equation in the wall layer and the number of grid points employed in the tests. Moreover, in comparison to the simulations of [17], the new wall layer model on the Cartesian grid predicts the separation point better than their case with the same resolution, and on the curvilinear grid it is even closer to the reference data than the WMLES of [17], but the reattachment point in their case is predicted closer to the reference. The other statistics obtained from the current simulation, especially the mean vertical velocity profile, are in better agreement with the wall-resolved LES than the simulations of [17].

**Table 3.** Resolution, separation point ($x_s$) and reattachment point ($x_r$) in resolved LES Fröhlich et al. [45], the present optimized IBM simulations and the wall model LES by Chen et al. [17].

| Case | Resolution | $x_s$ | $x_r$ |
|---|---|---|---|
| Resolved LES (Fröhlich et al.) | $196 \times 128 \times 186$ | $0.224h$ | $4.69h$ |
| IBM Cartesian ($n_2 = 64$) | $96 \times 64 \times 32$ | $0.54h$ | $4.20h$ |
| IBM curvilinear ($n_2 = 64$) | $96 \times 64 \times 32$ | $0.47h$ | $3.81h$ |
| WMLES (TBLE/LES at $y^+ = 30$) | $96 \times 64 \times 32$ | $0.65h$ | $4.00h$ |
| WMLES (TBLE/LES at $y^+ = 15$) | $192 \times 72 \times 48$ | $0.50h$ | $4.42h$ |

## 5. Summary and Conclusions

In the present paper, a wall layer modeling for massive separation has been developed within the LES context. Two main frameworks have been considered; namely, the case where the solid wall is reproduced using a body fitted curvilinear grid and the case where the solid surface is simulated with immersed boundaries. The research aimed at developing a simple wall layer model to be used in conjunction with LES for engineering applications where the near-wall resolution is generally coarse. The starting point was the equilibrium wall layer model, which gives good results in attached flows [47]. In the case of body fitted curvilinear grids, the equilibrium wall layer model was improved using the dynamic $\kappa$ procedure of Cabot and Moin [8] in a new fashion. The dynamic $\kappa$ was applied on the log law without using TBLE in order to deviate the wall shear stress and the two components of strain rate tensor from the standard log law.

When the solid surface was simulated with immersed boundaries, a different strategy was accomplished. The starting point was the equilibrium wall layer model of Roman et al. [34], in which the velocity at the near-wall nodes was reconstructed from a farther point (the projection point) based on the law of the wall, and a RANS-like eddy viscosity was given at the IB node. To reconstruct the fluctuations in the wall normal direction well, especially near wall, random stochastic forcing was applied to the first three near-wall cells. Moreover, the model was complemented with a procedure designed at reproducing massive separation without the need to use zonal approaches. Specifically, after removing the advective terms from the boundary layer equation, the tangential streamwise velocity at the IB node ($u_{IB}$) was obtained from the pressure gradient instead of calculating it from shear velocity at the projection point based on the law of the wall: When the pressure gradient was unfavourable and the IB node was not in the log region, the tangential velocity was obtained through a balance between the wall normal diffusion term and the streamwise pressure gradient. In some sense, this represents a simplified procedure with respect to the zonal approach where the boundary layer equation is solved within the near-wall layer. In order to check what grid topology is best suited for reproducing this kind of flow, both Cartesian and curvilinear grids were considered in conjunction with immersed boundaries.

The models were tested and validated over two two-dimensional geometries, both being well studied in the literature through physical and high-resolution LESs: First the isolated hill (case 69 of the ERCOFTAC classic database) was considered; then, the periodic hill which is case 81 of ERCOFTAC classic datasbe was considered.

The model for body fitted geometry was generally in good agreement with the reference data; the starting point of separation was delayed with respect to the wall-resolved LES and the experiment, but the reattachment point was closer to the experiment, while the wall-resolved LES displayed an early reattachment. Applying 32 and 20 cells in the vertical direction correspondent to $\Delta y^+ \approx 37$ and 60, respectively, the results were comparatively acceptable. However, decreasing the vertical resolution further would delay the starting point of separation more. The model developed for immersed boundaries gave a mean velocity profile and other statistics in acceptable agreement with the reference data on both isolated and periodic hills, also in consideration of the coarseness of the grids employed; for the isolated hill, 40 cells in the vertical direction were required to cover the hill with four cells applying the Cartesian grid, which was correspondent to $\Delta y^+ \approx 30$; higher resolution was not used to avoid the near-wall computational node being located in the viscous sub-layer, and lower resolution did not capture the flow separation.

For the periodic hill, 64 cells in the vertical direction gave $\Delta y^+ \approx 105$ applying the Cartesian grid. Lower resolution did not display flow separation because of an excessive under-resolution of the hill geometry.

**Funding:** The research has been supported by project COSMO "CFD open source per opera morta", the grant number CUP n. J94C14000090006, law number 3865, PAR FSC 2007-2013, Friuli Venezia Giulia.

**Acknowledgments:** The author wholeheartedly thanks Vincenzo Armenio from Università degli Studi di Trieste, Italia, and Federico Roman from IEFLUIDS S.r.l., Università degli Studi di Trieste, Italia, for their effective collaboration to accomplish this work.

**Conflicts of Interest:** The author declares no conflict of interest.

## References


1. Piomelli, U.; Balaras, E. Wall-Layer Models For Large-Eddy Simulations. *Annu. Rev. Fluid Mech.* **2002**, *34*, 349–374. [CrossRef]
2. Piomelli, U. Wall-Layer Models For Large-Eddy Simulations. *Prog. Aerosp. Sci.* **2008**, *44*, 437–446. [CrossRef]
3. Bae, H.J.; Lozano-Durán, A.; Bose, S.T.; Moin, P. Dynamic slip wall model for large-eddy simulation. *J. Fluid Mech.* **2019**, *859*, 40-432. [CrossRef] [PubMed]
4. Deardorff, J.W. A Numerical Study of Three-Dimensional Turbulent Channel Flow at Large Reynolds Numbers. *J. Fluid Mech.* **1970**, *41*, 453–480. [CrossRef]
5. Schumann, U. Subgrid-Scale Model For Finite Difference Simulation of Turbulent Flows in Plane Channels And Annuli. *J. Comput. Phys.* **1975**, *18*, 376–404. [CrossRef]
6. Balaras, E.; Benocci, C. *Sub-Grid Scale Models in Finite-Difference Simulations of Complex Wall Bounded Flows*; AGARD CP: Neuilly-Sur-Seine, France, 1994; Volume 551, pp. 2.1–2.5.
7. Cabot, W.H. Near-Wall Models in Large-Eddy Simulations of Flow Behind a Backward-Facing Step. In *Annual Research Briefs*—+; Center for Turbulence Research, Stanford University: Stanford, CA, USA, 1996; pp. 199–210.
8. Cabot, W.; Moin, P. Approximate Wall Boundary Conditions in The Large-Eddy Simulation of High Reynolds Number Flows. *Flow. Turbul. Combust.* **2000**, *63*, 269–291. [CrossRef]
9. Spalart, P.R.; Jou, W.H.; Strelets, M.; Allmaras, S.R. *Comments On The Feasibility of LES For Wings and on a Hybrid RANS/LES Approach*; Advances in DNS/LES; Liu, C., Liu, Z., Eds.; Greyden Press: Columbus, OH, USA, 1997; pp. 137–148.
10. Mockett, C.; Fuchs, M.; Thiele, F. Progress in DES For Wall-Modelled LES of Complex Internal Flows. *Comput. Fluids* **2012**, *65*, 44–55. [CrossRef]
11. Keating, A.; Piomelli, U. A Dynamic Stochastic Forcing Method As A Wall-Layer Model For Large-Eddy Simulation. *J. Turbul.* **2006**, *7* 1–24. [CrossRef]
12. Temmerman, L.; Hadziabdic, M.; Leschziner, M.A.; Hanjalic, K. A Hybrid Two-Layer URANS-LES Approach For Large Eddy Simulation at High Reynolds Numbers. *Int. J. Heat Fluid Flow* **2005**, *26*, 173–190. [CrossRef]
13. Mittal, R.; Iaccarino, G. Immersed Boundary Methods. *Annu. Rev. Fluid Mech.* **2005**, *37*, 239–261. [CrossRef]
14. Roman, F.; Napoli, E.; Milici, B.; Armenio, V. An Improved Immersed Boundary Method For Curvilinear Grids. *J. Comput. Fluids* **2009**, *38*, 1510–1527. [CrossRef]
15. Tessicini, F.; Iaccarino, G.; Fatica, M.; Wang, M.; Verzicco, R. Wall Modelling For Large-Eddy Simulation Using an Immersed Boundary Method. In *Annual Research Briefs*; NASA Ames Research Center/Stanford University Center of Turbulence Research: Stanford, CA, USA, 2005; pp. 181–187.

16. Posa, A.; Balaras, E. Model-Based Near-Wall Reconstructions For Immersed-Boundary Methods. *J. Theor. Comput. Fluid Dyn.* **2014**, *28*, 473–483. [CrossRef]

17. Chen, Z.L.; Hickel, S.; Devesa, A.; Berl, J.; Adams, N.A. Wall Modelling For Implicit Large-Eddy Simulation and Immersed-Interface Methods. *J. Theor. Comput. Fluid Dyn.* **1994**, *28*, 1–21.

18. Roman, F.; Stipcich, G.; Armenio, V.; Inghilesi, R.; Corsini, S. Large Eddy Simulation of Mixing in Coastal Areas. *J. Heat Fluid Flow* **2010**, *31*, 327–341. [CrossRef]

19. Galea, A.; Grifoll, M.; Roman, F.; Mestres, M.; Armenio, V.; Sanchez-Arcilla, A.; Mangion, L.Z. Numerical Simulation of Water Mixing And Renewals In The Barcelona Harbour Area. *Environ. Fluid Mech.* **2014**, *14*, 1405–1425. [CrossRef]

20. Smagorinsky, J. General Circulation Experiments with the Primitive Equations. *Mon. Weather Rev.* **1963**, *91*, 99–164. [CrossRef]

21. Lilly, D.K. The representation of small-scale turbulence in numerical simulation experiments. In Proceedings of the IBM Scientific Computing Symposium on Environmental Sciences, New York, NY, USA, 14–16 November 1967.

22. Blazek, J. *Computational Fluid Dynamics: Principles and Applications*, 3rd ed.; Butterworth-Heinemann: Oxford, UK, 2015.

23. Stocca, V. Development of a Large Eddy Simulation Model for the Study of pOllutant Dispersion in Urban Areas. Ph.D. Thesis, University of Trieste, Trieste, Italy, 2010.

24. Germano, M.; Piomelli, U.; Moin, P.; Cabot, W.H. A dynamic subgrid-scale eddy viscosity model. *Phys. Fluids* **1991**, *3*, 1760–1765. [CrossRef]

25. Zang, J.; Street, R.L.; Koseff, J.R. A Non-Staggered Grid, Fractional Step Method For Time-Dependent Incompressible Navier-Stokes Equations in Curvilinear Coordinates. *J. Comput. Phys.* **1994**, *14*, 459–486. [CrossRef]

26. Armenio, V.; Piomelli, U. A Lagrangian Mixed Subgrid-Scale Model in Generalized Coordinates. *J. Flow Turbul. Combust.* **2000**, *65* , 51–81. [CrossRef]

27. Petronio, A.; Roman, F.; Nasello, C.; Armenio, V. Large Eddy Simulation Model For Wind-Driven Sea Circulation In Coastal Areas. *Nonlin. Process. Geophys.* **2013**, *20*, 1095–1112. [CrossRef]

28. Fakhari, A. Wall-Layer Modelling of Massive Separation in Large Eddy Simulation of Coastal Flows. Ph.D. Thesis, University of Trieste, Trieste, Italy, 2015.

29. Flow over Isolated 2D Hill. Available online: http://cfd.mace.manchester.ac.uk/cgi-bin/cfddb/prpage.cgi?69& EXP&database/cases/case69/Case_data&database/cases/case69&cas69_head.html&cas69_desc.html&cas69_ meth.html&cas69_data.html&cas69_refs.html&cas69_rsol.html&1&0&0&0&0/ (accessed on 22 November 2019).

30. Khurshudyan, L.H.; Snyder, W.H.; Nekrasov, I.V. *Flow and Dispersion Of Pollutants over Two-Dimensional Hills*; Environment Protection Agency Report No EPA-600/4-81-067; Environment Protection Agency: Research Triangle Park, NC, USA, 1981.

31. Chaudhari, A.; Vuorinen, V.; Hämäläinen, J.; Hellsten, A. Large-eddy simulations for hill terrains: Validation with wind-tunnel and field measurements, *J. Comput. Appl. Math.* **2018**, *37*, 2017–2038. [CrossRef]

32. Chaudhari, A. Large-Eddy Simulation of Wind Flows over Complex Terrains for Wind Energy Applications. Ph.D. Thesis, Lappeenranta University of Technology, Lappeenranta, Finland, 2014.

33. Vinokur, M. On One-Dimensional Stretching Functions For Finite-Difference Calculations. *J. Comput. Phys.* **1983**, *50*, 215–234. [CrossRef]

34. Roman, F.; Armenio, V.; Fröhlich, J. A Simple Wall-Layer Model For Large Eddy Simulation With Immersed Boundary Method. *J. Phys. Fluids* **2009**, *21*, 101701. [CrossRef]

35. Roman, F. A Numerical tool for Large Eddy Simulations in Environmental and Industrial Processes. Ph.D. Thesis, University of Trieste, Trieste, Italy, 2009.

36. Kim, J.; Kim, D.; Choi, H. An Immersed-Boundary Finite-Volume Method for Simulations of Flow In Complex Geometries. *J. Comput. Phys.* **2001**, *171*, 132–150. [CrossRef]

37. Meyer, M.; Devesa, A.; Hickel, S.; Hu, X.Y.; Adams, N.A. A Conservative Immersed Interface Method For Large-Eddy Simulation of Incompressible Flows. *J. Comput. Phys.* **2010**, *229*, 6300–6317. [CrossRef]

38. Rapaka, N.R.; Sarkar, S. An Immersed Boundary Method For Direct And Large Eddy Simulation of Stratified Flows in Complex Geometry. *J. Comput. Phys.* **2016**, *322*, 511–534. [CrossRef]

39. Hoyas, S.; Jiménez, J. Reynolds Number Effects On The Reynolds-Stress Budgets In Turbulent Channels. *J. Phys. Fluids* **2008**, *20*, 101511. [CrossRef]

40. Taylor, J.; Sarkar, S. Near-Wall Modelling For LES of an Oceanic Bottom Boundary Layer. In Proceedings of the Fifth International Symposium on Environmental Hydraulics, Tempe, AZ, USA, 4–7 December 2007.

41. Ferziger, J.H.; Peric, M. *Computational Methods For Fluid Dynamics*, 3rd ed.; Springer: Berlin, Germany, 2002.

42. Almeida, G.P.; Durao, D.F.G.; Heitor, M.V. Wake Flows Behind Two Dimensional Model Hills. *Exp. Therm. Fluid Sci.* **1992**, *7*, 87–101. [CrossRef]

43. Flow over Periodic Hills. Available online: http://cfd.mace.manchester.ac.uk/cgi-bin/cfddb/prpage.cgi?81&LES&database/cases/case81/Case_data&database/cases/case81&cas81_head.html&cas81_desc.html&cas81_meth.html&cas81_data.html&cas81_refs.html&cas81_rsol.html&1&1&1&1&1&unknown (accessed on 22 November 2019).

44. Temmerman, L.; Leschziner, A. Large Eddy Simulation of Separated Flow in a Streamwise Periodic Channel Construction. In Proceedings of the International Symposium on Turbulence and Shear Flow Phenomena, Stockholm, Sweden, 27–29 June 2001.

45. Fröhlich, J.; Mellen, C.P.; Rodi, W.; Temmerman, L.; Leschziner, M.A. Highly Resolved Large-Eddy Simulation of Separated Flow In a Channel In a Channel With Streamwise Periodic Constructions. *J. Fluid Mech.* **2005**, *526*, 19–66. [CrossRef]

46. LES: 2-D Periodic Hill. Available online: https://turbmodels.larc.nasa.gov/Other_LES_Data/2dhill_periodic.html (accessed on 22 November 2019).

47. Fakhari, A.; Cintolesi, C.; Petronio, A.; Roman, F.; Armenio, V. Numerical simulation of hot smoke plumes from funnels. In *Technology and Science for the Ships of the Future, Proceedings of NAV 2018: 19th International Conference on Ship & Maritime Research, Trieste, Italy, 20–22 June 2018*; IOS Press: Amsterdam, The Netherlands, 2018; pp. 238–245.

*Article*

# Breaking the Kolmogorov Barrier in Model Reduction of Fluid Flows

**Shady E. Ahmed and Omer San ***

School of Mechanical and Aerospace Engineering, Oklahoma State University, Stillwater, OK 74078, USA;
shady.ahmed@okstate.edu
* Correspondence: osan@okstate.edu; Tel.: +1-405-744-2457; Fax: +1-405-744-7873

Received: 15 January 2020; Accepted: 14 February 2020; Published: 18 February 2020

**Abstract:** Turbulence modeling has been always a challenge, given the degree of underlying spatial and temporal complexity. In this paper, we propose the use of a partitioned reduced order modeling (ROM) approach for efficient and effective approximation of turbulent flows. A piecewise linear subspace is tailored to capture the fine flow details in addition to the larger scales. We test the partitioned ROM for a decaying two-dimensional (2D) turbulent flow, known as 2D Kraichnan turbulence. The flow is initiated using an array of random vortices, corresponding to an arbitrary energy spectrum. We show that partitioning produces more accurate and stable results than standard ROM based on a global application of modal decomposition techniques. We also demonstrate the predictive capability of partitioned ROM through an energy spectrum analysis, where the recovered energy spectrum significantly converges to the full order model's statistics with increased partitioning. Although the proposed approach incurs increased memory requirements to store the local basis functions for each partition, we emphasize that it permits the construction of more compact ROMs (i.e., of smaller dimension) with comparable accuracy, which in turn significantly reduces the online computational burden. Therefore, we consider that partitioning acts as a converter which reduces the cost of online deployment at the expense of offline and memory costs. Finally, we investigate the application of closure modeling to account for the effects of modal truncation on ROM dynamics. We illustrate that closure techniques can help to stabilize the results in the inertial range, but over-stabilization might take place in the dissipative range.

**Keywords:** reduced order modeling; Kolmogorov $n$-width; Galerkin projection; proper orthogonal decomposition; turbulent flows

---

## 1. Introduction

Reduced order modeling (ROM) has been a hot topic in response to the increasing complexity of engineering systems and the requirement of near real-time and multi-query responses along with the limited advancement in computational resources [1–5]. This is vital for informed decision-making in the context of digital twins [6–14], model-based control [15–22], data assimilation [23–33], parameter estimation [34–37], and uncertainty quantification [38–44]. In fluid systems, this has been feasible thanks to the existence of a few underlying structures (e.g., vortices) that dominate the flow dynamics as well as the tremendous availability of data, which enables us to identify those structures (also called modes or basis functions) in a data-driven fashion [45–57]. However, most of the ROM developments have been applied mainly on prototypical flow problems. In particular, there is a limited number of investigations regarding ROMs in turbulent flows (e.g., see [58–64]). This is basically due to the excessive complexity of turbulent flows and existence of so many scales that interact with each other and control the mass, momentum, and energy transfer.

Challenges in the development of ROM for turbulent flows are also related to the Kolmogorov $n$-width for these systems. The Kolmogorov $n$-width is a concept from approximation theory that determines the linear reducibility of a system [65,66]. Mathematically, it is defined as [66–68]

$$d_n(\mathcal{M}) := \inf_{\mathcal{S}_n} \sup_{q \in \mathcal{M}} \|q - \Pi_{\mathcal{S}_n} q\|, \tag{1}$$

where $\mathcal{S}_n$ is a linear $n$-dimensional subspace, $\mathcal{M}$ is the solution manifold, and $\Pi_{\mathcal{S}_n}$ is the orthogonal projector onto $\mathcal{S}_n$. In other words, $d_n(\mathcal{M})$ quantifies the maximum possible error that might arise from the projection of solution manifold onto the best-possible $n$-dimensional linear subspace. When the decay of $d_n(\mathcal{M})$ with increasing $n$ is fast, this means that the energy cascade and modal hierarchy allow the reduction of the system and the solution trajectory can be adequately approximated to evolve on a reduced linear subspace. However, for systems with strong nonlinearity, like turbulent flows, the intense interactions between different modes reduce the decay rate of the Kolmogorov $n$-width. Consequently, the linear reducibility is hindered. In ROM context, this is denoted as the "Kolmogorov barrier" because it necessitates involving more modes in ROM to guarantee sufficient approximation of the underlying dynamics. In short, turbulent flows are generally characterized by a slow decay of the Kolmogorov $n$-width, which raises the Kolmogorov barrier and the present work aims at addressing this issue.

In particular, we propose the use of a time domain partitioning approach, called principal interval decomposition (PID) [69–73] to break the Kolmogorov barrier of turbulent flows. PID has been investigated in previous studies and has been proven to be more accurate than standard ROMs based on the global basis construction (see [71,73] for example). Here, we aim at addressing a more complicated and involved flow problem characterizing two-dimensional turbulence. This allows us to investigate the effect of PID to capture the scales responsible for energy transfer and dissipation through an energy spectrum analysis. In addition, we relate the notion of the Kolmogorov barrier to the well-defined concept of relative information content in ROM context. Although turbulence is, by nature, three-dimensional, we investigate it in a simplified two-dimensional (2D) setting. It was shown that many aspects of idealized 2D turbulence are relevant for physical systems like geostrophic turbulence [74] and large-scale motions in the atmosphere and oceans [75,76]. The study of 2D turbulence idealizes geophysical phenomena and provides a starting point for their modeling. Specifically, we study ROM application on 2D Kraichnan turbulence. Moreover, we present a spectral study of the energy spectrum to emphasize the efficiency of PID approach for ROM in turbulence. Finally, we investigate the possibility of applying closure models for ROM stabilization. As a standard ROM technique, we employ proper orthogonal decomposition (POD) [46] for basis construction and Galerkin projection for dynamics approximation [58,77].

Fluid flow systems often include some sort of nonlinear interactions (e.g., corresponding to the convective effects). This implies that different modes (including the truncated ones) experience strong dynamical interactions between each others. Therefore, many modes need to be included for certain flows (especially for complex geometries and/or high Reynolds numbers). For example, Zhang and Stevens [78] stated that 2000 modes are necessary to capture 80% of the total energy in a neutral atmospheric boundary layer (ABL) process, while Shah and Bou-Zeid [79] found that about 500 POD modes are required to capture the same amount of energy by using a two-dimensional snapshot POD. Nonetheless, the computational cost of a Galerkin ROM usually scales with $R^3$ in case of quadratic nonlinearity, where $R$ is the number of retained/resolved modes. This necessitates the use of only a few modes for feasible ROM computations, especially for real-time control applications or multiple-query estimations. Other than the accuracy loss that would be incurred due to modal truncation, instabilities can easily occur. For instance, Östh et al. [80] showed a case for a flow around Ahmed body where the solution of a POD Galerkin model with 100 (or less) modes converges to infinity. Instabilities in standard ROMs have been attributed mainly to the truncation of higher modes, but Grimberg et al. [81] recently demonstrated that these instabilities are largely due to the standard Galerkin projection where the *left* and *right* set of basis functions match. To mitigate the instabilities encountered in Galerkin projection-based ROMs, Carlberg et al. [82] showed that least-squares Petrov–Galerkin can indeed build more accurate ROMs while Galerkin ROM becomes unstable for long time intervals. Alternatively, closure and stabilization models represent a feasible choice in the accuracy/stability-cost trade-off for Galerkin ROMs. Closure models have been popular in turbulence community, and found their way into ROM for fluid flows as well. Therefore, we also investigate the effect of a linear mode dependent eddy viscosity type closure model to stabilize the solution of ROM in case of 2D Kraichnan turbulence.

## 2. Two-Dimensional Kraichnan Turbulence

The two-dimensional (2D) turbulence cannot be realized in practice or experiments, but only in numerical simulations. Furthermore, 2D turbulence behaves differently from realistic 3D turbulence. In 3D turbulence, energy is transferred forward, from large scales to smaller scales, via vortex stretching. Conversely, in 2D turbulence, most of the forcing and dissipation energy will be transferred from smaller scales to larger scales and follows the Kraichnan–Batchelor–Leith (KBL) theory [83–85]. Despite the apparent simplicity in dealing with two spatial dimensions, 2D turbulence is very rich in its dynamics due to the inverse energy and forward enstrophy cascading mechanisms, which makes it a feasible testbed for a study of ROM in turbulence. Nonetheless, readers should be cautious while trying to extend the investigated framework into more realistic 3D turbulence cases.

The decaying homogeneous isotropic 2D Kraichnan turbulence follows the 2D Navier–Stokes equations, which can be written in vorticity-streamfunction formulation as below,

$$\frac{\partial \omega}{\partial t} + J(\omega, \psi) = \frac{1}{\text{Re}} \nabla^2 \omega, \tag{2}$$

where $\omega$ is the vorticity and $\psi$ is the streamfunction. Re is Reynolds number relating the inertial and viscous effects. $J(\omega, \psi)$ and $\nabla^2 \omega$ are the Jacobian and Laplacian operators, respectively, which can be defined as

$$J(\omega, \psi) = \frac{\partial \omega}{\partial x} \frac{\partial \psi}{\partial y} - \frac{\partial \omega}{\partial y} \frac{\partial \psi}{\partial x}, \tag{3}$$

$$\nabla^2 \omega = \frac{\partial^2 \omega}{\partial x^2} + \frac{\partial^2 \omega}{\partial y^2}. \tag{4}$$

The vorticity-streamfunction formulation enforces the incompressibility condition, where the vorticity and streamfunction fields are related by the following Poisson equation,

$$\nabla^2 \psi = -\omega. \tag{5}$$

According to the KBL theory for 2D turbulence, the inertial range in the energy spectrum is proportional to $k^{-3}$ in the inviscid limit. In our numerical experiments, the initial energy spectrum in Fourier space is given by

$$E(k) = \frac{4k^4}{3\sqrt{\pi}k_p^5} \exp\left[-\left(\frac{k}{k_p}\right)^2\right], \tag{6}$$

where $k = \sqrt{k_x^2 + k_y^2}$ and $k_p$ is the wavenumber at which the maximum value of initial energy spectrum occurs. During the time evolution process, due to the nonlinear interactions, this spectrum quickly approaches toward $k^{-3}$ spectrum as we will show later when we present our results. The magnitude of the vorticity Fourier coefficients is related to the energy spectrum as

$$|\tilde{\omega}(k)| = \sqrt{\frac{k}{\pi} E(k)}. \tag{7}$$

Then, the initial vorticity distribution (in Fourier space) is obtained by introducing a random phase. More details regarding derivation of the initial vorticity distribution from an assumed energy spectrum can be found in [86]. In the present study, we use a spatial computational domain of $(x, y) \in [0, 2\pi] \times [0, 2\pi]$ and a time domain of $t \in [0, 4]$. Periodic boundary conditions are applied in both $x$- and $y$-directions. A spatial grid of $1024^2$ and a timestep of $\Delta t = 0.001$ are used. A fourth-order accurate Arakawa scheme [87] is used for spatial discretization and a third-order total variation diminishing Runge–Kutta scheme (TVD-RK3) [88] is adopted for temporal discretization.

The energy spectrum for different values of Reynolds number is given in Figure 1 at $t = 2$ and $t = 4$. We can observe that the energy spectrum in the inertial range flattens toward the classical $k^{-3}$ scaling limit as Re increases, in agreement with the KBL theory for two-dimensional turbulence. It should be noted that the same initial vorticity distribution is enforced for all numerical experiments, by fixing the seed of the random numbers generator. In order to verify the validity of the used spatial resolution, Figure 2 shows the energy spectrum for Re = 16,000 (left) and Re = 32,000 (right) using two spatial resolutions (i.e., $1024^2$ and $2048^2$). A good match between the energy spectra corresponding to the two spatial resolutions can be seen, even for Re = 32,000. Nonetheless, we will consider Re = 16,000 with a spatial grid of $1024^2$ for the rest of the present study. In addition, results obtained from solving Equation (2) using the aforementioned numerical schemes and resolution will be denoted as the full order model (FOM) results.

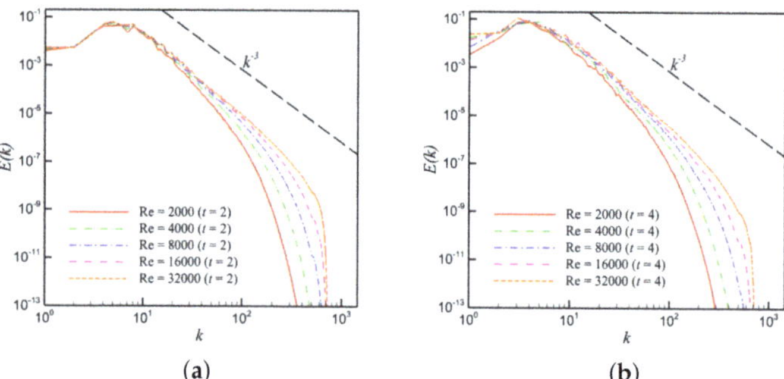

**Figure 1.** Evolution of energy spectra in 2D decaying Kraichnan turbulence for Re $\in \{2000, 4000, 8000, 16,000, 32,000 \}$ at $t = 2$ (**a**) and $t = 4$ (**b**). Notice the convergence of energy spectrum in the inertial range to the $k^{-3}$ scaling with increasing Reynolds number.

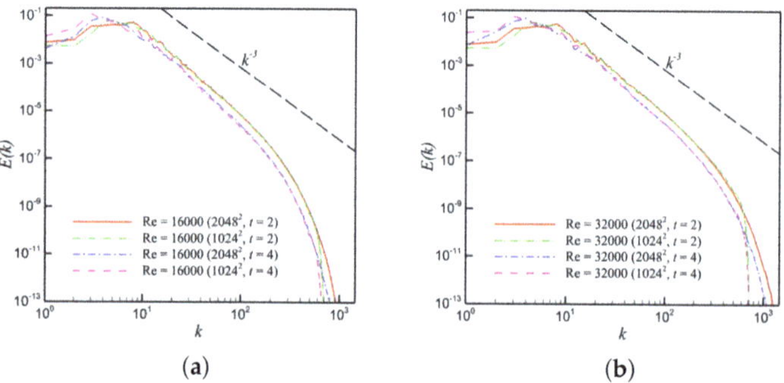

**Figure 2.** A mesh independence test for Re = 16,000 (**a**) and Re = 32,000 (**b**) using two spatial resolutions. An acceptable match is obtained from both resolutions.

## 3. Reduced Order Modeling

In order to build a Galerkin projection-based reduced order model (ROM) of a physical system, the system's state, $\mathbf{u}(\mathbf{x}, t)$, is assumed to live in a reduced trial subspace. Then, the high-dimensional operators are projected onto the same subspace, constraining the residual to be orthogonal to that subspace. Therefore, generating effective reduced subspaces that sufficiently contain the solution manifold is vital for the successful construction of Galerkin ROMs. One of the very early-developed and well-known approaches to build reduced subspaces is Fourier analysis. However, it assumes universal basis functions (or modes) which have no specific relation to the system in hand, except for boundary conditions. More efficiently, snapshot-based model reduction techniques can tailor a reduced subspace that best fits the problem by extracting the underlying coherent structures that control the major dynamical evolution we are interested in. Among those snapshot-based techniques, proper orthogonal decomposition (POD) is the most popular and well-established approach extracting the modes which contribute the most to the total system's energy [4,89]. The basic implementation of POD is presented in the following subsection.

### 3.1. Proper Orthogonal Decomposition

As mentioned before, proper orthogonal decomposition (POD) is a snapshot-based approach. Therefore, a collection of system's realizations (called snapshots) are required in advance. Assuming that a number of snapshots ($N_s$) of the 2D vorticity field, denoted as $w(x, y, t_n)$, are stored at consecutive times $t_n$ for $n = 1, 2, \ldots, N_s$, the time-averaged field, called "base flow", can be computed as

$$\bar{w}(x, y) = \frac{1}{N} \sum_{n=1}^{N} w(x, y, t_n). \tag{8}$$

The mean-subtracted snapshots, also called anomaly or fluctuation fields, are then computed as the difference between the instantaneous fields and the mean field,

$$w'(x, y, t_n) = w(x, y, t_n) - \bar{w}(x, y). \tag{9}$$

This mean-subtraction is common in ROM community, and it guarantees that ROM solution would satisfy the same boundary conditions as a full order model [90]. This anomaly field procedure is also interpreted as a mapping of snapshot data to its origin (i.e., centering), and will be used for constructing the POD basis functions. The POD approach basically relies on an eigenvalue decomposition of a snapshot correlation (or covariance) matrix. Since the number of snapshots ($O(10^2)$) is usually much less than the number of grid points ($O(10^6)$), we follow the method of snapshots [54] for efficient implementation of the POD algorithm.

An $N_s \times N_s$ temporal correlation data matrix $\mathbf{A} = [a_{ij}]$ is computed from the inner product of temporal anomaly fields as

$$a_{ij} = \langle w'(x, y, t_i); w'(x, y, t_j) \rangle, \tag{10}$$

where the angle-parenthesis denotes the inner product in Hilbert space defined as

$$\langle w'(x, y, t_i); w'(x, y, t_j) \rangle = \int_{\Omega} w'(x, y, t_i) w'(x, y, t_j) dx dy, \tag{11}$$

where $\Omega$ represent the physical (spatial) domain. Then, an eigenvalue decomposition of $\mathbf{A}$ is performed as

$$\mathbf{A}\mathbf{V} = \mathbf{V}\mathbf{\Lambda}, \tag{12}$$

where $\mathbf{\Lambda}$ is a diagonal matrix whose entries are the eigenvalues $\lambda_k$ of $\mathbf{A}$, and $\mathbf{V}$ is a matrix whose columns $\mathbf{v}_k$ are the corresponding eigenvectors. $\mathbf{A}$ is a Gramian matrix, a property which can be utilized for efficient solution of the eigenvalue problem (e.g., using Jacobi transformations [91]).

The eigenvalues $\lambda_k$ represent the amount of information contained in each mode; therefore, they should be arranged in a descending order (i.e., $\lambda_1 \geq \lambda_2 \geq \cdots \geq \lambda_N$), for proper selection of the POD modes. The POD modes $\phi_k$ are finally computed as

$$\phi_k(x,y) = \frac{1}{\sqrt{\lambda_k}} \sum_{n=1}^{N} v_k^n \omega'(x,y,t_n), \tag{13}$$

where $v_k^n$ is the $n$-th component of the eigenvector $\mathbf{v}_k$. It can be easily verified that the POD modes are orthonormal (i.e., $\langle \phi_i; \phi_j \rangle = \delta_{ij}$), which simplifies the derivation of ROM.

*3.2. Galerkin Projection*

With the POD algorithm implemented, a set of POD basis functions are obtained $\{\phi_k(x,y)\}_{k=1}^{R}$, where $R$ is the number of POD modes retained in ROM (i.e., the size of the reduced subspace). The solution is therefore approximated as a superposition of the contributions of these selected modes as follows:

$$w(x,y,t) = \bar{\omega}(x,y) + \sum_{k=1}^{R} \alpha_k(t)\phi_k(x,y), \tag{14}$$

where $\alpha_k(t)$ are the so-called temporal modal coefficients. Since the vorticity and streamfunction are related by the kinematic relationship given by Equation (5), the basis functions ($\theta_k(x,y)$) and mean field ($\bar{\psi}(x,y)$) corresponding to the streamfunction can be obtained from those of the vorticity as follows:

$$\nabla^2 \bar{\psi}(x,y) = -\bar{\omega}(x,y), \tag{15}$$

$$\nabla^2 \theta_k(x,y) = -\phi_k(x,y), \quad k = 1,2,\ldots,R, \tag{16}$$

which might result in a set of basis functions for the streamfunction that are not necessarily orthonormal. Moreover, the reduced order approximation of streamfunction can share the same temporal coefficients $\alpha_k(t)$,

$$\psi(x,y,t) = \bar{\psi}(x,y) + \sum_{k=1}^{R} \alpha_k(t)\theta_k(x,y). \tag{17}$$

Those approximations are then substituted in Equation (2), as follows:

$$\frac{\partial}{\partial t}\left[\bar{\omega}(x,y) + \sum_{k=1}^{R} \alpha_k(t)\phi_k(x,y)\right] = -J\left(\left[\bar{\omega}(x,y) + \sum_{k=1}^{R} \alpha_k(t)\phi_k(x,y)\right], \left[\bar{\psi}(x,y) + \sum_{k=1}^{R} \alpha_k(t)\theta_k(x,y)\right]\right)$$

$$+ \frac{1}{Re}\nabla^2\left[\bar{\omega}(x,y) + \sum_{k=1}^{R} \alpha_k(t)\phi_k(x,y)\right].$$

Then, we use the definition of Jacobian operator and expand the above expression as

$$\frac{\partial}{\partial t}\bar{\omega}(x,y) + \frac{\partial}{\partial t}\left[\sum_{k=1}^{R} \alpha_k(t)\phi_k(x,y)\right] = -J\left(\bar{\omega}(x,y), \bar{\psi}(x,y)\right) - J\left(\sum_{k=1}^{R} \alpha_k(t)\phi_k(x,y), \sum_{k=1}^{R} \alpha_k(t)\theta_k(x,y)\right)$$

$$- J\left(\bar{\omega}(x,y), \sum_{k=1}^{R} \alpha_k(t)\theta_k(x,y)\right) - J\left(\sum_{k=1}^{R} \alpha_k(t)\phi_k(x,y), \bar{\psi}(x,y)\right)$$

$$+ \frac{1}{Re}\nabla^2\bar{\omega}(x,y) + \frac{1}{Re}\nabla^2\left[\sum_{k=1}^{R} \alpha_k(t)\phi_k(x,y)\right].$$

Note that the first term $\frac{\partial}{\partial t}\bar{\omega}(x,y)$ is zero. In addition, the Jacobian and Laplacian operators are spatial functions; therefore, the following can be written (after change of indices)

$$\frac{\partial}{\partial t}\left[\sum_{i=1}^{R}\alpha_i(t)\phi_i(x,y)\right] = -J\left(\bar{\omega}(x,y),\bar{\psi}(x,y)\right) - \sum_{i=1}^{R}\sum_{j=1}^{R}\alpha_i(t)\alpha_j(t)J\left(\phi_i(x,y),\theta_j(x,y)\right)$$

$$-\sum_{i=1}^{R}\alpha_i(t)J\left(\bar{\omega}(x,y),\theta_i(x,y)\right) - \sum_{i=1}^{R}\alpha_i(t)J\left(\phi_i(x,y),\bar{\psi}(x,y)\right)$$

$$+\frac{1}{Re}\nabla^2\bar{\omega}(x,y) + \sum_{i=1}^{R}\alpha_i(t)\frac{1}{Re}\nabla^2\phi_i(x,y).$$

Then, we can collect the constant, linear, and nonlinear terms (in time) as follows:

$$\frac{\partial}{\partial t}\left[\sum_{i=1}^{R}\alpha_i(t)\phi_i(x,y)\right] = \left[-J\left(\bar{\omega}(x,y),\bar{\psi}(x,y)\right) + \frac{1}{Re}\nabla^2\bar{\omega}(x,y)\right]$$

$$+\sum_{i=1}^{R}\alpha_i(t)\left[-J\left(\bar{\omega}(x,y),\theta_i(x,y)\right) - J\left(\phi_i(x,y),\bar{\psi}(x,y)\right) + \frac{1}{Re}\nabla^2\phi_i(x,y)\right]$$

$$-\sum_{i=1}^{R}\sum_{j=1}^{R}\alpha_i(t)\alpha_j(t)J\left(\phi_i(x,y),\theta_j(x,y)\right).$$

Finally, an inner product with an arbitrary basis function $\phi_k(x,y)$ is performed, where we make use of the orthonormality of the vorticity basis functions. Therefore, a reduced order model is obtained, which can be simply represented by the following set of ODEs for $\alpha_k(t)$,

$$\frac{d\alpha_k}{dt} = \mathfrak{B}_k + \sum_{i=1}^{R}\mathfrak{L}_{i,k}\alpha_i + \sum_{i=1}^{R}\sum_{j=1}^{R}\mathfrak{N}_{i,j,k}\alpha_i\alpha_j, \quad k = 1,2,...,R, \tag{18}$$

where the predetermined model coefficients can be computed by the following numerical integration (offline computing)

$$\mathfrak{B}_k = \left\langle -J(\bar{\omega},\bar{\psi}) + \frac{1}{Re}\nabla^2\bar{\omega}; \phi_k\right\rangle,$$

$$\mathfrak{L}_{i,k} = \left\langle -J(\bar{\omega},\theta_i) - J(\phi_i,\bar{\psi}) + \frac{1}{Re}\nabla^2\phi_i; \phi_k\right\rangle,$$

$$\mathfrak{N}_{i,j,k} = \left\langle -J(\phi_i,\theta_j); \phi_k\right\rangle. \tag{19}$$

To complete the dynamical system given by Equation (18), the initial conditions for $\alpha_k(t)$ may be obtained by the following projection of the initial vorticity field onto the respective basis functions,

$$\alpha_k(t_0) = \left\langle \omega(x,y,t_0) - \bar{\omega}(x,y); \phi_k\right\rangle, \tag{20}$$

where $\omega(x,y,t_0)$ is the vorticity field specified at initial time $t_0$.

We note that the approach described here is called tensorial ROM, since the model predetermined terms (especially those corresponding to nonlinear terms) are computed in advance as tensors during an offline stage, while the online solution of Equation (18) scales with $R^3$. Alternatively, the model terms can be calculated online while minimizing the cost of computing the nonlinear terms through approximating approaches like discrete empirical interpolation method (DEIM) [92,93], gappy POD [60,94], and missing point estimation (MPE) [95,96]. An overview of these techniques can be found in [97,98].

## 4. Partitioned ROM

A representative solution subspace is a key factor in the construction of ROMs of physical systems and POD has been successfully offering a systematic approach for building such a subspace. However, POD in principle depends on a global approximation of the snapshot data. This has several consequences, a few of which are mentioned here. First, local temporary excursions in state space which contain small amounts of energy can be overlooked by POD because their contribution to the total energy may be negligible. These excursions can, however, be of interest and have significant impact on the dynamical evolution (see [99], for example). Second, the global nature of POD approach can result in overall deformation of the obtained modes for systems with fast variations in state. As a result, the constructed POD modes are smoothed out and cannot capture any dominant structure at all. This in turn causes an energy distribution across a large number of modes; therefore, the truncated modes contain a significant amount of system's energy, which causes inaccuracies and instabilities in the solution unless the size of the ROM is increased.

An alternative approach which preserves the optimality of POD, while giving care to the system's local characteristics, is called principal interval decomposition (PID) [69–73]. In PID, the time domain is divided into a few partitions, each characterizing a specific stage in the system's dynamics and evolution. Then, the same POD algorithm is applied within each partition to generate a set of basis functions that best fit the respective partition *locally*. In a sense, the PID approach can be viewed as decomposing the global POD subspace into a few of locally optimal subspaces. As a result, accurate partitioned ROMs with smaller sizes can be dedicated to each individual sub-interval. More details about the properties of PID can be found in [73]. This partitioning idea can also be performed in the physical domain, state space, or parameter space [64,100–102].

In brief, the time domain $\mathcal{T} = [t_0, t_f]$ is divided into a number $P$ of non-overlapping partitions (every two consecutive partitions share only the interface). Adaptive partitioning and clustering techniques can be used to tailor partitions with arbitrary widths if no a priori information is available about the dynamics of the systems. However, for simplicity and clarity of presentation, we assume equidistant decomposition of the time domain. In other words, the total number of snapshots $N_s$ is divided into $N_s/P$ sets. After that, local mean fields $\bar{\omega}^{(l)}(x,y)$ and $\bar{\psi}^{(l)}(x,y)$ are computed and the POD algorithm described in Section 3.1 is applied to each group of snapshots to generate local basis functions $\{\phi_k^{(l)}(x,y)\}_{k=1}^{R}$ and $\{\theta_k^{(l)}(x,y)\}_{k=1}^{R}$ for $l = 1, 2, \ldots, P$. Then, the Galerkin ROM equations are obtained for each interval as

$$\frac{d\alpha_k^{(l)}}{dt} = \mathfrak{B}_k^{(l)} + \sum_{i=1}^{R} \mathfrak{L}_{i,k}^{(l)}\alpha_i^{(l)} + \sum_{i=1}^{R}\sum_{j=1}^{R} \mathfrak{N}_{i,j,k}^{(l)}\alpha_i^{(l)}\alpha_j^{(l)}, \quad k = 1, 2, \ldots, R, \quad l = 1, 2, \ldots, P, \tag{21}$$

where the predetermined model coefficients can be computed partition-wise by the following numerical integrations:

$$\mathfrak{B}_k^{(l)} = \left\langle -J(\bar{\omega}^{(l)}, \bar{\psi}^{(l)}) + \frac{1}{Re}\nabla^2\bar{\omega}^{(l)}; \phi_k^{(l)}\right\rangle,$$

$$\mathfrak{L}_{i,k}^{(l)} = \left\langle -J(\bar{\omega}^{(l)}, \theta_i^{(l)}) - J(\phi_i^{(l)}, \bar{\psi}^{(l)}) + \frac{1}{Re}\nabla^2\phi_i^{(l)}; \phi_k^{(l)}\right\rangle,$$

$$\mathfrak{N}_{i,j,k}^{(l)} = \left\langle -J(\phi_i^{(l)}, \theta_j^{(l)}); \phi_k^{(l)}\right\rangle, \tag{22}$$

and the initial conditions for $\alpha_k^{(1)}(t)$ are obtained by the following projection:

$$\alpha_k^{(1)}(t_0) = \left\langle \omega(x,y,t_0) - \bar{\omega}^{(1)}(x,y); \phi_k^{(1)}\right\rangle. \tag{23}$$

It remains to enforce the interface constraints to enable solution transition from one partition to the next, if no access to the true field at the interface instant is available. We do so by imposing the following condition at the interface [103],

$$\langle \omega^{(l)}(x,y,t_l) - \omega^{(l+1)}(x,y,t_l); \phi_k^{(l+1)} \rangle = 0, \tag{24}$$

where $t_l$ represents the time instant at the interface between the interval $l$ and $l+1$. Here, $\omega^{(l)}(x,y,t_l)$ is the reconstructed vorticity field at the interface using the left interval $(l)$, while $\omega^{(l+1)}(x,y,t_l)$ is reconstructed at the same instant, but using the right interval $(l+1)$. Those can be rewritten as

$$\omega^{(l)}(x,y,t_l) = \bar{\omega}^{(l)}(x,y) + \sum_{k=1}^{R} \alpha_k^{(l)}(t_l)\phi_k^{(l)}(x,y), \tag{25}$$

$$\omega^{(l+1)}(x,y,t_l) = \bar{\omega}^{(l+1)}(x,y) + \sum_{k=1}^{R} \alpha_k^{(l+1)}(t_l)\phi_k^{(l+1)}(x,y). \tag{26}$$

This reduces to performing the following computation at the interface to re-initiate our solver at the first timestep of the new time interval,

$$\alpha_k^{(l+1)}(t_l) = \langle \sum_{i=1}^{R} \alpha_i^{(l)}(t_l)\phi_i^{(l)}(x,y) + \bar{\omega}^{(l)}(x,y) - \bar{\omega}^{(l+1)}(x,y); \phi_k^{(l+1)}(x,y) \rangle. \tag{27}$$

## 5. Results

As indicated in Section 2, the full order model is solved over a square domain with a side length of $2\pi$ using a spatial grid of $1024^2$ for $t \in [0,4]$ with a timestep of 0.001. These were shown to be sufficient resolutions to solve the problem at a Reynolds number of $16,000$. To build the reduced order subspace using the POD and PID algorithms, described in Sections 3.1 and 4, 800 snapshots of vorticity field were collected (i.e., every five timesteps). It was also shown that the sampling rate has minor effect on POD-based modal decomposition, as long as the Nyquist–Shannon sampling criterion is met [104]. Another factor we consider in our choice of sampling rate is to have adequate number of snapshots in each sub-interval (after partitioning the time domain).

The partitioned ROM approach is used to solve Equation (21) using different numbers of partitions. In particular, we present the results for $P \in \{1,4,8\}$. Note that, for $P = 1$, the approach reduces to the standard Galerkin-POD ROM with a single interval covering the whole time domain. In Figure 3, the energy spectrum at $t = 4$ is used to illustrate the effect of increasing the number of partitions on ROM accuracy and convergence to the FOM solution. This is shown for $P = 1$ (top), $P = 4$ (middle), and $P = 8$ (bottom), along with a compensated (for zooming-in) view of the energy spectrum at the inertial range on the right. We can easily observe that increasing the number of partitions improves the convergence of ROM energy spectrum to the true one for all values of $R$.

Although increasing the number of retained modes generally improves the ROM accuracy, this does not help a lot in the case with $P = 1$. This is shown in Figure 3, where we still see a large deviation of ROM results in the inertial range even with 32 modes. This might be attributed to the smoothing-out of constructed basis functions in such a way that information is distributed across a large number of modes. Consequently, the effect of modal truncation on system's dynamics is no more negligible. On the other hand, a partitioned ROM with $P = 8$ enables us to build accurate ROMs with relatively compact sizes (e.g., $R = 16$).

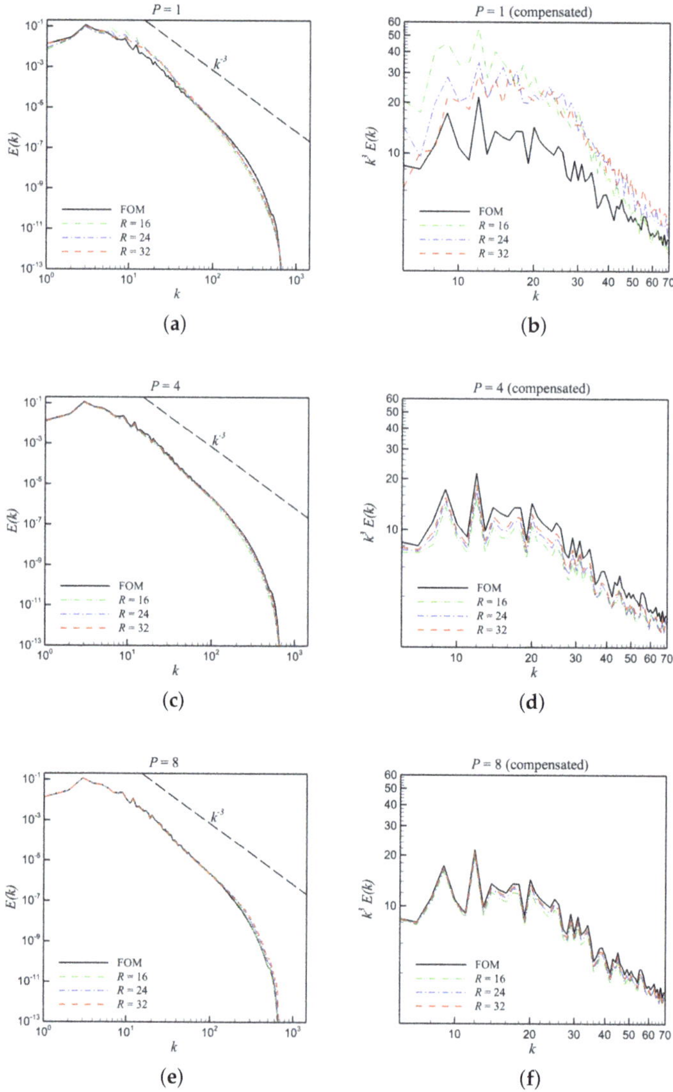

**Figure 3.** Energy spectra obtained from partitioned ROM with $P = 1$ (**a,b**), $P = 4$ (**c,d**), and $P = 8$ (**e,f**). Note that $P = 1$ corresponds to the standard Galerkin-based reduced order model (ROM) with global proper orthogonal decomposition (POD) modes.

To provide more insights about the scales resolved with partitioned ROM, we plot the evolution of the temporal coefficients of the first and eighth modes in Figure 4. We can notice that a ROM with global POD modes (i.e., $P = 1$) can adequately capture the evolution of the first mode's contribution (with increasing deviations toward the end of the time domain). On the other hand, a phase shift and amplification of the temporal evolution of the $8^{th}$ modal coefficient are spotted after a small initial period. From a physical point of view, the first modes correspond to the large dominating scales, while the last ones correspond to the small (or fine) scales. This indicates that a global POD application provides ROMs that moderately capture the large scales, but can fail to capture the finer details. Moreover, we can deduce from Figure 4a that the dynamics at different time instants is different (where all the snapshots are projected onto the same set of basis functions). This justifies the use of paritioned ROM to tailor unique bases for each sub-interval. On the other hand, with partitioned ROMs (i.e., $P = 4$ and $P = 8$), the evolution of both $\alpha_1$ and $\alpha_8$ is tracked accurately, implying that the presented partitioning approach improves the ROM's capabilities to represent the fine scales. Moreover, the accuracy of $\alpha_1$ prediction is improved compared to $P = 1$. Although the evolution of temporal coefficients in different sub-intervals looks similar, this is because the fields in each region are projected onto their *respective* bases. Therefore, once reconstructed, the dynamics in the full order space will be different. In addition, we can note the "jump" at the interfaces (e.g., at $t \in \{1, 2, 3, 4\}$ for $P = 4$) corresponding to updating the solution manifold. The true modal coefficients (denoted as FOM) is obtained from projecting the vorticity field at different times on the corresponding basis function as

$$\alpha_k^{(l)}(t) = \left\langle \omega(x, y, t) - \bar{\omega}^{(l)}(x, y); \phi_k^{(l)}(x, y) \right\rangle. \tag{28}$$

More visualizations of ROM's predictability are provided in Figures 5 and 6, showing the vorticity (in color) and streamfunction (in height) at the final time (i.e., $t = 4$), compared to the FOM results. In Figure 5, we illustrate that ROM with $P = 1$ fails to correctly predict the final fields at different values of $R$. At $R = 16$, even the large scales are not captured accurately (recall the larger deviation of $\alpha_1$ at final time from Figure 4). Conversely, both the large fine scales are predicted reliably using the partitioned ROM approach as shown in Figure 6 with $R = 16$.

In order to obtain more insights about the scales resolved by partitioned ROM, compared to standard POD-based ROM, we compare PID ROM results for $P = 8$ and $R = 16$, with standard ROM (i.e., $P = 1$) and increased number of modes. The root-mean-squares errors between Galerkin ROMs and FOM are shown in Table 1. We find that at least around $R = 128$ modes are needed for standard Galerkin ROM to achieve similar accuracy as partitioned ROM, which suggests that the relevant scales represented in those global 128 modes are almost encapsulated in 16 local modes if eight partitions are used. The computational gain of the PID approach can be easily seen from this analysis since online computational cost scales with $O(R^3)$.

**Table 1.** Root-mean-squares error at time $t = 4$ between Galerkin ROM (without any additional stabilization $\kappa = 0$) and FOM at Re = 16,000.

| Model | $\|\psi^{ROM} - \psi^{FOM}\|$ | $\|\omega^{ROM} - \omega^{FOM}\|$ |
|---|---|---|
| $P = 8$ $(R = 16)$ | $7.47 \times 10^{-3}$ | 1.2774 |
| $P = 1$ $(R = 16)$ | $1.14 \times 10^{-1}$ | 9.7322 |
| $P = 1$ $(R = 24)$ | $7.43 \times 10^{-2}$ | 8.8779 |
| $P = 1$ $(R = 32)$ | $4.48 \times 10^{-2}$ | 7.6098 |
| $P = 1$ $(R = 48)$ | $1.77 \times 10^{-2}$ | 5.2468 |
| $P = 1$ $(R = 64)$ | $1.12 \times 10^{-2}$ | 4.1695 |
| $P = 1$ $(R = 96)$ | $5.99 \times 10^{-3}$ | 2.7428 |
| $P = 1$ $(R = 128)$ | $3.86 \times 10^{-3}$ | 1.9213 |

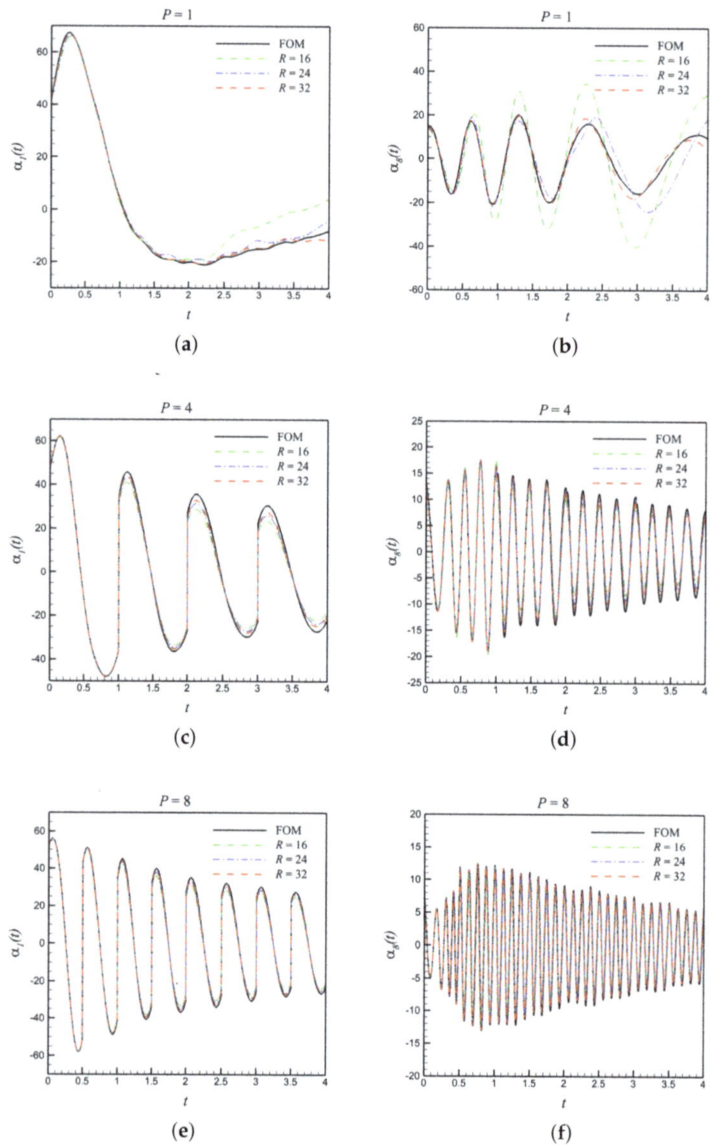

**Figure 4.** Temporal evolution of the first and eighth modal coefficients using different numbers of partitions, compared to the true values obtained from the projection of FOM vorticity field on the corresponding basis function. (**a**) shows the first modal coefficient with $P = 1$, while the eighth is given in (**b**). Results for $P = 4$ are shown in (**c**,**d**), while those for $P = 8$ are given in (**e**,**f**).

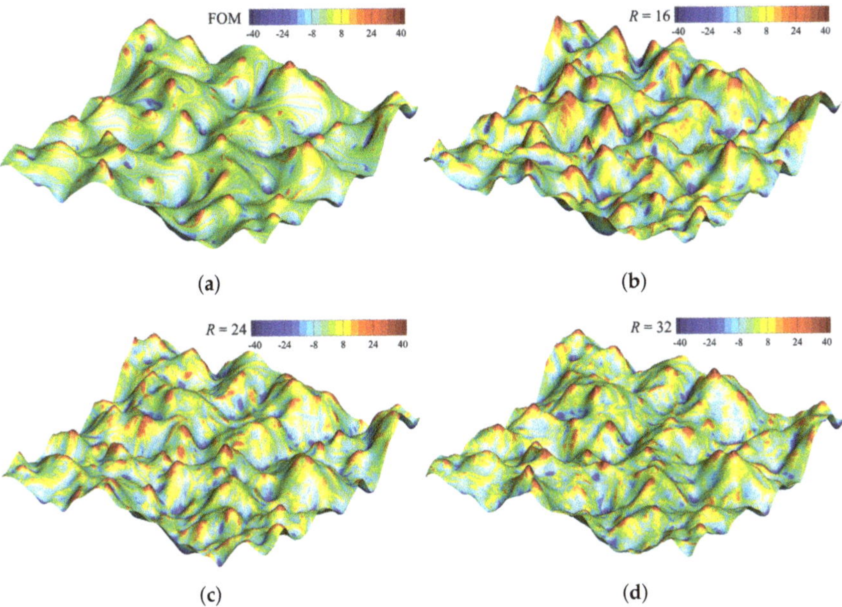

**Figure 5.** Predicted results at time $t = 4$ showing vorticity contours in color with the level of streamfunction as height. (**a**) Full order model (FOM), (**b**) ROM ($R = 16$) with $P = 1$, (**c**) ROM ($R = 24$) with $P = 1$, and (**d**) ROM ($R = 32$) with $P = 1$. Note that $P = 1$ refers to the standard Galerkin ROM approach using a single partitioning zone.

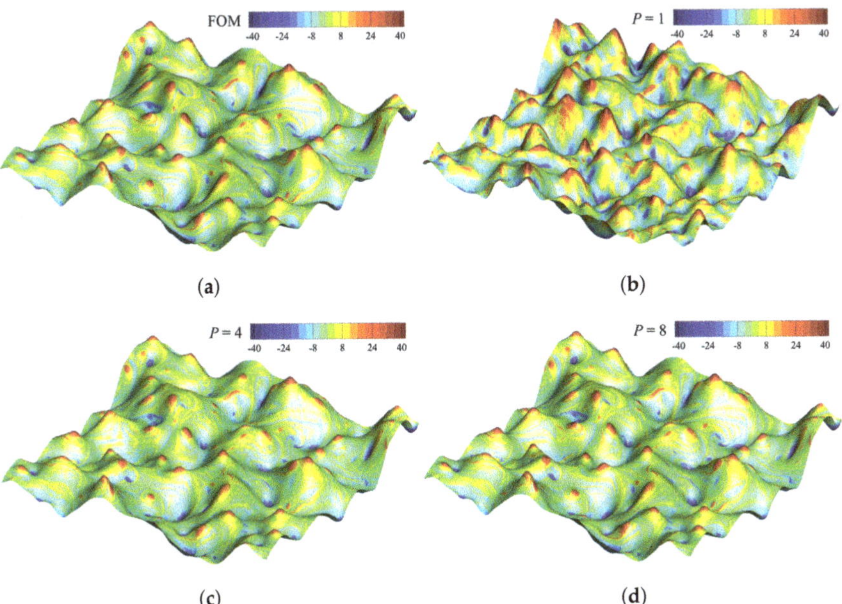

**Figure 6.** Predicted results at time $t = 4$ showing vorticity contours in color with the level of streamfunction as height. (**a**) Full order model (FOM), (**b**) ROM ($R = 16$) with $P = 1$, (**c**) ROM ($R = 16$) with $P = 4$, and (**d**) ROM ($R = 16$) with $P = 8$.

### 5.1. Kolmogorov n-Width and Relative Information Content

The Kologorov $n$-width provides a measure of the system's reducibility, and in POD/PID context, it can be considered as a measure of how well a linear superposition of POD modes might represent the underlying dynamics. Similarly, a relative information content (RIC) of the snapshot matrix can be used as a simple quantitative metric to understand the Kolmogorov $n$-width of the system. The following RIC formula [105] is used to compute the percentage modal energy,

$$\text{RIC}^{(l)}(k) = \left( \frac{\sum_{j=1}^{k} \lambda_j^{(l)}}{\sum_{j=1}^{N_s/P} \lambda_j^{(l)}} \right) \times 100. \tag{29}$$

In other words, $\text{RIC}(k)$ represents the fraction of information (variance) of the snapshot matrix that can be recovered using a specific number ($k$) of basis functions. Figure 7 shows the growth of RIC values with increasing $k$ for different values of $P$. We can observe that, in case of $P = 1$, the increase of the fraction of preserved information is relatively slow and more modes are required to exceed a 90% RIC. On the other hand, with the use of a partitioned ROM approach, the slope of RIC curve increases significantly. For example, less than 10 modes are enough to hit the 90% RIC value. In Kolmogorov $n$-width context, a faster increase in RIC corresponds to a faster decrease of the Kolmogorov $n$-width. Therefore, we consider that partitioned ROM can, in turn, help to break the Kolmogorov barrier and increase the system's reducibility.

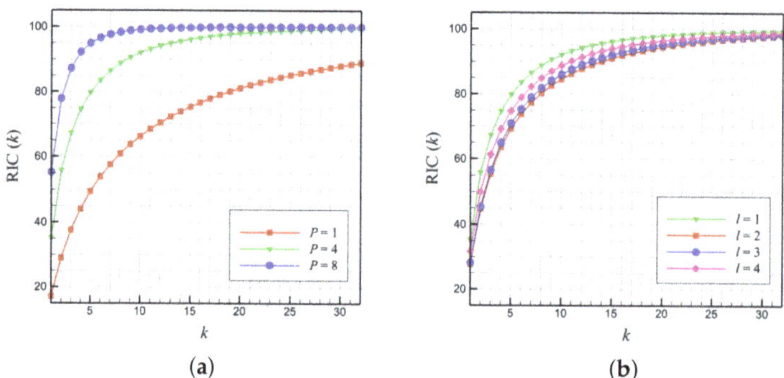

**Figure 7.** Relative information content (RIC) of the snapshot matrix. (**a**) varying the number of partitions $P$, and (**b**) its sensitivity with respect to different zones for $P = 4$ case.

We note that, in Figure 7a, we only show the RIC in the first sub-interval (i.e., $l = 1$) for the sake of clarity. In Figure 7b, we plot the RIC curves for $P = 4$ in the four sub-intervals, where they almost match each other. An interesting observation from Figure 7b is that the slopes of RIC in the second and third sub-intervals are the lowest, compared to those in the first and last sub-intervals. This might be justified by the fact that flow is just initiated in the first sub-interval with relatively smooth dynamics. After that, most of the system's vortical evolution and energy exchange take place within the following sub-intervals, decaying to quasi steady-state in the last sub-interval. For fast evolution and strong dynamics, more modes are required to capture sufficient amount of information. Alternatively, constructing local modes (through further partitioning) can help to accurately capture those dynamics. Since we use fixed number of modes in each sub-interval, this might correspond to different RIC values. That is the ROM approximation in some regions might be more or less accurate than the approximation in other regions. This might be crucial for flow problems with stronger and more complicated dynamics, where that difference in RIC values can cause inconsistencies of approximations. Alternatively, varying values of $R$ can be used to satisfy pre-determined level of RIC in each partition.

## 5.2. Closure Modeling

It turns out that the partitioning approach helps with building more accurate and stable ROM due to several effects. First, it helps to construct more compact and localized subspaces which capture fine scales efficiently. Second, the energy (information) is concentrated in fewer modes. As a result of these two effects, modal truncation works effectively as the discarded modes have negligible effects on the system's dynamics. However, as shown in previous discussions, a large number of partitions might be needed to provide acceptable accuracy. Other than the increased storage requirement to store basis function for each sub-interval, a projection error is accumulated due to interface treatment. If the true field data are available at the interface, it can be projected onto the respective basis functions to obtain the corresponding modal coefficients at the beginning of the sub-interval. Since this is not generally available, the update of ROM solution manifold is performed using Equation (27), where the obtained reduced order approximation of the field from the previous partition is projected onto the basis functions of the following partition. Since that approximation depends on the accuracy of the ROM in the current partition, and lives in a reduced subspace different than that of the following subspace, discontinuities can easily occur at the interface. In Figure 8, we show the modal coefficients at the interfaces for $P = 4$ obtained from ROMs with different $R$ dimensions. Moreover, the mismatch between initial conditions at each interface increases in subsequent partitions, due to the successive interface treatment, resulting in amplification of the error. However, we highlight that the projection error generated at the interface is bounded and overall accuracy gain due to the localization in the PID approach surpasses that error.

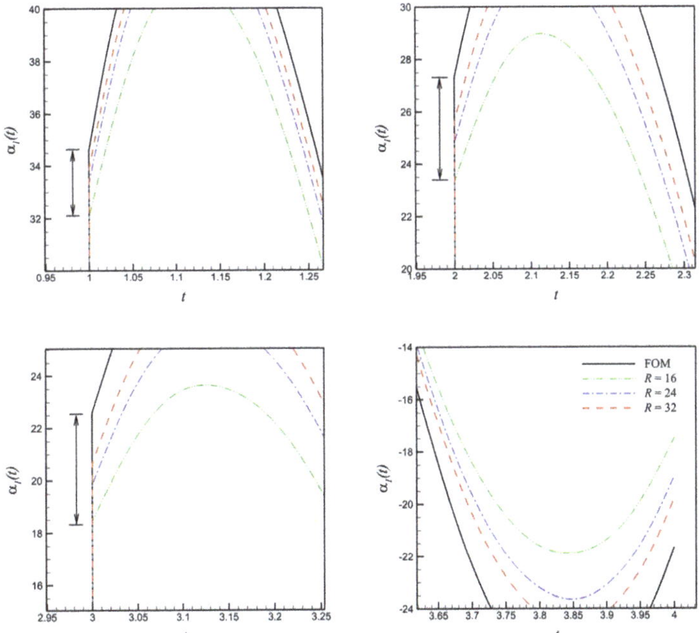


**Figure 8.** Close-up views of temporal coefficients at the interfaces for $P = 4$ case.

Conversely, for ROM with global POD application without any partitioning, the truncated modes might include substantial amount of energy and/or have strong interactions with the retained modes. Hence, inaccuracies and instabilities can easily take place in the produced ROM. Therefore, a compromise between accuracy gain from localization and projection error from interfaces should be considered while selecting the number of intervals. To enhance the stability of ROMs, closure models have been effectively incorporated (inspired from large eddy simulation (LES) studies). Here, we investigate the effect of closure modeling to stabilize and correct ROMs based on global application of POD (i.e., $P = 1$). We utilize a linear mode length eddy viscosity closure based on Rempfer's model [72,106,107], where the 2D Navier–Stokes equation (given in Equation (2)) is re-written as

$$\frac{\partial \omega}{\partial t} + J(\omega, \psi) = \frac{1}{\mathrm{Re}} \left( 1 + \kappa \frac{k}{R} \right) \nabla^2 \omega, \tag{30}$$

where $\kappa$ is a free-parameter controlling the amount of stabilization added to the model, $R$ is the number of modes, and $k$ is the mode index (i.e., $k = 1, 2, \ldots, R$). More recently, several ideas from a plethora of predictive turbulence models have been used in ROMs in the spirit of closure modeling to account for the effects of the truncated scales [108–115]. In the present study, we test the effect of the linear mode dependent eddy viscosity closure model for $\kappa \in \{0, 4, 8\}$, where $\kappa = 0$ corresponds to the standard Galerkin ROM without stabilization.

In Figure 9, we present the temporal evolution of $\alpha_1$ and $\alpha_8$ for ROM with 16 modes and a single global interval (i.e., $P = 1$). We find that the closure significantly improves the predictive capabilities of the ROM, and with increasing $\kappa$ (i.e., adding more stabilization), results converge to the true values of the modal coefficients.

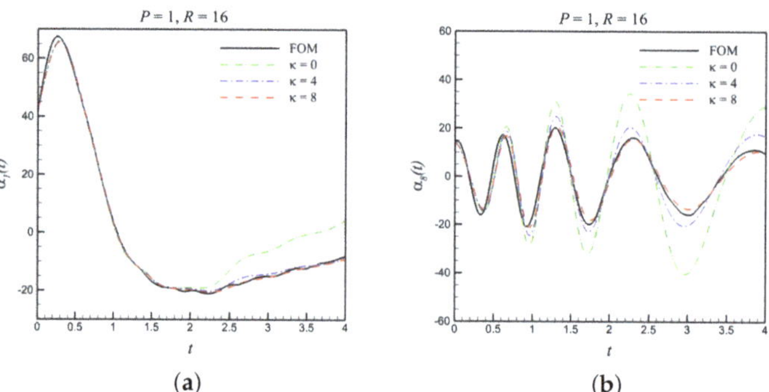

**Figure 9.** Temporal evolution of the first (**a**) and eighth (**b**) modal coefficients in the presence of closure modeling.

In addition, the final vorticity and streamfunction fields at $t = 4$ are shown in Figure 10. We can notice tremendous improvement of predicted results with closure modeling. With close investigation of surface topology, we can observe that Figure 10d captures the correct large scales as Figure 10a. However, we can see that smaller and finer details are overlooked and smoothed-out. This is to be expected given the fact that closure models try to compensate the effect of the truncated modes on the dynamics of the retained modes. In other words, it helps to improve our predictions of the dynamics of these retained modes, but does not add the contribution of the truncated modes into our reduced order approximation. Therefore, our ROM prediction will always be restricted to the scales represented by the retained modes. In case of global POD modes, only the large scales are captured and the fine scales are smoothed-out, which justifies Figure 10d.

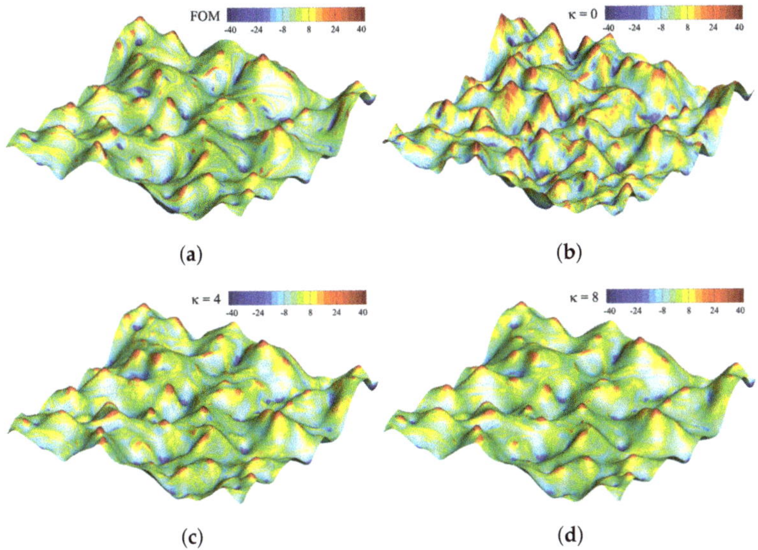

(a)  (b)

(c)  (d)

**Figure 10.** Predicted results in the presence of closure modeling at time $t = 4$ showing vorticity contours in color with the level of streamfunction as height. (**a**) full order model (FOM), (**b**) ROM with $\kappa = 0$, (**c**) ROM with $\kappa = 4$, and (**d**) ROM with $\kappa = 8$. For (**b**–**d**): $R = 16$ and $P = 1$.

Finally, the energy spectrum at final time is given in Figure 11, where we can see the influence of closure modeling in improving ROM results. A very nice observation from this figure is that stabilization helps a lot in the inertial regime of energy spectrum, but it causes over-stabilization in the dissipative regime.

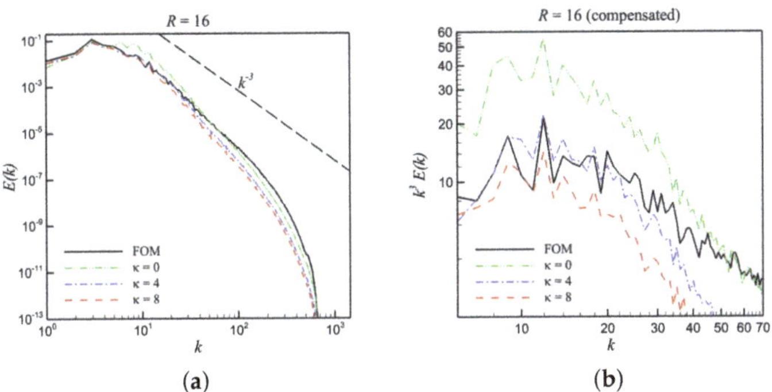

(a)  (b)

**Figure 11.** Energy spectra at final time predicted by ROMs in the presence of stabilization, compared to the FOM spectrum and $k^{-3}$ scaling (**a**), with a compensated view in (**b**).

The root-mean-squares errors (RMSE) between Galerkin ROM approximation (with $P = 1$ and $R = 16$), in the presence of stabilization, and FOM solution are shown in Table 2, compared to the base case without closure. It is clear that the addition of stabilization to Galerkin ROM helps to improve the predictions and reduces the approximation error. However, compared to Table 1, we can deduce that the accuracy gain due to partitioning is much more superior than that of closure. In addition, the RMSE is sensitive to the value of the stabilization parameter. Therefore, a dynamic selection of this parameter might be needed to guarantee sufficient correction.

Moreover, it should be noted that there might exist situations where the gain from only closure is minimum, even in case of "optimal" closure [73]. Hence, a combination of both partitioning and stabilization seems to be a feasible choice for building accurate, stable, and compact ROMs for turbulent flows. Indeed, Ahmed and San [72] demonstrated that an eddy viscosity based closure in conjunction with PID yields significant improvements in accuracy for capturing strong moving discontinuities with a negligibly small computational overhead. In addition, Table 3 shows the effect of stabilization through closure modeling in the partitioned ROM (with $P =$ and $R = 16$). It is clear that the closure slightly improves the RMSE of vorticity field approximation. In addition, we can see that further increase in $\kappa$ can actually increase the RMSE due to over stabilization. It can be also observed that RMSE in streamfunction predictions in case of closure modeling with PID is generally larger than that in case of PID without closure. This can be attributed to the nature of streamfunction being inverse of the Laplacian operator of vorticity, making it a smoother field by construction. Therefore, closure with PID might cause over-stabilization and excessive smoothing of the streamfunction fields predictions. This again suggests the development of a dynamical strategy to estimate optimal values of eddy viscosity coefficient.

**Table 2.** Root-mean-squares error at time $t = 4$ between the standard Galerkin ROM ($P = 1$, $R = 16$) and FOM in order to demonstrate the sensitivity with respect to the additional stabilization through $\kappa$ at Re = 16,000.

| Model with 16 Modes ($P = 1$) | $\|\psi^{ROM} - \psi^{FOM}\|$ | $\|\omega^{rom} - \omega^{FOM}\|$ |
| --- | --- | --- |
| $\kappa = 0$ | $1.14 \times 10^{-1}$ | 9.7322 |
| $\kappa = 0.1$ | $1.11 \times 10^{-1}$ | 9.6010 |
| $\kappa = 0.25$ | $1.07 \times 10^{-1}$ | 9.4103 |
| $\kappa = 1.0$ | $8.97 \times 10^{-2}$ | 8.5579 |
| $\kappa = 2.0$ | $7.24 \times 10^{-2}$ | 7.6032 |
| $\kappa = 4.0$ | $5.39 \times 10^{-2}$ | 6.3506 |
| $\kappa = 8.0$ | $5.27 \times 10^{-2}$ | 5.6756 |
| $\kappa = 16.0$ | $6.41 \times 10^{-2}$ | 5.7574 |

**Table 3.** Root-mean-squares error at time $t = 4$ between the partitioned ROM ($P = 8$, $R = 16$) and FOM in order to demonstrate the sensitivity with respect to the additional stabilization through $\kappa$ at Re = 16,000.

| Model with 16 Modes ($P = 8$) | $\|\psi^{ROM} - \psi^{FOM}\|$ | $\|\omega^{ROM} - \omega^{FOM}\|$ |
| --- | --- | --- |
| $\kappa = 0$ | $7.47 \times 10^{-3}$ | 1.2774 |
| $\kappa = 0.1$ | $7.62 \times 10^{-3}$ | 1.2627 |
| $\kappa = 0.25$ | $7.85 \times 10^{-3}$ | 1.2445 |
| $\kappa = 1.0$ | $8.89 \times 10^{-3}$ | 1.2058 |
| $\kappa = 2.0$ | $1.01 \times 10^{-2}$ | 1.2322 |
| $\kappa = 4.0$ | $1.18 \times 10^{-2}$ | 1.3574 |
| $\kappa = 8.0$ | $1.40 \times 10^{-2}$ | 1.5752 |
| $\kappa = 16.0$ | $1.61 \times 10^{-2}$ | 1.8108 |

*5.3. Computational Cost*

In addition to the accuracy gains through the partitioned ROM approach, presented in previous subsections, a comparison in terms of computational cost is required for fair assessment. Here, we consider two phases of ROM computations: offline and online. By offline, we refer to the evaluation of the vectors, matrices, and tensors corresponding to the constant, linear, and nonlinear predetermined model coefficients (i.e., $\mathcal{B}$, $\mathcal{L}$, and $\mathcal{R}$). Meanwhile, we denote the actual prediction and solution of Equation (21) as the online stage. Table 4 provides the CPU times in the offline and online stages for different number of partitions and modes. From this table, we can see that increasing the number of modes increases the CPU times for both the offline (due to computation of larger vectors, matrices, and tensors) and online (due to solving an increased number of equations) stages. Interestingly, we can observe that the computational cost of the online stage scales cubically with increasing $R$. This conforms to the quadratic nonlinearity in the investigated system, making the online cost an order of $O(R^3)$.

On the other hand, increasing the number of partitions has a negligible effect on the online CPU time. Therefore, the partitioned ROM approach provides more accurate results with similar online costs as standard ROM with the same number of modes. However, increasing $P$ (with fixed $R$) results in an increase in the offline CPU time. This corresponds to the computation of $\mathcal{B}$, $\mathcal{L}$, and $\mathcal{R}$ for each individual partition. That is why the offline CPU time scales linearly with increasing $P$. Nonetheless, it was shown in [73] that the CPU times for basis construction are largely reduced in the PID framework due to the solution of smaller eigenvalue problems rather than solving a large one. Finally, it is noteworthy that the partitioned ROM approach requires more memory to store the local basis functions and model predetermined coefficients in each partition. Therefore, we can consider that partitioned ROM converts the offline and storage costs into a speed-up of the online deployment.

Table 4. CPU time (in seconds) for ROM prediction and predetermined model coefficients computations, denoted as online and offline stages, respectively. We note that the CPU time assessments documented in this table are based on Fortran executions.

|  | $R = 16$ | $R = 24$ | $R = 32$ |
|---|---|---|---|
| $P = 1$ | | | |
| online | 0.3889 | 1.1448 | 2.6925 |
| offline | 153.1237 | 393.5841 | 745.8216 |
| $P = 4$ | | | |
| online | 0.3719 | 1.1568 | 2.6026 |
| offline | 619.4428 | 1507.8947 | 3032.1081 |
| $P = 8$ | | | |
| online | 0.3699 | 1.1368 | 2.6036 |
| offline | 1142.9432 | 2882.4478 | 5776.0589 |

## 6. Conclusions and Future Works

We utilize a partitioning approach in the current study in order to build a compact reduced order models of 2D turbulent flows. We demonstrate the framework using the two-dimensional decaying Kraichnan turbulence. It is shown that this partitioned ROM helps to break the Kolmogorov $n$-width barrier, in the context of a relative information content criterion. An energy spectrum analysis is performed to emphasize the capabilities of the presented framework to sufficiently capture both the large and fine scales and recover the energy spectra within the inertial and dissipative ranges. The partitioned ROM produces significantly more accurate predictions than standard ROM without partitioning. A computational CPU time analysis reveals that the partitioned ROM minimizes the online cost at the expense of offline cost and increased memory to build reduced order approximation with acceptable accuracy. In addition, we investigate the use of a linear mode dependent eddy viscosity to stabilize ROM prediction. We find that closure improves the results in the inertial kinetic energy regime, corresponding to the large scales. However, it presents over-stabilization in the dissipative regime, which smoothens out the field finer details. This might advocate a strategy to develop optimal (or near optimal) estimates for the eddy viscosity coefficients to be posed. Therefore, we suggest the use of partitioning along with closure as an enabler for more efficient ROMs turbulent flow systems. Incorporating PID with closure would provide the flexibility of constructing ROMs with less modes (and same number of partitions), or with less partitions (and same number of modes). The former would provide higher speed-ups during online deployment, while the latter would reduce the offline cost.

To this end, the partitioned ROM has been shown to improve the predictions for the investigated 2D Kraichnan turbulence case. However, scaling this framework into a 3D setting is still required to fully assess the validity of the framework to address realistic turbulent flows with strong anisotropy. In addition, the PID has been shown to capture finer details than standard global POD. Application of this approach onto flow situations where global basis fails drastically and results in completely unstable and inaccurate ROMs can be an interesting extension to the current study. Also, the adopted PID approach assumes evenly-spaced partitions with an equal number of modes. Despite the resulting simplicity of implementation, this fixed partitioning might be less meaningful for more challenging flow problems where dynamics change rapidly. Therefore, an adaptive partitioning with varying number of modes can be another extension to generalize the present work. The assumption of non-overlapping partitions might also be relaxed to allow overlapping sub-intervals with an indicator for smooth transition between partitions based on the actual flow conditions. Finally, the development of a strategy to dynamically update the basis functions using data assimilation techniques would enable the extrapolation of ROM in time while reflecting local phenomena.

**Author Contributions:** Data curation, S.E.A. and O.S.; Supervision, O.S.; Writing—original draft, S.E.A.; and Writing—review and editing, S.E.A. and O.S. All authors have read and agreed to the published version of the manuscript.

**Funding:** U.S. Department of Energy, Office of Science, Office of Advanced Scientific Computing Research, Award Number: DE-SC0019290.

**Acknowledgments:** This material is based upon work supported by the U.S. Department of Energy, Office of Science, Office of Advanced Scientific Computing Research under Award Number DE-SC0019290. Omer San gratefully acknowledges their support. Disclaimer: This report was prepared as an account of work sponsored by an agency of the United States Government. Neither the United States Government nor any agency thereof, nor any of their employees, makes any warranty, express or implied, or assumes any legal liability or responsibility for the accuracy, completeness, or usefulness of any information, apparatus, product, or process disclosed, or represents that its use would not infringe privately owned rights. Reference herein to any specific commercial product, process, or service by trade name, trademark, manufacturer, or otherwise does not necessarily constitute or imply its endorsement, recommendation, or favoring by the United States Government or any agency thereof. The views and opinions of authors expressed herein do not necessarily state or reflect those of the United States Government or any agency thereof.

**Conflicts of Interest:** The authors declare no conflict of interest.

## References

1. Powell, J.R. The quantum limit to Moore's law. *Proc. IEEE* **2008**, *96*, 1247–1248. [CrossRef]
2. Waldrop, M.M. The chips are down for Moore's law. *Nat. News* **2016**, *530*, 144. [CrossRef] [PubMed]
3. Kumar, S. The end of Moore's law and reinventing computing. In *High-Speed and Lower Power Technologies: Electronics and Photonics*; Choi, J.H., Ed.; CRC Press: Boca Raton, FL, USA, 2018.
4. Taira, K.; Brunton, S.L.; Dawson, S.T.; Rowley, C.W.; Colonius, T.; McKeon, B.J.; Schmidt, O.T.; Gordeyev, S.; Theofilis, V.; Ukeiley, L.S. Modal analysis of fluid flows: An overview. *AIAA J.* **2017**, *55*, 4013–4041. [CrossRef]

5.  Puzyrev, V.; Ghommem, M.; Meka, S. pyROM: A computational framework for reduced order modeling. *J. Comput. Sci.* **2019**, *30*, 157–173. [CrossRef]

6.  Rasheed, A.; San, O.; Kvamsdal, T. Digital Twin: Values, Challenges and Enablers. *arXiv* **2019**, arXiv:1910.01719.

7.  Grieves, M.; Vickers, J. Digital twin: Mitigating unpredictable, undesirable emergent behavior in complex systems. In *Transdisciplinary Perspectives on Complex Systems*; Springer: Berlin/Heidelberg, Germany, 2017; pp. 85–113.

8.  Boschert, S.; Rosen, R. Digital twin—The simulation aspect. In *Mechatronic Futures*; Springer: Berlin/Heidelberg, Germany, 2016; pp. 59–74.

9.  Tao, F.; Cheng, J.; Qi, Q.; Zhang, M.; Zhang, H.; Sui, F. Digital twin-driven product design, manufacturing and service with big data. *Int. J. Adv. Manuf. Technol.* **2018**, *94*, 3563–3576. [CrossRef]

10.  Uhlemann, T.H.J.; Lehmann, C.; Steinhilper, R. The digital twin: Realizing the cyber-physical production system for industry 4.0. *Procedia CIRP* **2017**, *61*, 335–340. [CrossRef]

11.  Glaessgen, E.; Stargel, D. The digital twin paradigm for future NASA and US Air Force vehicles. In Proceedings of the 53rd AIAA/ASME/ASCE/AHS/ASC Structures, Structural Dynamics and Materials Conference 20th AIAA/ASME/AHS Adaptive Structures Conference 14th AIAA, Honolulu, HI, USA, 23–26 April 2012; p. 1818.

12.  Kapteyn, M.G.; Willcox, K.; Knezevic, D.J. Toward predictive digital twins via component-based reduced-order models and interpretable machine learning. In Proceedings of the AIAA Scitech 2020 Forum, Orlando, FL, USA, 6–10 January 2020; p. 0418.

13.  Singh, V.; Willcox, K.E. Engineering Design with Digital Thread. *AIAA J.* **2018**, *56*, 4515–4528. [CrossRef]

14.  Hartmann, D.; Herz, M.; Wever, U. Model order reduction a key technology for digital twins. In *Reduced-Order Modeling (ROM) for Simulation and Optimization*; Springer: Berlin/Heidelberg, Germany, 2018; pp. 167–179.

15.  Noack, B.R.; Morzynski, M.; Tadmor, G. *Reduced-Order Modelling for Flow Control*; Springer: Berlin/Heidelberg, Germany, 2011; Volume 528.

16.  Ito, K.; Ravindran, S.S. Reduced basis method for optimal control of unsteady viscous flows. *Int. J. Comput. Fluid Dyn.* **2001**, *15*, 97–113. [CrossRef]

17.  McNamara, A.; Treuille, A.; Popović, Z.; Stam, J. Fluid control using the adjoint method. *ACM Trans. Graph.* **2004**, *23*, 449–456. [CrossRef]

18.  Bergmann, M.; Cordier, L. Optimal control of the cylinder wake in the laminar regime by trust-region methods and POD reduced-order models. *J. Comput. Phys.* **2008**, *227*, 7813–7840. [CrossRef]

19.  Ravindran, S.S. A reduced-order approach for optimal control of fluids using proper orthogonal decomposition. *Int. J. Numer. Methods Fluids* **2000**, *34*, 425–448. [CrossRef]

20.  Graham, W.; Peraire, J.; Tang, K. Optimal control of vortex shedding using low-order models. Part I–open-loop model development. *Int. J. Numer. Methods Eng.* **1999**, *44*, 945–972. [CrossRef]

21.  Graham, W.; Peraire, J.; Tang, K. Optimal control of vortex shedding using low-order models. Part II–model-based control. *Int. J. Numer. Methods Eng.* **1999**, *44*, 973–990. [CrossRef]

22.  Brunton, S.L.; Noack, B.R. Closed-loop turbulence control: Progress and challenges. *Appl. Mech. Rev.* **2015**, *67*, 050801. [CrossRef]

23.  Daescu, D.N.; Navon, I.M. Efficiency of a POD-based reduced second-order adjoint model in 4D-Var data assimilation. *Int. J. Numer. Methods Fluids* **2007**, *53*, 985–1004. [CrossRef]

24.  Navon, I.M. Data assimilation for numerical weather prediction: A review. In *Data Assimilation for Atmospheric, Oceanic and Hydrologic Applications*; Park, S.K., Xu, L., Eds.; Springer: New York, NY, USA, 2009; pp. 21–65.

25.  Cao, Y.; Zhu, J.; Navon, I.M.; Luo, Z. A reduced-order approach to four-dimensional variational data assimilation using proper orthogonal decomposition. *Int. J. Numer. Methods Fluids* **2007**, *53*, 1571–1583. [CrossRef]

26.  He, J.; Sarma, P.; Durlofsky, L.J. Use of reduced-order models for improved data assimilation within an EnKF context. In Proceedings of the SPE Reservoir Simulation Symposium, The Woodlands, TX, USA, 21–23 February 2011; Society of Petroleum Engineers: Richardson, TX, USA, 2011.

27.  Houtekamer, P.L.; Mitchell, H.L. Data assimilation using an ensemble Kalman filter technique. *Mon. Weather. Rev.* **1998**, *126*, 796–811. [CrossRef]

28. Houtekamer, P.L.; Mitchell, H.L. A sequential ensemble Kalman filter for atmospheric data assimilation. *Mon. Weather. Rev.* **2001**, *129*, 123–137. [CrossRef]

29. Bennett, A.F. *Inverse Modeling of the Ocean and Atmosphere*; Cambridge University Press: New York, NY, USA, 2005.

30. Evensen, G. *Data Assimilation: The Ensemble Kalman Filter*; Springer: Berlin/Heidelberg, Germany, 2009.

31. Law, K.J.; Stuart, A.M. Evaluating data assimilation algorithms. *Mon. Weather. Rev.* **2012**, *140*, 3757–3782. [CrossRef]

32. Buljak, V. *Inverse Analyses with Model Reduction: Proper Orthogonal Decomposition in Structural Mechanics*; Springer: Berlin/Heidelberg, Germany, 2011.

33. Xiao, D.; Du, J.; Fang, F.; Pain, C.; Li, J. Parameterised non-intrusive reduced order methods for ensemble Kalman filter data assimilation. *Comput. Fluids* **2018**, *177*, 69–77. [CrossRef]

34. Navon, I. Practical and theoretical aspects of adjoint parameter estimation and identifiability in meteorology and oceanography. *Dyn. Atmos. Ocean.* **1998**, *27*, 55–79. [CrossRef]

35. Fu, H.; Wang, H.; Wang, Z. POD/DEIM reduced-order modeling of time-fractional partial differential equations with applications in parameter identification. *J. Sci. Comput.* **2018**, *74*, 220–243. [CrossRef]

36. Boulakia, M.; Schenone, E.; Gerbeau, J.F. Reduced-order modeling for cardiac electrophysiology. Application to parameter identification. *Int. J. Numer. Methods Biomed. Eng.* **2012**, *28*, 727–744. [CrossRef] [PubMed]

37. Kramer, B.; Peherstorfer, B.; Willcox, K. Feedback control for systems with uncertain parameters using online-adaptive reduced models. *SIAM J. Appl. Dyn. Syst.* **2017**, *16*, 1563–1586. [CrossRef]

38. Galbally, D.; Fidkowski, K.; Willcox, K.; Ghattas, O. Non-linear model reduction for uncertainty quantification in large-scale inverse problems. *Int. J. Numer. Methods Eng.* **2010**, *81*, 1581–1608. [CrossRef]

39. Biegler, L.; Biros, G.; Ghattas, O.; Heinkenschloss, M.; Keyes, D.; Mallick, B.; Tenorio, L.; Waanders, B.V.B.; Willcox, K.; Marzouk, Y. *Large-Scale Inverse Problems and Quantification of Uncertainty*; Wiley Online Library: Chichester, UK, 2011.

40. Mathelin, L.; Hussaini, M.Y.; Zang, T.A. Stochastic approaches to uncertainty quantification in CFD simulations. *Numer. Algorithms* **2005**, *38*, 209–236. [CrossRef]

41. Smith, R.C. *Uncertainty Quantification: Theory, Implementation, and Applications*; SIAM: Philadelphia, PA, USA, 2014.

42. Najm, H.N. Uncertainty quantification and polynomial chaos techniques in computational fluid dynamics. *Annu. Rev. Fluid Mech.* **2009**, *41*, 35–52. [CrossRef]

43. Zahr, M.J.; Carlberg, K.T.; Kouri, D.P. An efficient, globally convergent method for optimization under uncertainty using adaptive model reduction and sparse grids. *SIAM/ASA J. Uncertain. Quantif.* **2019**, *7*, 877–912. [CrossRef]

44. Fairbanks, H.R.; Jofre, L.; Geraci, G.; Iaccarino, G.; Doostan, A. Bi-fidelity approximation for uncertainty quantification and sensitivity analysis of irradiated particle-laden turbulence. *J. Comput. Phys.* **2020**, *402*, 108996. [CrossRef]

45. Kevlahan, N.R.; Hunt, J.; Vassilicos, J. A comparison of different analytical techniques for identifying structures in turbulence. *Appl. Sci. Res.* **1994**, *53*, 339–355. [CrossRef]

46. Holmes, P.; Lumley, J.L.; Berkooz, G. *Turbulence, Coherent Structures, Dynamical Systems and Symmetry*; Cambridge University Press: New York, NY, USA, 1998.

47. Rowley, C.W.; Colonius, T.; Murray, R.M. Model reduction for compressible flows using POD and Galerkin projection. *Phys. D Nonlinear Phenom.* **2004**, *189*, 115–129. [CrossRef]

48. Ma, Z.; Ahuja, S.; Rowley, C.W. Reduced-order models for control of fluids using the eigensystem realization algorithm. *Theor. Comput. Fluid Dyn.* **2011**, *25*, 233–247. [CrossRef]

49. Rowley, C.W. Model reduction for fluids, using balanced proper orthogonal decomposition. *Int. J. Bifurc. Chaos* **2005**, *15*, 997–1013. [CrossRef]

50. Willcox, K.; Peraire, J. Balanced model reduction via the proper orthogonal decomposition. *AIAA J.* **2002**, *40*, 2323–2330. [CrossRef]

51. Schmid, P.J. Dynamic mode decomposition of numerical and experimental data. *J. Fluid Mech.* **2010**, *656*, 5–28. [CrossRef]

52. Williams, M.O.; Kevrekidis, I.G.; Rowley, C.W. A data–driven approximation of the koopman operator: Extending dynamic mode decomposition. *J. Nonlinear Sci.* **2015**, *25*, 1307–1346. [CrossRef]

53. Peterson, J.S. The reduced basis method for incompressible viscous flow calculations. *SIAM J. Sci. Stat. Comput.* **1989**, *10*, 777–786. [CrossRef]

54. Sirovich, L. Turbulence and the dynamics of coherent structures. I. Coherent structures. *Q. Appl. Math.* **1987**, *45*, 561–571. [CrossRef]

55. Rim, D.; Peherstorfer, B.; Mandli, K.T. Manifold Approximations via Transported Subspaces: Model reduction for transport-dominated problems. *arXiv* **2019**, arXiv:1912.13024.

56. Lai, X.; Wang, X.; Nie, Y.; Li, Q. Characterizing complex flows via adaptive sparse dynamic mode decomposition with error approximation. *Int. J. Numer. Methods Fluids* **2019**. [CrossRef]

57. Kastian, S.; Reese, S. An Adaptive Way of Choosing Significant Snapshots for Proper Orthogonal Decomposition. In Proceedings of the IUTAM Symposium on Model Order Reduction of Coupled Systems, Stuttgart, Germany, 22–25 May 2018; Springer: Berlin/Heidelberg, Germany, 2020; pp. 67–79.

58. Aubry, N.; Holmes, P.; Lumley, J.L.; Stone, E. The dynamics of coherent structures in the wall region of a turbulent boundary layer. *J. Fluid Mech.* **1988**, *192*, 115–173. [CrossRef]

59. Wang, Z.; Akhtar, I.; Borggaard, J.; Iliescu, T. Proper orthogonal decomposition closure models for turbulent flows: A numerical comparison. *Comput. Methods Appl. Mech. Eng.* **2012**, *237*, 10–26. [CrossRef]

60. Carlberg, K.; Farhat, C.; Cortial, J.; Amsallem, D. The GNAT method for nonlinear model reduction: Effective implementation and application to computational fluid dynamics and turbulent flows. *J. Comput. Phys.* **2013**, *242*, 623–647. [CrossRef]

61. Protas, B.; Noack, B.R.; Östh, J. Optimal nonlinear eddy viscosity in Galerkin models of turbulent flows. *J. Fluid Mech.* **2015**, *766*, 337–367. [CrossRef]

62. Stabile, G.; Ballarin, F.; Zuccarino, G.; Rozza, G. A reduced order variational multiscale approach for turbulent flows. *Adv. Comput. Math.* **2019**, *45*, 2349–2368. [CrossRef]

63. Xiao, D.; Heaney, C.; Mottet, L.; Fang, F.; Lin, W.; Navon, I.; Guo, Y.; Matar, O.; Robins, A.; Pain, C. A reduced order model for turbulent flows in the urban environment using machine learning. *Build. Environ.* **2019**, *148*, 323–337. [CrossRef]

64. Xiao, D.; Heaney, C.; Fang, F.; Mottet, L.; Hu, R.; Bistrian, D.; Aristodemou, E.; Navon, I.; Pain, C. A domain decomposition non-intrusive reduced order model for turbulent flows. *Comput. Fluids* **2019**, *182*, 15–27. [CrossRef]

65. Kolmogoroff, A. Über die beste Annäherung von Funktionen einer gegebenen Funktionenklasse. *Ann. Math.* **1936**, *37*, 107–110. [CrossRef]

66. Pinkus, A. *N-Widths in Approximation Theory*; Springer Science & Business Media: Berlin/Heidelberg, Germany, 2012; Volume 7.

67. Taddei, T. A registration method for model order reduction: Data compression and geometry reduction. *arXiv* **2019**, arXiv:1906.11008.

68. Greif, C.; Urban, K. Decay of the Kolmogorov N-width for wave problems. *Appl. Math. Lett.* **2019**, *96*, 216–222. [CrossRef]

69. IJzerman, W. Signal Representation and Modeling of Spatial Structures in Fluids. Ph.D. Thesis, University of Twente, Enschede, The Netherlands, 2000.

70. Borggaard, J.; Hay, A.; Pelletier, D. Interval-based reduced order models for unsteady fluid flow. *Int. J. Numer. Anal. Model.* **2007**, *4*, 353–367.

71. San, O.; Borggaard, J. Principal interval decomposition framework for POD reduced-order modeling of convective Boussinesq flows. *Int. J. Numer. Methods Fluids* **2015**, *78*, 37–62. [CrossRef]

72. Ahmed, M.; San, O. Stabilized principal interval decomposition method for model reduction of nonlinear convective systems with moving shocks. *Comput. Appl. Math.* **2018**, *37*, 6870–6902. [CrossRef]

73. Ahmed, S.E.; Rahman, S.M.; San, O.; Rasheed, A.; Navon, I.M. Memory embedded non-intrusive reduced order modeling of non-ergodic flows. *Phys. Fluids* **2019**, *31*, 126602. [CrossRef]

74. Charney, J.G. Geostrophic turbulence. *J. Atmos. Sci.* **1971**, *28*, 1087–1095. [CrossRef]

75. Kraichnan, R.H.; Montgomery, D. Two-dimensional turbulence. *Rep. Prog. Phys.* **1980**, *43*, 547. [CrossRef]

76. Boffetta, G.; Ecke, R.E. Two-dimensional turbulence. *Annu. Rev. Fluid Mech.* **2012**, *44*, 427–451. [CrossRef]

77. Rempfer, D. On low-dimensional Galerkin models for fluid flow. *Theor. Comput. Fluid Dyn.* **2000**, *14*, 75–88. [CrossRef]

78. Zhang, M.; Stevens, R.J. Characterizing the Coherent Structures Within and Above Large Wind Farms. *Bound.-Layer Meteorol.* **2019**, 1–20. [CrossRef]

79. Shah, S.; Bou-Zeid, E. Very-large-scale motions in the atmospheric boundary layer educed by snapshot proper orthogonal decomposition. *Bound.-Layer Meteorol.* **2014**, *153*, 355–387. [CrossRef]

80. Östh, J.; Noack, B.R.; Krajnović, S.; Barros, D.; Borée, J. On the need for a nonlinear subscale turbulence term in POD models as exemplified for a high-Reynolds-number flow over an Ahmed body. *J. Fluid Mech.* **2014**, *747*, 518–544. [CrossRef]

81. Grimberg, S.; Farhat, C.; Youkilis, N. On the stability of projection-based model order reduction for convection-dominated laminar and turbulent flows. *arXiv* **2020**, arXiv:2001.10110.

82. Carlberg, K.; Barone, M.; Antil, H. Galerkin v. least-squares Petrov–Galerkin projection in nonlinear model reduction. *J. Comput. Phys.* **2017**, *330*, 693–734. [CrossRef]

83. Kraichnan, R.H. Inertial ranges in two-dimensional turbulence. *Phys. Fluids* **1967**, *10*, 1417–1423. [CrossRef]

84. Batchelor, G.K. Computation of the energy spectrum in homogeneous two-dimensional turbulence. *Phys. Fluids* **1969**, *12*, II–233. [CrossRef]

85. Leith, C. Atmospheric predictability and two-dimensional turbulence. *J. Atmos. Sci.* **1971**, *28*, 145–161. [CrossRef]

86. San, O.; Staples, A.E. High-order methods for decaying two-dimensional homogeneous isotropic turbulence. *Comput. Fluids* **2012**, *63*, 105–127. [CrossRef]

87. Arakawa, A. Computational design for long-term numerical integration of the equations of fluid motion: Two-dimensional incompressible flow. Part I. *J. Comput. Phys.* **1966**, *1*, 119–143. [CrossRef]

88. Gottlieb, S.; Shu, C.W. Total variation diminishing Runge-Kutta schemes. *Math. Comput.* **1998**, *67*, 73–85. [CrossRef]

89. Rowley, C.W.; Dawson, S.T. Model reduction for flow analysis and control. *Annu. Rev. Fluid Mech.* **2017**, *49*, 387–417. [CrossRef]

90. Chen, K.K.; Tu, J.H.; Rowley, C.W. Variants of dynamic mode decomposition: Boundary condition, Koopman, and Fourier analyses. *J. Nonlinear Sci.* **2012**, *22*, 887–915. [CrossRef]

91. Press, W.H.; Flannery, B.P.; Teukolsky, S.A.; Vetterling, W.T. *Numerical Recipes in FORTRAN*; Cambridge University Press: New York, NY, USA, 1992.

92. Chaturantabut, S.; Sorensen, D.C. Discrete empirical interpolation for nonlinear model reduction. In Proceedings of the 48th IEEE Conference on Decision and Control, Shanghai, China, 16–18 December 2009; pp. 4316–4321.

93. Chaturantabut, S.; Sorensen, D.C. Nonlinear model reduction via discrete empirical interpolation. *SIAM J. Sci. Comput.* **2010**, *32*, 2737–2764. [CrossRef]

94. Everson, R.; Sirovich, L. Karhunen–Loeve procedure for gappy data. *JOSA A* **1995**, *12*, 1657–1664. [CrossRef]

95. Astrid, P.; Weiland, S.; Willcox, K.; Backx, T. Missing point estimation in models described by proper orthogonal decomposition. *IEEE Trans. Autom. Control.* **2008**, *53*, 2237–2251. [CrossRef]

96. Zimmermann, R.; Willcox, K. An accelerated greedy missing point estimation procedure. *SIAM J. Sci. Comput.* **2016**, *38*, A2827–A2850. [CrossRef]

97. Ştefănescu, R.; Sandu, A.; Navon, I.M. Comparison of POD reduced order strategies for the nonlinear 2D shallow water equations. *Int. J. Numer. Methods Fluids* **2014**, *76*, 497–521. [CrossRef]

98. Dimitriu, G.; Ştefănescu, R.; Navon, I.M. Comparative numerical analysis using reduced-order modeling strategies for nonlinear large-scale systems. *J. Comput. Appl. Math.* **2017**, *310*, 32–43. [CrossRef]

99. Cazemier, W.; Verstappen, R.; Veldman, A. Proper orthogonal decomposition and low-dimensional models for driven cavity flows. *Phys. Fluids* **1998**, *10*, 1685–1699. [CrossRef]

100. Amsallem, D.; Zahr, M.J.; Farhat, C. Nonlinear model order reduction based on local reduced-order bases. *Int. J. Numer. Methods Eng.* **2012**, *92*, 891–916. [CrossRef]

101. Peherstorfer, B.; Willcox, K. Online adaptive model reduction for nonlinear systems via low-rank updates. *SIAM J. Sci. Comput.* **2015**, *37*, A2123–A2150. [CrossRef]

102. Taddei, T.; Perotto, S.; Quarteroni, A. Reduced basis techniques for nonlinear conservation laws. *ESAIM Math. Model. Numer. Anal.* **2015**, *49*, 787–814. [CrossRef]

103. Dihlmann, M.; Drohmann, M.; Haasdonk, B. Model reduction of parametrized evolution problems using the reduced basis method with adaptive time-partitioning. In Proceedings of the International Conference on Adaptive Modeling and Simulation (ADMOS 2011), Paris, France, 6–8 June 2011; pp. 156–167.

104. Ahmed, S.E.; San, O.; Bistrian, D.A.; Navon, I.M. Sampling and resolution characteristics in reduced order models of shallow water equations: Intrusive vs non-intrusive. *arXiv* **2019**, arXiv:1910.07654,

105. Gunzburger, M.D. *Flow Control*; Springer: Berlin, Gemany, 2012.
106. Rempfer, D. Kohärente Strukturen und Chaos beim Laminar-Turbulenten Grenzschichtumschlag. Ph.D. Thesis, University of Stuttgart, Stuttgart, Germany, 1997.
107. San, O.; Iliescu, T. Proper orthogonal decomposition closure models for fluid flows: Burgers equation. *INternational J. Numer. Anal. Model. Ser. B* **2014**, *5*, 217–237.
108. Wang, Z.; Akhtar, I.; Borggaard, J.; Iliescu, T. Two-level discretizations of nonlinear closure models for proper orthogonal decomposition. *J. Comput. Phys.* **2011**, *230*, 126–146. [CrossRef]
109. Borggaard, J.; Iliescu, T.; Wang, Z. Artificial viscosity proper orthogonal decomposition. *Math. Comput. Model.* **2011**, *53*, 269–279. [CrossRef]
110. Akhtar, I.; Wang, Z.; Borggaard, J.; Iliescu, T. A new closure strategy for proper orthogonal decomposition reduced-order models. *J. Comput. Nonlinear Dyn.* **2012**, *7*, 034503. [CrossRef]
111. Iliescu, T.; Wang, Z. Variational multiscale proper orthogonal decomposition: Navier-stokes equations. *Numer. Methods Part. Differ. Equations* **2014**, *30*, 641–663. [CrossRef]
112. Xie, X.; Wells, D.; Wang, Z.; Iliescu, T. Approximate deconvolution reduced order modeling. *Comput. Methods Appl. Mech. Eng.* **2017**, *313*, 512–534. [CrossRef]
113. Xie, X.; Wells, D.; Wang, Z.; Iliescu, T. Numerical analysis of the Leray reduced order model. *J. Comput. Appl. Math.* **2018**, *328*, 12–29. [CrossRef]
114. Xie, X.; Mohebujjaman, M.; Rebholz, L.G.; Iliescu, T. Data-driven filtered reduced order modeling of fluid flows. *SIAM J. Sci. Comput.* **2018**, *40*, B834–B857. [CrossRef]
115. Rahman, S.M.; Ahmed, S.E.; San, O. A dynamic closure modeling framework for model order reduction of geophysical flows. *Phys. Fluids* **2019**, *31*, 046602. [CrossRef]

*Article*

# Closure Learning for Nonlinear Model Reduction Using Deep Residual Neural Network

**Xuping Xie [1], Clayton Webster [2] and Traian Iliescu [3,***

[1]   Courant Institute of Mathematical Sciences, New York University, New York, NY 10012, USA; xxie@nyu.edu
[2]   Department of Mathematics, University of Tennessee, Knoxville, TN 37996, USA; cwebst13@utk.edu
[3]   Department of Mathematics, Virginia Tech, Blacksburg, VA 24061, USA
*   Correspondence: iliescu@vt.edu; Tel.: +1-540-231-5296

Received: 18 March 2020; Accepted: 19 March 2020; Published: 23 March 2020

**Abstract:** Developing accurate, efficient, and robust closure models is essential in the construction of reduced order models (ROMs) for realistic nonlinear systems, which generally require drastic ROM mode truncations. We propose a deep residual neural network (ResNet) closure learning framework for ROMs of nonlinear systems. The novel ResNet-ROM framework consists of two steps: (i) In the first step, we use ROM projection to filter the given nonlinear system and construct a spatially filtered ROM. This filtered ROM is low-dimensional, but is not closed. (ii) In the second step, we use ResNet to close the filtered ROM, i.e., to model the interaction between the resolved and unresolved ROM modes. We emphasize that in the new ResNet-ROM framework, data is used only to complement classical physical modeling (i.e., only in the closure modeling component), not to completely replace it. We also note that the new ResNet-ROM is built on general ideas of spatial filtering and deep learning and is independent of (restrictive) phenomenological arguments, e.g., of eddy viscosity type. The numerical experiments for the 1D Burgers equation show that the ResNet-ROM is significantly more accurate than the standard projection ROM. The new ResNet-ROM is also more accurate and significantly more efficient than other modern ROM closure models.

**Keywords:** reduced order model; closure model; variational multiscale method; deep residual neural network

---

## 1. Introduction

Many scientific and engineering applications, such as weather forecasting, ocean modeling, and cardiovascuar flow simulation, can often be represented by multiscale systems of ordinary differential equations (ODE) in high-dimensional modal spaces. The analysis and high-fidelity simulation of such systems can be very expensive even on high-performance computing systems. Consequently, using the full order model (FOM) for such simulations can be impractical for time critical applications, such as flow control and parameter estimation. To alleviate the computation burden of the FOM simulation, reduced order models (ROMs) have been successfully used.

ROMs seek a low-dimensional approximation of the FOM with orders of magnitude reduction in computational cost. The classical projection based ROM approach is first to construct a low dimensional space using data-driven reduction methods, such as proper orthogonal decomposition (POD) or dynamical modal decomposition (DMD). The ROM dynamics can be obtained via Galerkin projection of the FOM onto the reduced space [1–5].

The Galerkin projection reduced order model (GP-ROM) can be very efficient and relatively accurate for many systems [3,4]. There are, however, systems (e.g., convection-dominated fluid flows) for which the GP-ROM can generate inaccurate approximations. There are several approaches to address this inaccuracy (see, e.g., [6–9]). In this paper, we focus on one of the main reasons for this inaccuracy: the ROM closure problem, i.e., modeling the interaction between the GP-ROM modes and the discarded modes. Indeed, due to the inherently drastic mode truncation required in realistic settings, the dimension of the GP-ROM space is too low to resolve the complex nonlinear interactions of the fluid system [1,2,10,11]. GP-ROMs that do not include a ROM closure model can yield inaccurate results, often in the form of spurious numerical oscillations [1,2,12,13]. Endowing GP-ROM with closure models could extend the applicability of GP-ROM in many fluid mechanics applications, such as flow control, climate modeling, and weather forecasting [1,2].

ROM closure models for nonlinear systems have been proposed in, e.g., [14–24]. The vast majority of the current ROM closure models aim at mitigating the numerical instability observed in GP-ROMs that do not include a closure model. Some of these ROM closure models use stabilization techniques that have been developed in standard discretization methods (e.g., in the finite element community) [25–27]. Other ROM closure models have imported ideas developed in standard CFD methodologies, e.g., large eddy simulation (LES) [1,24,28,29]. The overwhelming majority of the current ROM closure models can be categorized as stabilization techniques (for a notable exception, see the approximate deconvolution ROM closure model [28] that uses a mathematical framework inspired from image processing).

This is in stark contrast with classical LES, where a plethora of closure models have been proposed over the years. The main difference between ROM closure and LES closure is that the latter has been entirely built around physical insight from the statistical theory of turbulence (e.g., energy cascade and Kolmogorov's theory), which is generally posed in the Fourier setting [30,31]. This physical insight is generally not available in the ROM setting (see, however, [32] for initial steps). Thus, current ROM closure models have generally been deprived of this powerful methodology that represents the core of most LES closure models.

We believe that machine learning represents a natural solution for ROM closure modeling. Indeed, since physical insight cannot be easily extended to the ROM setting, available data and machine learning can be utilized instead to develop ROM closure models.

We propose a novel ROM closure modeling framework that is constructed by using available data and deep residual neural network (ResNet) (for details, see, e.g., [33–37]). The resulting ROM, which we call the *residual neural network ROM (ResNet-ROM)*, is schematically illustrated in (1) and Figure 1 (see Section 3 for details). We emphasize that, in the new ResNet-ROM framework, data is used only to *complement* classical physical modeling (i.e., only in the closure modeling component) [29,38], not to completely replace it [39]. Thus, the resulting ResNet-ROM combines the strengths of both physical and data-driven modeling.

$$\text{FOM} \xrightarrow{\text{filtering}} \text{GP-ROM} \xrightarrow{\text{learning closure}} \text{ROM} \tag{1}$$

The main contributions of this paper can be summarized as follows:

- A novel ROM closure learning framework centered around deep neural networks.
- A hybrid framework that synthesizes the strengths of physical modeling and data-driven modeling.
- Very good performance in numerical tests, in both the reconstructive and the predictive regime.
- Significant improvement in numerical accuracy compared with state of the art ROM closure models.

Machine learning has recently been utilized to construct ROM closure models (see, e.g., [22,38,40–42]). The ROM closure model proposed in [22] is similar to the new ResNet-ROM. There are, however, two major differences between the two ROM closure models: The first difference is that the ROM closure model in [22] uses FOM data to model both the linear and the nonlinear terms in the underlying equations (see Section 5.1 in [22]). In contrast, in the LES spirit [31], the ResNet-ROM models only the nonlinear terms (see Algorithm 1). One motivation for modeling only the nonlinear terms is given in the theoretical and numerical investigation in [43], where it was shown that the contribution of the linear terms to the ROM closure term (i.e., the commutation error) is not significant in convection-dominated problems, such as those we consider in this study. The second major difference between the ROM closure model in [22] and the ResNet-ROM is the machine learning approach used to construct the ROM closure model: Bayesian regularization and extreme learning machine approaches are used in [22], whereas a deep residual neural network (ResNet) is used to construct the ResNet-ROM.

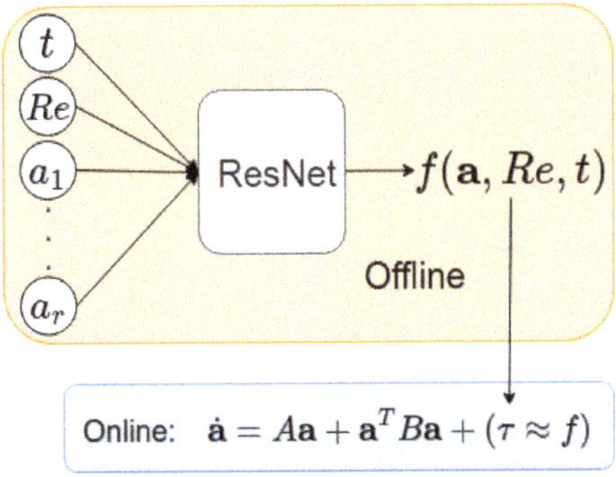


**Figure 1.** Flow chart of the new ResNet-ROM.

## 2. Reduced Order Model

To construct the novel ResNet-ROM framework, we use the Navier-Stokes equations (NSE) for incompressible fluid flows:

$$
\begin{cases}
\dfrac{\partial u}{\partial t} - Re^{-1}\,\Delta u + (u \cdot \nabla)\,u + \nabla p = 0, \\[2mm]
\hspace{3cm} \nabla \cdot u = 0,
\end{cases}
\tag{2}
$$

which are defined on a spatial domain $\Omega$ and on the time interval $[0, T]$. In the NSE (2), $u$ is the velocity, $p$ the pressure, and $Re$ the Reynolds number. The ROM basis $\{\varphi_1, \ldots, \varphi_r\}$, where $r$ is small, represents the large, energy containing structures in the flow, and is obtained by using available numerical or experimental data and, e.g., the POD or DMD methods [1,3,4]. The ROM velocity approximation is defined as

$$
u_r(\mathbf{x}, t) \equiv \sum_{j=1}^{r} a_j(t)\varphi_j(\mathbf{x}),
\tag{3}
$$

where $\{a_j(t)\}_{j=1}^{r}$ are the sought time-varying coefficients. To determine these coefficients, we use a Galerkin procedure. We replace $u$ in (2) with the ROM approximation (3), we take the $L^2$ inner product between the resulting system and each ROM basis function $\{\varphi_1, \ldots, \varphi_r\}$, and then we integrate by parts: $\forall\, i = 1, \ldots, r,$

$$
\left(\frac{\partial u_r}{\partial t}, \varphi_i\right) + \frac{1}{Re}\left(\nabla u_r, \nabla \varphi_i\right) + \left((u_r \cdot \nabla)\,u_r, \varphi_i\right) = 0.
$$

To derive the above equation, we assumed that the ROM velocity modes are perpendicular to the discrete pressure space, which is the case if standard mixed FEs (e.g., Taylor-Hood) are used for the snapshot creation [1,29]. Using (3), we obtain the standard *Galerkin projection ROM (GP-ROM)*:

$$
\dot{a} = A\,a + a^{\top} B\,a,
\tag{4}
$$

which can be written componentwise as follows: $\forall\, i = 1 \ldots r,$

$$\dot{a}_i = \sum_{m=1}^{r} A_{im}\, a_m(t) + \sum_{m=1}^{r} \sum_{n=1}^{r} B_{imn}\, a_n(t)\, a_m(t)\,, \tag{5}$$

where $A_{im} = -Re^{-1}\,(\nabla\boldsymbol{\varphi}_m, \nabla\boldsymbol{\varphi}_i)$ and $B_{imn} = -(\boldsymbol{\varphi}_m \cdot \nabla\boldsymbol{\varphi}_n, \boldsymbol{\varphi}_i)$.

## 3. Closure Learning

### 3.1. Residual Neural Network (ResNet)

Deep residual neural network (ResNet) has been first introduced for image recognition in [33]. ResNet has been widely studied and applied in many supervised learning tasks. Recent mathematical understanding of deep ResNet has been achieved in the ODE representation of ResNet; for a comprehensive introduction, see, e.g., [35–37,44].

To construct the novel ResNet-ROM framework, we consider the ResNet model, which is illustrated in Figure 2.

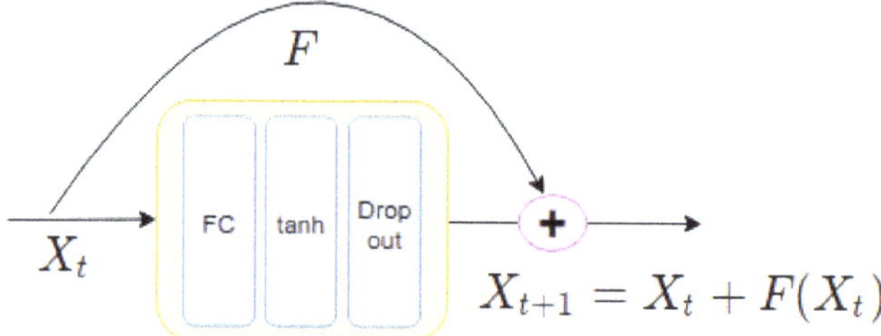


**Figure 2.** ResNet block used in our training. Each ResNet block consists of a flatten fully connected layer with a tanh activation function, followed by a drop out layer to prevent overfitting. 128 neurons and 5% drop out rate are used.

The input values of forward propagation in the ResNet are given by

$$X_{t+1} = X_t + \tanh(W_t X_t + b_t), \quad t = 1, \ldots, N-1, \tag{6}$$

where $N$ is the number of layers in the network architecture, and $X_t \in \mathbb{R}^s$ is the output from each ResNet block at time $t$. $W_t$ and $b_t$ are the weight matrix and bias at each layer, respectively. The ResNet propagation starts from step $t = 1$ with the nonlinear activation function tanh. The initial input layer for the network is $X_0 = [Re, t, a_1, \ldots, a_r]^T$. For a standard ResNet in image classification, a convolution layer (CNN) is often included in a residual block. In our work, we use a simplified version of the ResNet that does not include CNN.

### 3.2. ROM Closure Modeling

In realistic nonlinear systems (e.g., convection-dominated flows), current GP-ROMs of the form (4) generally yield inaccurate results, often in the form on numerical instabilities. This inaccurate behavior is due to the fact that, in order to maintain a low computational cost, GP-ROMs are constructed with a drastically truncated ROM basis $\{\boldsymbol{\varphi}_1, \ldots, \boldsymbol{\varphi}_r\}$, where $r$ is generally small (e.g., $r = \mathcal{O}(10)$). In realistic nonlinear systems, this extreme truncation cannot capture the complex nonlinear interactions among the various degrees of freedom of the system. Thus, current GP-ROMs are computationally efficient, but numerically inaccurate. To alleviate this inaccurate behavior, GP-ROMs are often supplemented with a ROM closure model [1,24], i.e., a model for the important interactions between the ROM basis

$\{\boldsymbol{\varphi}_1,\ldots,\boldsymbol{\varphi}_r\}$ and the discarded ROM modes $\{\boldsymbol{\varphi}_{r+1},\ldots,\boldsymbol{\varphi}_R\}$, where $R$ is the dimension of the input data. For example, for the NSE, the standard GP-ROM (4) is generally modified as follows:

$$\dot{a} = A\,a + a^\top B\,a + \tau, \tag{7}$$

where $\tau$ is a *ROM closure model* that represents the interactions between the ROM modes and the discarded modes.

We emphasize that the same closure problem needs to be addressed when classical numerical discretization schemes (e.g., finite element or spectral methods) are used in the numerical simulation of turbulent flows. In those settings, classical discretizations schemes are used in inherently under-resolved regimes (i.e., they use coarse meshes or too few spectral modes). To ensure the relative accuracy of these under-resolved simulations, various types of closure models for the unresolved (e.g., subgrid-scale) information are generally utilized. These closure models are central to, e.g., large eddy simulation (LES) [31], one of the main approaches to the numerical simulation of turbulent flows. The vast majority of LES closure models have been constructed by using physical insight from Kolmogorov's statistical theory of turbulence. The concept of energy cascade is central in the development of LES closure models. The energy cascade states that energy enters the system at the large scales, is transferred to smaller and smaller scales through nonlinear interactions, and is dissipated at the smallest scale (i.e., the Kolmogorov's scale). Thus, most LES closure models (e.g., of eddy viscosity type) aim at recovering the energy cascade displayed by the original system (i.e., the NSE).

This physical insight cannot be easily extended to the ROM setting (see, however, [32] for a preliminary numerical investigation). Thus, current ROM closure models have generally been deprived of this powerful methodology that represents the main tool in the development of most LES closure models.

### 3.3. ROM Closure Learning

Our vision is that machine learning represents a natural approach for constructing ROM closure modeling. Indeed, since physical insight cannot generally be used in a ROM setting, data and machine learning can be utilized instead to develop ROM closure models. Furthermore, data is used only to construct the ROM closure model; the other ROM operators are built by using the classical Galerkin projection. Thus, data and machine learning complement (instead of replace) physical based modeling, yielding a hybrid framework that synthesizes the strengths of both approaches [17,18,22,27,29,42,45–48].

In this section, we propose a novel ROM closure modeling framework that is constructed by using available data and the ResNet approach described in Section 3.1. The resulting ROM, which we call the *residual neural network ROM (ResNet-ROM)*, is schematically illustrated in (1) and Figure 1, and is summarized in Algorithm 1.

To obtain the explicit formula (11) in Algorithm 1, we develop a *large eddy simulation ROM (LES-ROM)* framework [24,28]: First, we filter the high-resolution ($R$-dimensional) ROM approximation of the NSE with a low-pass ROM spatial filter, denoted by overbar in (11). In this paper, as a ROM spatial filter, we use the ROM projection onto the space spanned by the first $r$ ROM modes [29,48] (see [28] for an alternative ROM spatial filter): For a given $u \in \text{span}\{\boldsymbol{\varphi}_1,\ldots,\boldsymbol{\varphi}_R\}$, the ROM projection seeks $\overline{u} \in \text{span}\{\boldsymbol{\varphi}_1,\ldots,\boldsymbol{\varphi}_r\}$ such that

$$(\overline{u}, \varphi_i) = (u, \varphi_i) \qquad \forall\, i = 1,\ldots,r, \tag{8}$$

where $(\cdot,\cdot)$ denotes the $L^2$ inner product. The resulting filtered equations can be cast within a *variational multiscale (VMS)* framework [49], yielding a two-scale VMS-ROM [50]. Equation (11) represents the explicit VMS-ROM formula for the ROM closure term $\tau$ in (10). In the two-scale VMS-ROM [50], the ROM closure model is developed in two steps. First, an ansatz is used to approximate the ROM closure term: $\tau \approx \tilde{A}\,a + a^\top \tilde{B}\,a$. Then, FOM data for $\tau$ in the training time interval is used to solve a least squares problem to determine the optimal entries in $\tilde{A}$ and $\tilde{B}$. In the new ResNet-ROM, we construct the ROM closure model in a fundamentally different way. Instead of making an ansatz on the structural form of the ROM closure term $\tau$ (as in the two-scale VMS-ROM [50]), in Algorithm 1 we consider a general structural formulation and use the ResNet approach to construct the ROM closure model. To this end, in (6), we make the following choices: The initial layer contains the Reynolds number ($Re$), the current time ($t$), and the current ROM coefficients in (3) ($a_1, \ldots, a_r$), i.e., $X_0 = [Re, t, a_1, \ldots, a_r] \in \mathbb{R}^{(s+2)}$. The final output layer provides an approximation for the closure term, i.e., $X_N \approx \tau^{FOM}$. The optimization problem associated with this ResNet is given by

$$\min \|\tau^{ansatz} - \tau^{FOM}\|_F^2 + \lambda R(W, b), \tag{9}$$

where the regularizer $R$ penalizes undesirable parameters and can prevent overfitting, $\tau^{ansatz} = f(a, Re, t_j)$ is the output from the neural network, $W$ and $b$ are weights in the network, $\|\cdot\|_F$ is the Frobenius norm, and $\lambda$ is a hyperparameter for the $L^2$ regularization to prevent overfitting [36,37]. The minimization problem (9) is over the parameters defining the structural form of the function $f$ used to define $\tau^{ansatz}$.

---

**Algorithm 1** ResNet-ROM

---

1: Consider the ROM closure model

$$\dot{a} = A\,a + a^\top B\,a + \tau. \tag{10}$$

2: Use snapshot data to compute the true vector $\tau$ in (10), $\tau^{FOM}$:

$$\tau_i^{FOM}(t_j) \;=\; -\Bigg( \overline{\left(u_R^{FOM}(t_j) \cdot \nabla\right) u_R^{FOM}(t_j)}^r$$

$$-\left(u_r^{FOM}(t_j) \cdot \nabla\right) u_r^{FOM}(t_j), \varphi_i \Bigg), \tag{11}$$

where an overbar indicates spatial filtering with ROM projection.

3: Use snapshot data and (11) to define the approximation function $\tau$ in (10), $\tau^{ansatz}$:

$$\tau^{ansatz}(t_j) = f(a, Re, t_j), \tag{12}$$

where $f$ is a generic function that needs to be determined.

4: Use ResNet to train the closure term, i.e., to find the form of $f$ in (12) that is optimal with respect to the minimization problem (9).

5: The novel ResNet-ROM has the following form:

$$\dot{a} = A\,a + a^\top B\,a + f(a, Re, t). \tag{13}$$

---

## 4. Numerical Experiments

### 4.1. Implementation

As a test problem for our new approach, we use the 1D Burgers equation, which has been used to test new ROM ideas in simplified settings [29,51,52]:

$$\begin{cases} u_t - Re^{-1}u_{xx} + uu_x = 0 & x \in \Omega, \\ u(x,0) = u_0(x) & x \in \Omega, \\ u(x,t) = 0 & x \in \partial\Omega, \end{cases} \tag{14}$$

where, for consistency with the notation used for the NSE, the diffusion parameter is denoted as $Re^{-1}$. In our numerical tests, $\Omega = [0,1]$ is the computational domain and the time domain is $[0,1]$. We use the same initial conditions as those utilized in [29,51,52]: $u_0(x) = 1, x \in (0,1/2], u_0(x) = 0, x \in (1/2,1]$. These initial conditions yield a steep internal layer that is difficult to capture in the convection-dominated regime that we consider [29,51,52]. We first use a piecewise linear finite element discretization to generate the FOM solution. To this end, we utilize a uniform mesh with $N = 1024$ grid points (which yields a meshsize $h = 1/1024$) and the forward Euler method with $\Delta t = 10^{-4}$ for the time discretization. To construct the ROM basis, we collect 101 snapshots sampled from $[0,1]$. We build the neural network in PyTorch and we train the ROM closure model with a 6 block ResNet with the common Adam optimizer [53]. We perform all the computational experiments on a Linux system laptop with Nvidia Geforce GTX GPU hardware.

### 4.2. Reconstruction

In this section, we consider the reconstructive regime, i.e., we test the ROMs at the same $Re$ as the $Re$ at which the ROMs are constructed. We choose $Re = 1000$ in (14) and we use $r = 6$ basis functions in all ROMs. In Figure 3, we plot the solutions of the FOM (top left), GP-ROM (top right), and ResNet-ROM (bottom left). When compared with the FOM data, the ResNet-ROM solution is significantly more accurate than the standard GP-ROM solution. In Figure 3, we also plot the FOM, GP-ROM, and ResNet-ROM solutions at the final time step (i.e., at $t = 1$). This plot shows that the closure term in the ResNet-ROM plays an important role in stabilizing the ROM approximation. Indeed, the GP-ROM solution displays large, spurious numerical oscillations. These oscillations are dramatically decreased in the ResNet-ROM solution.

In Figure 4, we plot the the time evolution of the ROM coefficients $a$ for the FOM, GP-ROM, and ResNet-ROM. The plots show that the ResNet-ROM is significantly more accurate than the standard GP-ROM.

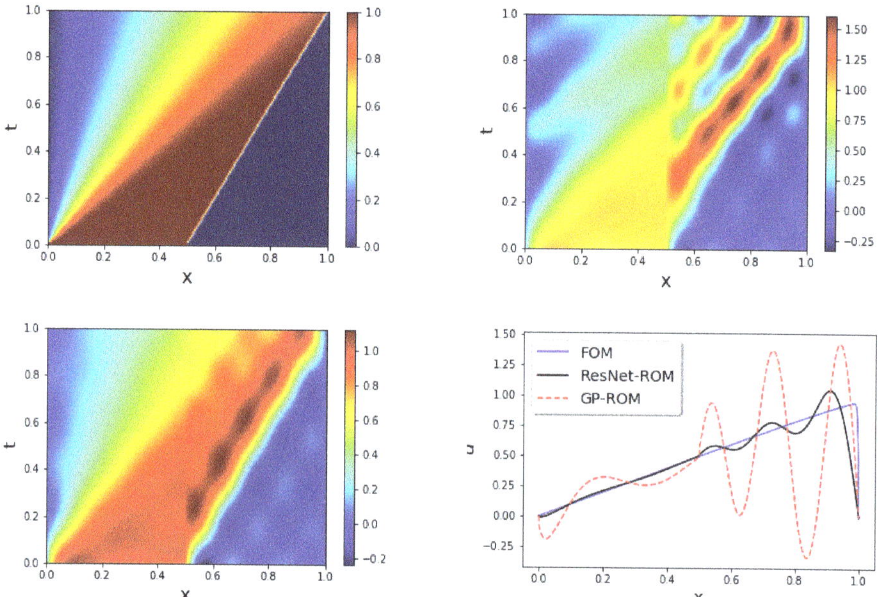

**Figure 3.** Reconstructive regime, $Re = 1000$: FOM (**top left**), GP-ROM (**top right**), ResNet-ROM (**bottom left**), and final time solution for all three simulations (**bottom right**). ResNet-ROM yields the most accurate solution.

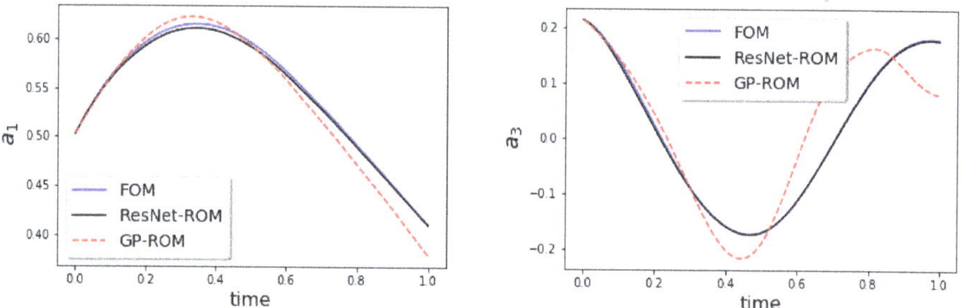

**Figure 4.** Reconstructive regime, $Re = 1000$: Time evolution of ROM coefficients $a_1$ and $a_3$ for FOM, GP-ROM, and ResNet-ROM. ResNet-ROM yields the most accurate solution.

### 4.3. Prediction

To study the robustness of the new ResNet-ROM, we test its predictability, i.e., we train the ResNet-ROM closure term on data from multiple $Re$ and we then test the ResNet-ROM to predict the ROM dynamics at different $Re$. The training data space is sampled at $Re = [20, 50, 100, 200, 500, 800, 1000]$, and the test data contains $Re = [30, 80, 300, 1200]$. Note that the solution of the Burgers equation is affected by $Re$. Small $Re$ values yield a slow movement of the sharp internal layer, while large $Re$ values can speed up this movement.

In Figure 5, for $Re = 30, 80$, and 1200 (which are different from the training $Re$ values), we plot the solutions for the DNS (first column), GP-ROM (second column), ResNet-ROM (third column), and final time solution for all three simulations (fourth column). These plots show that the ResNet-ROM is consistently the most accurate ROM, especially for the largest $Re$ value. In Figure 5, we also plot the

FOM, GP-ROM, and ResNet-ROM solutions at the final time step (i.e., at $t = 1$). These plots show that the closure term in the ResNet-ROM plays an important role in stabilizing the ROM approximation.

In Figure 6, we plot the the time evolution of the ROM coefficients $a$ for the FOM, GP-ROM, and ResNet-ROM. These plots show that the ResNet-ROM is significantly more accurate than the standard GP-ROM.

Overall, we draw the same conclusion as in the reconstructive regime (Section 4.2): In all cases, the ResNet-ROM is significantly more accurate than the standard GP-ROM in the predictive regime.

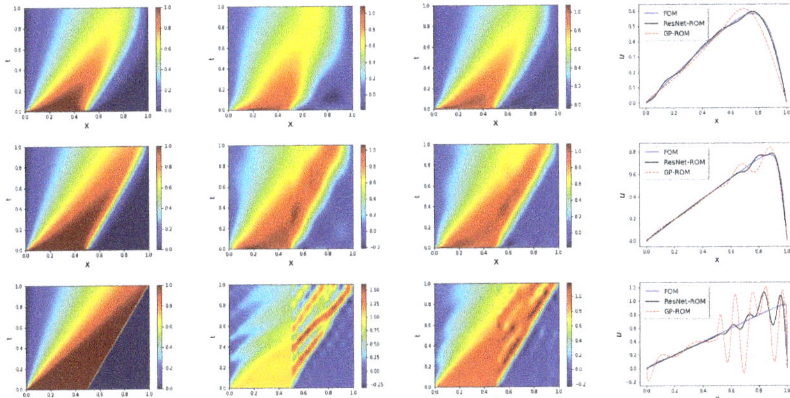

**Figure 5.** Predictive regime: ROMs are trained on data from $Re = [20, 50, 100, 200, 500, 800, 1000]$ and are tested at $Re = 30$ (**first row**), $Re = 80$ (**second row**), $Re = 1200$ (**third row**). Results presented for FOM (**first column**), GP-ROM (**second column**), ResNet-ROM (**third column**), and final time solution for all three simulations (**fourth column**).

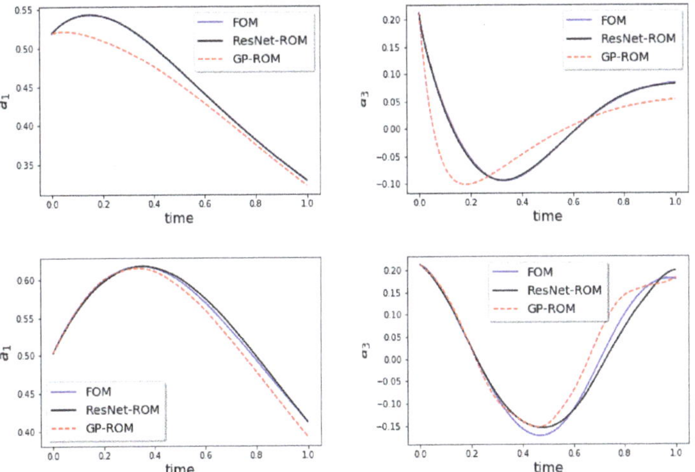

**Figure 6.** Predictive regime: Time evolution of ROM coefficients $a_1$ and $a_3$ for FOM, GP-ROM, and ResNet-ROM. Results for $Re = 30$ (**top**) and $Re = 1200$ (**bottom**). ResNet-ROM yields the most accurate solution.

*4.4. Comparison*

In this section, in the numerical simulation of the Burgers equation, we also make a numerical comparison between the new ResNet-ROM and several modern ROM closure models: the POD artificial viscosity model (POD-AV) [51], the POD Smagorinsky model (POD-L) [54], the evolve-then-filter ROM (EF-ROM) [55], the approximate deconvolution ROM (AD-ROM) [28], and the data-driven filtered ROM (DDF-ROM) [29].

In Table 1, we list the $L^2$ errors of the ResNet-ROM and the other closure models. These results show that the ResNet-ROM error is at least *an order of magnitude lower* than the errors of the other ROM closure models.

**Table 1.** $L^2$ errors of the new ResNet-ROM and other closure models.

| Model | $L^2$ Error |
| --- | --- |
| ResNet-ROM | $4.32 \times 10^{-4}$ |
| POD-AV [51] | $9.01 \times 10^{-3}$ |
| POD-L [54] | $1.73 \times 10^{-2}$ |
| EF-ROM [55] | $6.99 \times 10^{-2}$ |
| AD-ROM [28] | $6.33 \times 10^{-2}$ |
| DDF-ROM [29] | $6.27 \times 10^{-2}$ |

For the online ResNet-ROM integration, we use the *scipy* package with the built-in integration function *odeint*. In the online stage, the new ResNet-ROM is more efficient than the other ROM closure models: The online CPU time of the ResNet-ROM is 3.89 s, whereas the online CPU times of the EF-ROM, AD-ROM, and DDF-ROM are 6.91, 7.26, and 4.42 s, respectively [28,29,55].

The CPU time of the offline training of the new ResNnet-ROM is 122.74 s for the current dataset with 10,000 epochs (iterations). Thus, even though the online cost for ResNet-ROM is lower than the online cost of the other closure models that are compared in this paper, the neural network training cost of ResNet-ROM is much higher than the offline training cost of the other ROM closure models [28,29,55].

*4.5. Sensitivity*

In this section, we perform a sensitivity study of the new ResNet-ROM with respect to two parameters: (i) the hyperparameter $\lambda$ used in the regularization of the neural network training; and (ii) the meshsize $h$ used in the snapshot generation. We also investigate the potential improvement in GP-ROM accuracy when the number of snapshots and the dimension $r$ of the ResNet-ROM are increased.

The parameter $\lambda$ is an $L^2$ regularization parameter used in the neural network training to prevent overfitting. In our numerical simulations, we pick the parameter $\lambda$ based on the validation error performance in the training phase. In our training dataset for $Re = [20, 50, 100, 200, 500, 800, 1000]$, we test the following parameter values: $\lambda = 0, 1, 10^{-1}, 10^{-2}, 10^{-3}$. In Figure 7, we plot the ResNet-ROM error for different parameter values. The plot in Figure 7 shows that the ResNet-ROM is not very sensitive to the hyperparameter $\lambda$. Thus, in our numerical tests, we fix $\lambda = 0.01$ with 10,000 epochs (iterations) in the Resnet-ROM training.

We also perform a sensitivity study for the new ResNet-ROM with respect to the meshsize $h$. In Figure 8, for the reconstructive regime and $Re = 1000$, we plot the FOM, GP-ROM, and ResNet-ROM results for three coarse meshsize values. The plots in Figure 8 show that the ResNet-ROM is still significantly more accurate than the GP-ROM. In the predictive regime, however, both the ResNet-ROM and the GP-ROM yield inaccurate results for these three coarse meshes. The relationship between the FOM mesh utilized to generate the snapshots and the ROM accuracy is subject of current research (see, e.g., [56,57]) and should be further investigated for the new ResNet-ROM.

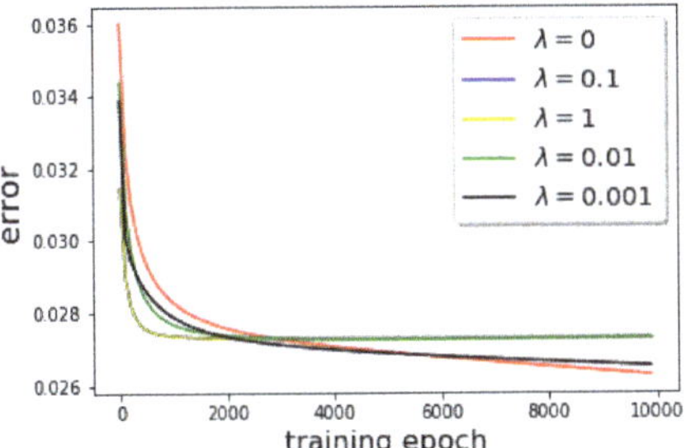

**Figure 7.** Validation error performance for different values of the regularization parameter $\lambda$.

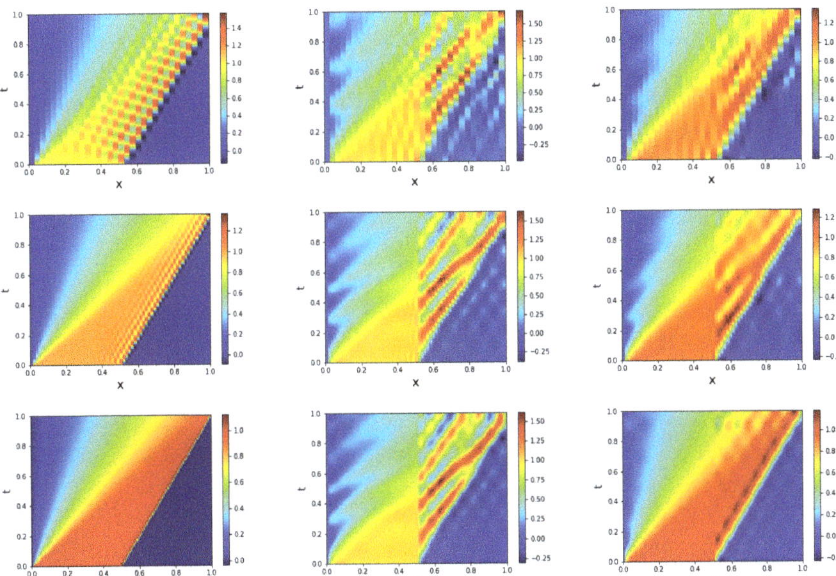

**Figure 8.** Reconstructive regime, $Re = 1000$: FOM (**first column**), GP-ROM (**second column**), and ResNet-ROM (**third column**), for meshsizes $h = 1/64$ (**first row**), $h = 1/128$ (**second row**), and $h = 1/256$ (**third row**). ResNet-ROM yields the most accurate solution.

We also investigate the ResNet-ROM rate of convergence with respect to the meshsize $h$. To this end, we fixed the ResNet-ROM dimension at $r = 30$ and the time-step size at $\Delta t = 10^{-3}$. The plot in Figure 9 shows that the rate of convergence (obtained with a least squares fit) is about $h^{1.70}$, which is an acceptable approximation to the theoretical rate of convergence of $h^2$.

Finally, we investigate the potential improvement in GP-ROM accuracy when the number of snapshots and the dimension $r$ of the ResNet-ROM are increased.

First, we collect the maximum number of snapshots that are available from the FOM simulations in the training interval. That is, we collect 1001 snapshots, which yield a snapshot matrix whose rank is 251. Comparing the left plot in Figure 10 (for the highest number of snapshots) with the top right plot in Figure 3 (for the lower number of snapshots), we conclude that increasing the number of snapshots does not seem to improve the ResNet-ROM accuracy.

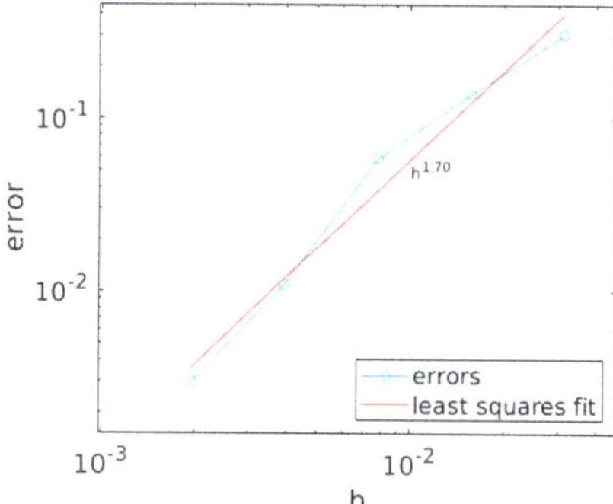



**Figure 9.** The ResNet-ROM rate of convergence with respect to the meshsize $h$.

Next, we increase the GP-ROM dimension. In Figure 10, we plot the GP-ROM results for three $r$ values: $r = 10$, $r = 20$, and $r = 30$. As expected, the GP-ROM accuracy improves as we increase $r$. We emphasize, however, that the ROM closure models (such as that used in the new ResNet-ROM) are designed to improve the GP-ROM accuracy in under-resolved numerical simulations, i.e., when only a few ROM basis functions can be used, which is often the case in realistic settings [1,2,14,16–18,20–23,25,27–29,32,38,40–42,46–48,50,51,54,58]. As shown earlier, for $r = 10$, the ResNet-ROM is significantly more accurate than the GP-ROM both in the reconstructive and the predictive regimes.

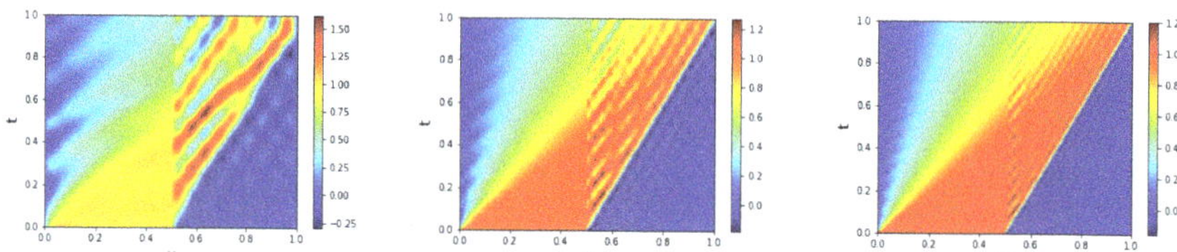

**Figure 10.** Reconstructive regime, $Re = 1000$, GP-ROM results: $r = 10$ (**left**), $r = 20$ (**middle**), and $r = 30$ (**right**).

## 5. Conclusions

In this paper, we used available data and deep residual neural networks (ResNet) to construct a novel reduced order model (ROM) closure for complex nonlinear settings. We emphasize that the ResNet-ROM closure terms are much more general than the phenomenological ansatzes generally used in ROM closure modeling [29]. We tested the novel ResNet-ROM in the numerical simulation of the Burgers equation. For comparison purposes, we investigated the standard Galerkin projection ROM (GP-ROM) and the full order model (FOM). We considered two settings: (i) a reconstructive regime, in which the Reynolds number $Re$ is the same in the training and testing stages; and (ii) a predictive regime, in which $Re$ used in the testing regime is different from the $Re$ used in the training regime. In both regimes, the new ResNet-ROM was consistently more accurate than the standard GP-ROM. Furthermore, the ResNet-ROM was also dramatically more accurate than several other ROM closure models from the literature.

There are several research directions that we plan to pursue: We will test the novel ResNet-ROM on realistic test problems (e.g., 3D turbulent flows) and we will compare it with state of the art closure models. We will also investigate alternative approaches to develop the ROM closure term $\tau$. Indeed, in this paper we used the ROM projection as a spatial filter in the construction of the ROM closure term $\tau$. We plan to investigate different ROM spatial filters, such as the ROM differential filter [55].

This alternative ROM filter will yield a different ROM closure term $\tau$ and, therefore, a different ResNet-ROM. Finally, a numerical investigation of alternative machine learning approaches (see, e.g., [22]) could yield improved ROM closure models. Indeed, for the one-dimensional Burgers equation test case considered in this paper, the computational cost of training the ROM closure model with ResNet was acceptable. For more complex settings, however, this computational cost could be much higher. In those cases, neural networks with a lower computational cost could be more effective.

**Author Contributions:** Conceptualization, X.X., C.W. and T.I.; methodology, X.X., C.W. and T.I.; writing—original draft preparation, X.X.; writing—review and editing, C.W., T.I. All authors have read and agreed to the published version of the manuscript.

**Funding:** The work of the second author was supported by U.S. Department of Energy, Office of Science, Office of Advanced Scientific Computing Research. The work of the third author was supported by National Science Foundation grant DMS-1821145.

**Conflicts of Interest:** The authors declare no conflict of interest.

## Abbreviations

The following abbreviations are used in this manuscript:

| | |
|---|---|
| ResNet | Residual Neural Network |
| ROM | Reduced Order Modeling |
| GP-ROM | Galerkin Projection Reduced Order Model |
| POD | Proper Orthogonal Decomposition |
| FOM | Full Order Model |
| LES | Large Eddy Simulation |
| VMS | Variational Multiscale |
| NSE | Navier-Stokes Equations |

## References


1. Holmes, P.; Lumley, J.L.; Berkooz, G. *Turbulence, Coherent Structures, Dynamical Systems and Symmetry*; Cambridge University Press: Cambridge, UK, 1996.
2. Noack, B.R.; Morzynski, M.; Tadmor, G. *Reduced-Order Modelling for Flow Control*; Springer: Berlin/Heidelberg, Germany, 2011; Volume 528.
3. Hesthaven, J.S.; Rozza, G.; Stamm, B. *Certified Reduced Basis Methods for Parametrized Partial Differential Equations*; Springer: Berlin/Heidelberg, Germany, 2015.
4. Quarteroni, A.; Manzoni, A.; Negri, F. *Reduced Basis Methods for Partial Differential Equations: An Introduction*; Springer: Berlin/Heidelberg, Germany, 2015; Volume 92.
5. Mohebujjaman, M.; Rebholz, L.G.; Xie, X.; Iliescu, T. Energy balance and mass conservation in reduced order models of fluid flows. *J. Comput. Phys.* **2017**, *346*, 262–277. [CrossRef]
6. Akkari, N.; Casenave, F.; Moureau, V. Time stable reduced order modeling by an enhanced reduced order basis of the turbulent and incompressible 3D Navier–Stokes equations. *Math. Comput. Appl.* **2019**, *24*, 45. [CrossRef]
7. Cagniart, N.; Maday, Y.; Stamm, B. Model order reduction for problems with large convection effects. In *Contributions to Partial Differential Equations and Applications*; Springer: Berlin, Germany, 2019; pp. 131–150.
8. Iollo, A.; Lanteri, S.; Désidéri, J.A. Stability properties of POD–Galerkin approximations for the compressible Navier–Stokes equations. *Theoret. Comput. Fluid Dyn.* **2000**, *13*, 377–396. [CrossRef]
9. Stabile, G.; Hijazi, S.; Mola, A.; Lorenzi, S.; Rozza, G. POD-Galerkin reduced order methods for CFD using finite volume discretisation: Vortex shedding around a circular cylinder. *Commun. Appl. Ind. Math.* **2017**, *8*, 210–236. [CrossRef]
10. Loiseau, J.C.; Brunton, S.L. Constrained sparse Galerkin regression. *J. Fluid Mech.* **2018**, *838*, 42–67. [CrossRef]
11. Noack, B.R.; Stankiewicz, W.; Morzyński, M.; Schmid, P.J. Recursive dynamic mode decomposition of transient and post-transient wake flows. *J. Fluid Mech.* **2016**, *809*, 843–872. [CrossRef]

12. Amsallem, D.; Farhat, C. Stabilization of projection-based reduced-order models. *Int. J. Num. Meth. Eng.* **2012**, *91*, 358–377. [CrossRef]
13. Carlberg, K.; Farhat, C.; Cortial, J.; Amsallem, D. The GNAT method for nonlinear model reduction: Effective implementation and application to computational fluid dynamics and turbulent flows. *J. Comput. Phys.* **2013**, *242*, 623–647. [CrossRef]
14. Baiges, J.; Codina, R.; Idelsohn, S. Reduced-order subscales for POD models. *Comput. Methods Appl. Mech. Eng.* **2015**, *291*, 173–196. [CrossRef]
15. Feppon, F.; Lermusiaux, P.F.J. Dynamically orthogonal numerical schemes for efficient stochastic advection and Lagrangian transport. *SIAM Rev.* **2018**, *60*, 595–625. [CrossRef]
16. Fick, L.; Maday, Y.; Patera, A.T.; Taddei, T. A stabilized POD model for turbulent flows over a range of Reynolds numbers: Optimal parameter sampling and constrained projection. *J. Comput. Phys.* **2018**, *371*, 214–243. [CrossRef]
17. Hijazi, S.; Stabile, G.; Mola, A.; Rozza, G. Data-driven POD-Galerkin reduced order model for turbulent flows. *arXiv* **2019**, arXiv:1907.09909.
18. Lu, F.; Lin, K.K.; Chorin, A.J. Data-based stochastic model reduction for the Kuramoto–Sivashinsky equation. *Phys. D* **2017**, *340*, 46–57. [CrossRef]
19. Majda, A.J.; Harlim, J. Physics constrained nonlinear regression models for time series. *Nonlinearity* **2012**, *26*, 201. [CrossRef]
20. Rebollo, T.C.; Ávila, E.D.; Mármol, M.G.; Ballarin, F.; Rozza, G. On a certified Smagorinsky reduced basis turbulence model. *SIAM J. Numer. Anal.* **2017**, *55*, 3047–3067. [CrossRef]
21. San, O.; Iliescu, T. A stabilized proper orthogonal decomposition reduced-order model for large scale quasigeostrophic ocean circulation. *Adv. Comput. Math.* **2015**, *41*, 1289–1319. [CrossRef]
22. San, O.; Maulik, R. Neural network closures for nonlinear model order reduction. *Adv. Comput. Math.* **2018**, *44*, 1–34. [CrossRef]
23. Stabile, G.; Ballarin, F.; Zuccarino, G.; Rozza, G. A reduced order variational multiscale approach for turbulent flows. *Adv. Comput. Math.* **2019**, *45*, 2349–2368. [CrossRef]
24. Wang, Z.; Akhtar, I.; Borggaard, J.; Iliescu, T. Proper orthogonal decomposition closure models for turbulent flows: A numerical comparison. *Comput. Methods Appl. Mech. Eng.* **2012**, *237–240*, 10–26. [CrossRef]
25. Bergmann, M.; Bruneau, C.H.; Iollo, A. Enablers for robust POD models. *J. Comput. Phys.* **2009**, *228*, 516–538. [CrossRef]
26. Carlberg, K.; Bou-Mosleh, C.; Farhat, C. Efficient non-linear model reduction via a least-squares Petrov–Galerkin projection and compressive tensor approximations. *Int. J. Num. Meth. Eng.* **2011**, *86*, 155–181. [CrossRef]
27. Parish, E.J.; Wentland, C.; Duraisamy, K. The adjoint Petrov-Galerkin method for non-linear model reduction. *arXiv* **2018**, arXiv:1810.03455.
28. Xie, X.; Wells, D.; Wang, Z.; Iliescu, T. Approximate deconvolution reduced order modeling. *Comput. Methods Appl. Mech. Eng.* **2017**, *313*, 512–534. [CrossRef]
29. Xie, X.; Mohebujjaman, M.; Rebholz, L.; Iliescu, T. Data-driven filtered reduced order modeling of fluid flows. *SIAM J. Sci. Comput.* **2018**, *40*, B834–B857. [CrossRef]
30. Pope, S. *Turbulent Flows*; Cambridge University Press: Cambridge, UK, 2000.
31. Sagaut, P. *Large Eddy Simulation for Incompressible Flows*, 3rd ed.; Scientific Computation; Springer: Berlin, Germany, 2006.
32. Couplet, M.; Sagaut, P.; Basdevant, C. Intermodal energy transfers in a proper orthogonal decomposition—Galerkin representation of a turbulent separated flow. *J. Fluid Mech.* **2003**, *491*, 275–284. [CrossRef]
33. He, K.; Zhang, X.; Ren, S.; Sun, J. Deep residual learning for image recognition. In Proceedings of the IEEE Conference on Computer Vision and Pattern Recognition, Las Vegas, NV, USA, 27 June 2016; pp. 770–778.
34. Szegedy, C.; Ioffe, S.; Vanhoucke, V.; Alemi, A.A. Inception-v4, inception-resnet and the impact of residual connections on learning. In Proceedings of the Thirty-First AAAI Conference on Artificial Intelligence, San Francisco, CA, USA, 4–9 February 2017.
35. Lu, Y.; Zhong, A.; Li, Q.; Dong, B. Beyond finite layer neural networks: Bridging deep architectures and numerical differential equations. *arXiv* **2017**, arXiv:1710.10121.
36. Chang, B.; Meng, L.; Haber, E.; Tung, F.; Begert, D. Multi-level residual networks from dynamical systems view. *arXiv* **2017**, arXiv:1710.10348.

37. Chang, B.; Meng, L.; Haber, E.; Ruthotto, L.; Begert, D.; Holtham, E. Reversible architectures for arbitrarily deep residual neural networks. In Proceedings of the Thirty-Second AAAI Conference on Artificial Intelligence, New Orleans, LO, USA, 2 February 2018.
38. Maulik, R.; Mohan, A.; Lusch, B.; Madireddy, S.; Balaprakash, P. Time-series learning of latent-space dynamics for reduced-order model closure. *arXiv* **2019**, arxiv:1906.07815.
39. Rahman, S.M.; Pawar, S.; San, O.; Rasheed, A.; Iliescu, T. A non-intrusive reduced order modeling framework for quasi-geostrophic turbulence. *arXiv* **2019**, arxiv:11906.11617.
40. Ahmed, S.E.; San, O.; Rasheed, A.; Iliescu, T. A long short-term memory embedding for hybrid uplifted reduced order models. *arXiv* **2019**, arXiv:1912.06756.
41. Maulik, R.; Lusch, B.; Balaprakash, P. Reduced-order modeling of advection-dominated systems with recurrent neural networks and convolutional autoencoders. *arXiv* **2020**, arXiv:2002.00470.
42. San, O.; Maulik, R. Machine learning closures for model order reduction of thermal fluids. *Appl. Math. Model.* **2018**, *60*, 681–710. [CrossRef]
43. Koc, B.; Mohebujjaman, M.; Mou, C.; Iliescu, T. Commutation error in reduced order modeling of fluid flows. *Adv. Comput. Math.* **2019**, *45*, 2587–2621. [CrossRef]
44. Chen, T.Q.; Rubanova, Y.; Bettencourt, J.; Duvenaud, D. Neural ordinary differential equations. *arXiv* **2018**, arXiv:1806.07366.
45. Baiges, J.; Codina, R.; Castanar, I.; Castillo, E. A finite element reduced order model based on adaptive mesh refinement and artificial neural networks. *Int. J. Numer. Methods Eng.* **2019**, *121*, 588–601. [CrossRef]
46. Chekroun, M.D.; Liu, H.; McWilliams, J.C. Variational approach to closure of nonlinear dynamical systems: Autonomous case. *J. Stat. Phys.* **2019**, 1–88. [CrossRef]
47. Lin, K.K.; Lu, F. Data-driven model reduction, Wiener projections, and the Mori-Zwanzig formalism. *arXiv* **2019**, arXiv:1908.07725.
48. Mohebujjaman, M.; Rebholz, L.G.; Iliescu, T. Physically-constrained data-driven correction for reduced order modeling of fluid flows. *Int. J. Num. Methods Fluids* **2019**, *89*, 103–122. [CrossRef]
49. John, V. *Finite Element Methods for Incompressible Flow Problems*; Springer: Berlin, Germany, 2016.
50. Mou, C.; Koc, B.; San, O.; Iliescu, T. Data-driven variational multiscale reduced order models. *arXiv* **2020**, arXiv:2002.06457.
51. Borggaard, J.; Iliescu, T.; Wang, Z. Artificial viscosity proper orthogonal decomposition. *Math. Comput. Model.* **2011**, *53*, 269–279. [CrossRef]
52. Kunisch, K.; Volkwein, S. Galerkin proper orthogonal decomposition methods for parabolic problems. *Numer. Math.* **2001**, *90*, 117–148. [CrossRef]
53. Kingma, D.P.; Ba, J. Adam: A method for stochastic optimization. *arXiv* **2014**, arXiv:1412.6980.
54. Akhtar, I.; Wang, Z.; Borggaard, J.; Iliescu, T. A new closure strategy for proper orthogonal decomposition reduced-order models. *J. Comput. Nonlinear Dyn.* **2012**, *7*, 39–54. [CrossRef]
55. Wells, D.; Wang, Z.; Xie, X.; Iliescu, T. An evolve-then-filter regularized reduced order model for convection-dominated flows. *Int. J. Num. Methods Fluids* **2017**, *84*, 598–615. [CrossRef]
56. Caiazzo, A.; Iliescu, T.; John, V.; Schyschlowa, S. A numerical investigation of velocity-pressure reduced order models for incompressible flows. *J. Comput. Phys.* **2014**, *259*, 598–616. [CrossRef]
57. Giere, S.; Iliescu, T.; John, V.; Wells, D. SUPG reduced order models for convection-dominated convection-diffusion-reaction equations. *Comput. Methods Appl. Mech. Eng.* **2015**, *289*, 454–474. [CrossRef]
58. Wang, Z. Reduced-Order Modeling of Complex Engineering and Geophysical Flows: Analysis and Computations. Ph.D. Thesis, Virginia Tech, Blacksburg, VA, USA, 2012.

*Article*

# Liquid-Cooling System of an Aircraft Compression Ignition Engine: A CFD Analysis

**Alessandro Coclite [1,*], Maria Faruoli [1], Annarita Viggiano [1], Paolo Caso [2] and Vinicio Magi [1,3]**

[1]   School of Engineering, University of Basilicata, 85100 Potenza, Italy; maria.faruoli@unibas.it (M.F.); annarita.viggiano@unibas.it (A.V.); vinicio.magi@unibas.it (V.M.)

[2]   Costruzioni Motori Diesel CMD S.p.A., 81020 San Nicola La Strada (CE), Italy; paolo.caso@cmdengine.com

[3]   Department of Mechanical Engineering, San Diego State University, San Diego, CA 92182, USA

*   Correspondence: alessandro.coclite@unibas.it

Received: 7 April 2020; Accepted: 9 May 2020; Published: 13 May 2020

**Abstract:** The present work deals with an analysis of the cooling system for a two-stroke aircraft engine with compression ignition. This analysis is carried out by means of a 3D finite-volume RANS equations solver with $k$-$\epsilon$ closure. Three different cooling system geometries are critically compared with a discussion on the capabilities and limitations of each technical solution. A first configuration of such a system is considered and analyzed by evaluating the pressure loss across the system as a function of the inlet mass-flow rate. Moreover, the velocity and vorticity patterns are analyzed to highlight the features of the flow structure. Thermal effects on the engine structure are also taken into account and the cooling system performance is assessed as a function of both the inlet mass-flow rate and the cylinder jackets temperatures. Then, by considering the main thermo-fluid dynamics features obtained in the case of the first configuration, two geometrical modifications are proposed to improve the efficiency of the system. As regards the first modification, the fluid intake is split in two manifolds by keeping the same total mass-flow rate. As regards the second configuration, a new single-inlet geometry is designed by inserting restrictions and enlargements within the cooling system to constrain the coolant flow through the cylinder jackets and by moving downstream the outflow section. It is shown that the second geometry modification achieves the best performances by improving the overall transferred heat of about 20% with respect to the first one, while keeping the three cylinders only slightly unevenly cooled. However, an increase of the flow characteristic loads occurs due to the geometrical restrictions and enlargements of the cooling system.

**Keywords:** internal combustion engines; liquid-cooling system; heat transfer; computational fluid dynamics

---

## 1. Introduction

It is well known that internal combustion engines for aeronautical applications operate within a specific temperature range to avoid structural damages, detonations and loss of efficiency of the combustion process. The heat released by combustion is only partially converted to work: a certain amount is within exhaust gases, while the remaining part warms-up the engine structure. This heat needs to be some how dissipated by keeping the engine within its optimal temperature range as to ensure reliability and long service life [1]. This temperature range (roughly from 350 to 380 K) ensures that the engine is efficiently working from the thermodynamic point of view. Indeed, a high cooling would reduce the engine thermal efficiency whereas if the engine temperature exceeds this range, the engine may over-heat and this over-heating is likely to damage the engine. The engine over-heating can be prevented by using efficient cooling systems to assist the engine running at its optimal performance. Hence, there is a need to look at the total heat balance and control system for the aircraft in order to search for performance optimization [2–4].

Generally, for piston aircraft engines, either air or liquid cooling systems are employed; only a few modern engines implement a combination of both, all with their own advantages and disadvantages [5]. Inline four, six or eight cylinders engines are almost exclusively air-cooled, except for the Rotax, Viking and Subaru engines and some aero diesels [6]. Such a cooling technique represents a good compromise between the structure design and the low weight achieved when compared to liquid cooling circuits. Nevertheless, air-cooled cylinders present a large number of fins cast around the heads increasing the total exposed surface. Indeed, the engine may be shock-cooled at high altitudes [7]. On the other hand, liquid-cooling systems have a weight penalty with respect to air-based systems while they are able to keep the temperature of all of the cylinders quite even and the engine cannot be shock-cooled during high speed or low power descends. Moreover, the coolant liquid is usually thermostatically controlled so that the engine is quickly warmed up at the start-up and works all the time around the optimal operating temperature [8].

In this context, a detailed analysis of such cooling systems is required by the designer to optimize the heat distribution of the engine structure. Computational Fluid Dynamics (CFD) simulations provide a means to optimize cooling circuit configurations by employing sensitivity analysis to the components of such circuits. Indeed, simulation models are very useful for engineers not only to support but also to reduce the amount of testing required during the design of the engine. Numerical models and simulations can greatly enhance development efforts by predicting performance trends and trade-offs and will, therefore, result in more efficient and better-optimized heating and cooling systems for high-performing engines [9–14]. Specifically, due to the wide spatial and temporal scales involved in internal combustion engine simulations, steady and unsteady Reynolds-averaged Navier–Stokes (RANS) solvers are largely employed supporting the rational design of such engines. Moreover, recently, hybrid approaches involving both unsteady RANS and Large Eddy Simulations (LES) techniques have also been proposed thus achieving the required resolutions needed for a comprehensive analysis of such engines [15–17].

In this work, a liquid cooling system for a specific aircraft engine is analyzed by means of a 3D finite-volume RANS equations solver with $k$-$\epsilon$ closure. Specifically, the RANS equations for an incompressible thermal fluid is used to analyze the cooling system performance as a function of the inlet mass-flow rate $G_{In}$. Such analysis is carried out by varying the cylinder jackets temperature and by comparing the outflow temperature and the transferred thermal power as a function of coolant mass-flow rate. The CFD results enable us to notice the drawbacks of the cooling system geometry and suggest some improvements that can be used to enhance the heat transfer. Two geometrical modifications to the cooling system are proposed. The first modification consists of two coolant inlets while keeping the same total mass-flow rate; the second configuration consists of a new single-inlet geometry with new restrictions and enlargements within the cooling system to constrain the coolant flow through the cylinder jackets and also by positioning downstream the outflow manifold. Capabilities and limitations of these three cooling system geometries are critically discussed and compared providing useful insights for the internal combustion engine community.

## 2. The Model

### 2.1. Mathematical Model

The steady three-dimensional Reynolds-Averaged Navier-Stokes (RANS) system of equations with the $k$–$\epsilon$ closure reads:

$$\frac{\partial(E - E_v)}{\partial x} + \frac{\partial(F - F_v)}{\partial y} + \frac{\partial(G - G_v)}{\partial z} = S, \tag{1}$$

where $E$, $F$ and $G$ are the inviscid flux vectors:

$$E = \begin{pmatrix} \overline{u} \\ \rho\overline{u}^2 + p_t \\ \rho\overline{uv} \\ \rho\overline{uw} \\ \rho c_p \overline{Tu} \\ \rho\overline{u}k \\ \rho\overline{u}\epsilon \end{pmatrix} , F = \begin{pmatrix} \overline{v} \\ \rho\overline{uv} \\ \rho\overline{v}^2 + p_t \\ \rho\overline{vw} \\ \rho c_p \overline{Tv} \\ \rho\overline{v}k \\ \rho\overline{v}\epsilon \end{pmatrix} , G = \begin{pmatrix} \overline{w} \\ \rho\overline{uw} \\ \rho\overline{vw} \\ \rho\overline{w}^2 + p_t \\ \rho c_p \overline{Tw} \\ \rho\overline{w}k \\ \rho\overline{w}\epsilon \end{pmatrix} ; \tag{2}$$

where overbar indicates Reynolds-averaging of the physical variable; $\rho$ is the density; $(\overline{u}, \overline{v}, \overline{w})$ and $\overline{T}$ are the Reynolds-averaged velocity components and temperature, respectively; $\overline{p}_t = \overline{p} + \frac{2}{3}k$ being $\overline{p}$ the Reynolds-averaged pressure; $k$ and $\epsilon$ are the turbulent kinetic energy and its dissipation rate. $E_v$, $F_v$ and $G_v$ represent the viscous flux vectors, reading:

$$E_v = \begin{pmatrix} 0 \\ \mu\frac{\partial \overline{u}}{\partial x} \\ \mu\frac{\partial \overline{v}}{\partial x} \\ \mu\frac{\partial \overline{w}}{\partial x} \\ \lambda\frac{\partial \overline{T}}{\partial x} + \mu[2(\frac{\partial \overline{u}}{\partial x})^2 + (\frac{\partial \overline{v}}{\partial x} + \frac{\partial \overline{u}}{\partial y})^2] \\ \mu_t\frac{\partial k}{\partial x} \\ \mu_t\frac{\partial \epsilon}{\partial x} \end{pmatrix} , F_v = \begin{pmatrix} 0 \\ \mu\frac{\partial \overline{u}}{\partial y} \\ \mu\frac{\partial \overline{v}}{\partial y} \\ \mu\frac{\partial \overline{w}}{\partial y} \\ \lambda\frac{\partial \overline{T}}{\partial y} + \mu[2(\frac{\partial \overline{v}}{\partial y})^2 + (\frac{\partial \overline{w}}{\partial y} + \frac{\partial \overline{v}}{\partial z})^2] \\ \mu_t\frac{\partial k}{\partial y} \\ \mu_t\frac{\partial \epsilon}{\partial y} \end{pmatrix} ,$$

$$G_v = \begin{pmatrix} 0 \\ \mu\frac{\partial \overline{u}}{\partial z} \\ \mu\frac{\partial \overline{v}}{\partial z} \\ \mu\frac{\partial \overline{w}}{\partial z} \\ \lambda\frac{\partial \overline{T}}{\partial z} + \mu[2(\frac{\partial \overline{w}}{\partial z})^2 + (\frac{\partial \overline{u}}{\partial z} + \frac{\partial \overline{w}}{\partial x})^2] \\ \mu_t\frac{\partial k}{\partial z} \\ \mu_t\frac{\partial \epsilon}{\partial z} \end{pmatrix} ; \tag{3}$$

being $\mu$, $\mu_t$ and $\lambda$ the molecular viscosity, turbulent viscosity and thermal diffusivity, respectively. Finally, $S$ corresponds to the source term vector,

$$S = \begin{pmatrix} 0 \\ 0 \\ 0 \\ 0 \\ 0 \\ S_k \\ S_\epsilon \end{pmatrix} , \tag{4}$$

where $S_k$ and $S_\epsilon$ are the source terms related to $k$ and $\epsilon$, respectively.

The steady-state RANS equations with $k$–$\epsilon$ turbulence closure are solved by means of a cell-centered finite volume numerical scheme. The computational domain is discretized by a multi-block unstructured mesh. A pressure-based solver is employed where the equations are solved segregated one to each other, i.e., each equation is decoupled from all the others. Indeed, this segregated algorithm is memory-efficient, since the discretized equations need only to be located in the memory one at a time [18]. Specifically, the SIMPLE solver by Patankar and Spalding [19] is employed. Viscous terms are discretized by a second-order-accurate least-squares cell-centered scheme [20], while convective terms are treated with a second-order upwind scheme [21]. The pressure and velocity fields are solved by using a SIMPLE-type pressure-velocity coupling in which the momentum equation is solved using an estimated initial pressure field. Then, the SIMPLE correction for the pressure field is employed to satisfy the continuity equation. The pressure-correction equation is solved by using the algebraic multigrid (AMG) method by Hutchinson and Raithby [22]. In the present work, only steady flows are considered; therefore, the algorithm is iterated in a pseudo-time until a residual drop of at least four orders of magnitude for all of the conservation-law equations is achieved. No-slip conditions are imposed along the engine walls while the near-wall region of the flow field is treated with the standard wall-functions approach proposed by Launder and Spalding [23].

### 2.2. Computational Domain and Boundary Conditions

In this work, a six-cylinder aircraft compression ignition engine with two opposite banks is considered. Specifically, due to the specular geometries of the banks, only one of the two cooling system banks is considered (Figure 1a). The bank is characterized by an inlet section and an outlet section and presents three slots corresponding to the three cylinders jackets. Moreover, two modifications of this geometry are proposed and discussed to provide improvements in terms of mass-averaged outflow temperature and cylinder specific and total transferred thermal power (see Figure 1b,c). These geometries are discretized by means of about 8,000,000 tetrahedral elements with an average cell size of about 1.5 mm and a global minimum size of about 0.05 mm. The spatial discretization used in the computations for the single-inlet geometry (depicted in Figure 1a) is shown in Figure 2. Three sections are considered along with close-ups to show the quality of the discretization in the near-wall regions. Specifically, two transversal sections are located at y = −0.250 m and y = −0.325 m and the longitudinal section corresponds to the valve plane at z = −0.01 m. The minimum and maximum $y^+$ for this numerical discretization are 20 and 97, respectively. These values are well confined in the range of $10 \leq y^+ \leq 300$, which represents a constraint for the employed $k$–$\epsilon$ closure joined with the use of standard wall functions.

The comparison between the three geometries is provided in terms of both the thermodynamic and kinematic flow fields, i.e., velocity, vorticity and temperature distributions, and of the evaluation of the pressure loss through the bank, the outflow temperature, as well as the cylinder specific and total transferred thermal power. To this end, two sets of simulations with five inlet mass-flow rates, namely $G_{in} = 0.5, 1.0, 1.5, 2.0,$ and 2.5 kg/s are considered. At first instance, thermal effects are neglected, i.e., fluid temperature is constant through the bank. This analysis is mainly carried out to provide a picture of the overall flow structure and to identify flow features under such an assumption. The working fluid is assumed to be water at 353 K with density and viscosity equal to 972 kg/s and $3.54 \times 10^{-4}$ Pa·s, respectively. The second set of computations includes heat transfer between fluid and walls with the inlet coolant water temperature equal to 343 K and the crankcase external temperature equal to 380 K, whereas the cylinder jacket temperature ranges from 400 K to 460 K. For such thermal analysis, the water density and viscosity are computed as functions of the water temperature by employing a second-order polynomial fitting [24]. For both sets of computations, the mass-flow rate is set at the inlet section with an outlet pressure $p_{out} = 1$ bar. As regards turbulence, the inlet turbulence intensity is $u'_{in} = 0.25\, u_{in}$ and the turbulence characteristic length is equal to $0.3\, d_{in}$, being $d_{in}$ and $u_{in}$ the inlet diameter and the inlet mass-averaged fluid velocity, respectively.

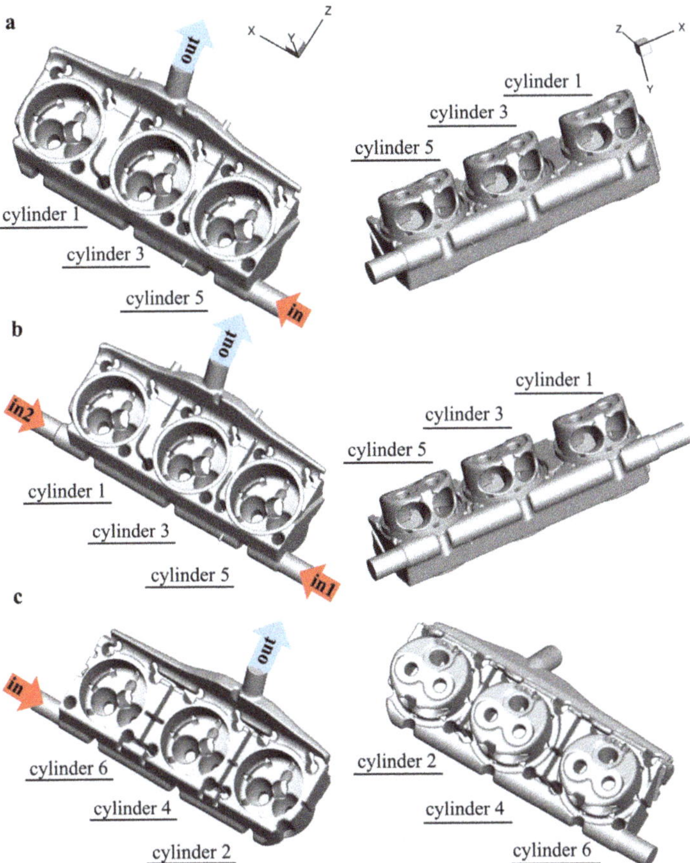



**Figure 1.** Geometry of the cooling system for the three configurations. (**a**) Sketch of the right bank of the cooling system. (**b**) Sketch of the right bank of the cooling system with a second intake duct (In2) on the opposite side with respect to the old one (In1). (**c**) Sketch of a new single inlet geometry (left bank) obtained by inserting enlargements and restrictions to the old one and by moving downstream the outlet duct. All of the graphics are represented from an internal (left column) and external (right column) point of view.

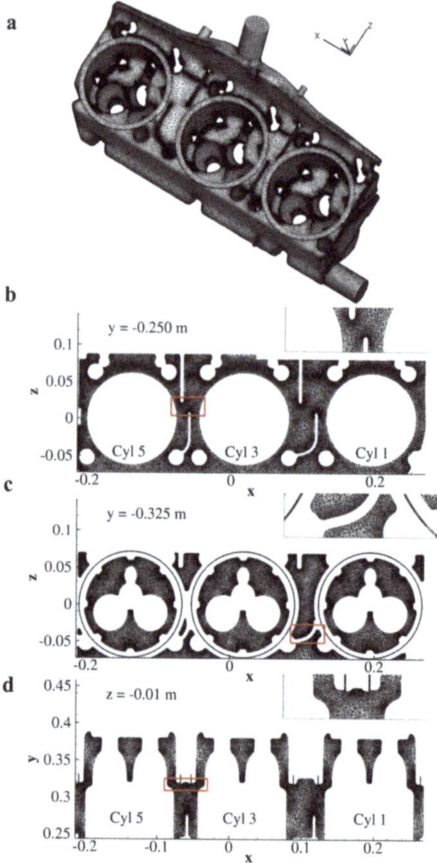

**Figure 2.** Spatial discretization of the cooling system for the single-inlet configuration. (**a**) Computational mesh of the right bank of the cooling system. (**b**,**c**) Transversal sections of the computational mesh at (**b**) y = −0.250 m and (**c**) y = −0.325 m. (**d**) Longitudinal section of the computational mesh at the valve plane z = −0.01 m.

## 3. Results and Discussion

### 3.1. Analysis of the Single-Inlet Geometry

At first instance, the flow field of the single-inlet geometry (see Figure 1a) is analyzed by neglecting thermal effects. The pressure drop across the cooling system is computed as a function of the inlet mass-flow rate and shown in Figure 3. The head losses are computed as the difference between the inlet ($< p_{in} >$) and the outlet ($< p_{out} >$) surface-averaged pressure, $\Delta p =< p_{in} >_{A_{in}} - < p_{out} >_{A_{out}}$, where $< p_{in/out} >_{A_{in/out}}$ correspond to the surface-averaged pressure related to the inlet (outlet) sections. The diagram shows that $\Delta p$ vs. $G_{in}$ follows a quadratic trend due to the fully developed turbulent regime of the flow field. The figure shows that $\Delta p$ is of the order of $10^{-2}$ bar for all the investigated inlet mass-flow rates.

The velocity and vorticity magnitude contour plots, given in Figure 4, show that a large fraction of the flow pattern sweeps over the cylinder that is closer to the inlet section, i.e., cylinder 5. Then, the flow is almost totally short-circuited from the inlet directly to the outlet for all values of $G_{in}$. Indeed, the cylinder in between, i.e., cylinder 3, is directly affected by the fluid flow, whereas cylinder 1 results poorly cooled down even for large values of $G_{in}$, as shown in Figure 4c. On the other hand, the small values of $\Delta p$ ($O(10^{-2})$ bar) reflect that the flow is somehow short-circuited so that the path actually performed by the fluid is relatively short and without much cross-sectional area variations. An interesting view of the flow structure is given in Figure 5 for $G_{In} = 0.5$, 1.5, and 2.5 kg/s, where the velocity magnitude colored streamlines are drawn to provide a qualitative picture of the coolant internal flow structure.

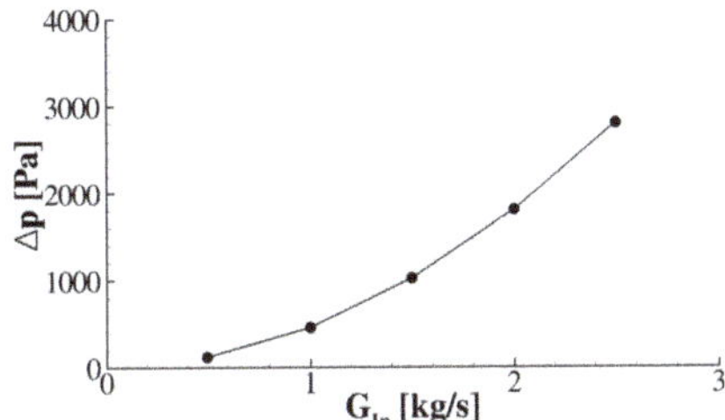

**Figure 3.** Single-inlet cooling system characteristic loads. Pressure loss as a function of the inlet mass-flow rate.

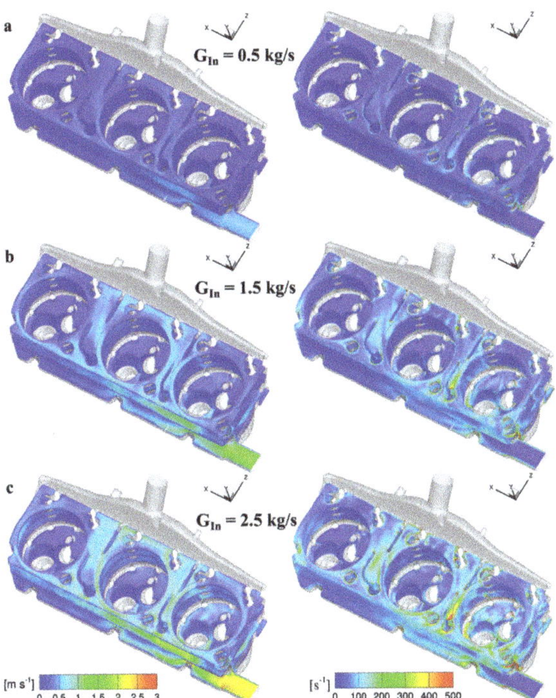



**Figure 4.** Velocity and vorticity contour plots for the single-inlet geometry. Distribution of velocity (left) and vorticity (right) magnitude for $G_{In} = 0.5$ (**a**), 1.5 (**b**), and 2.5 (**c**) kg/s.

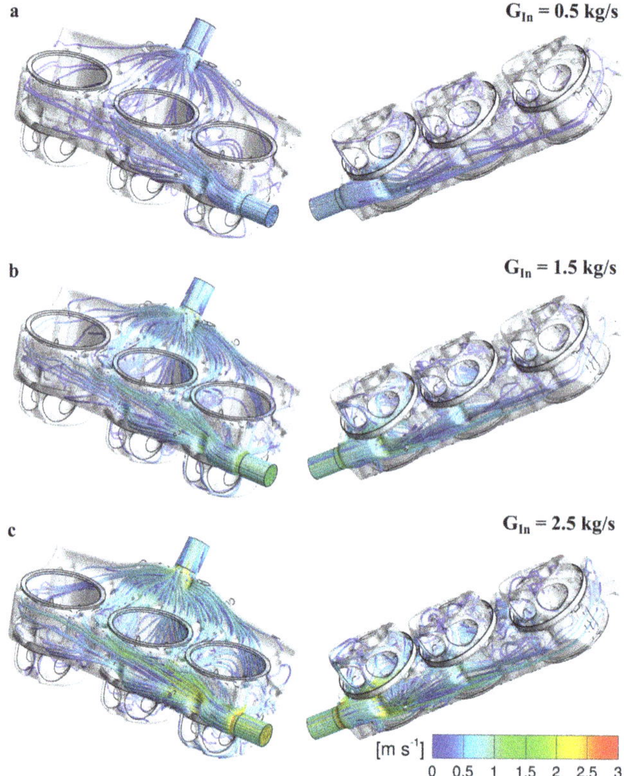



**Figure 5.** Velocity magnitude colored streamlines for the single-inlet geometry. Velocity magnitude colored streamlines, starting from the inlet section, for $G_{In}$ = 0.5 (**a**), 1.5 (**b**), and 2.5 (**c**) kg/s. Internal (left) and external (right) point of view.

In order to consider thermal effects, the temperatures of the inlet water and of the crankcase are kept constant and equal to 343 K and 380 K, respectively. Several simulations have been performed by changing the cylinder jackets temperatures ($T_{Cyls}$) from 400 to 460 K with a 20 K step. For the sake of conciseness, only the temperature contour plots for the case $T_{Cyls}$ = 440 K are reported in Figure 6. Like the velocity and the vorticity patterns, the fluid temperature field clearly shows that the highest temperature is located around cylinder 1, regardless of $G_{in}$. The water, flowing through cylinder 5 jacket, results colder than that through cylinder 3 jacket which is colder than that through cylinder 1 jacket. As regards the mass-averaged outflow temperature, Table 1 shows that $T_{Out}$ decreases, by increasing $G_{In}$, with a smaller rate with respect to the mass flow rates (see Figure 7a). Besides, Table 1 shows the cylinder-specific and the total heat transferred. The cylinder-specific heat is computed by considering only the numerical cells that surround each cylinder, whereas the total heat is the sum over all the cells of the computational domain. The values of Table 1 clearly show that the cooling of the cylinder reduces as the distance between the cylinders and the inlet section increases. Specifically, the cylinder closest to the inlet transfers to the fluid a thermal power that is 2 to 3 times that of the furthest cylinder. $\dot{Q}_{Tot}$ increases linearly with $G_{In}$ as depicted in Figure 7b. The distributions of $T_{Out}$ for $T_{Cyls}$ = 400, 420, 440, and 460 K are summarized in Figure 7c. Indeed, the functional form of these distributions is roughly constant with respect to the parameter $T_{Cyls}$, while the distributions themselves are displaced with an offset which varies between 1.3 and 1.7 K. Two convenient thresholds for both the mass-averaged outflow temperature and the total transferred heat flux have been selected, which are superimposed to Figure 7c,d. $T_{Out}$ has been supposed to be lower than 363 K since too high values would represent an inefficient cooling for the engine itself. Besides, since the output mechanical power of the six-cylinder engine ranges from 200 to 250 kW, the total transferred heat flux for each bank should range from 100 to 120 kW. The second condition is more stringent than the first and returns only some of the couples ($T_{Cyl}$, $G_{In}$) which are acceptable in the parameter space spanned with the simulations. The physically acceptable parameters are enlighten in Table 2.

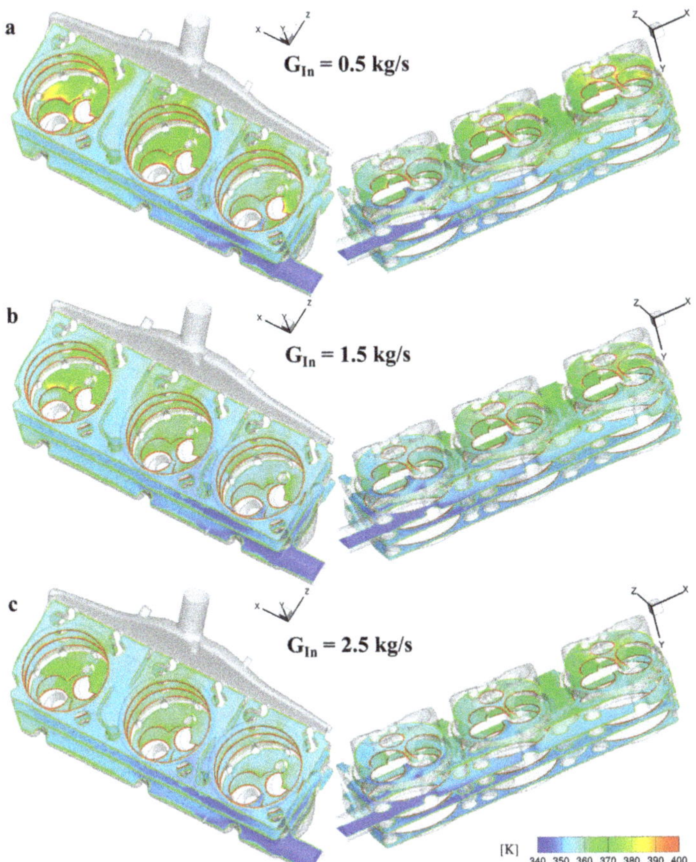


**Figure 6.** Temperature patterns for $T_{Cyls} = 440$ K for the single-inlet geometry. Temperature contour plots for $T_{Cyls} = 440$ K and $G_{In} = 0.5$ (**a**), 1.5 (**b**), and 2.5 (**c**) kg/s.

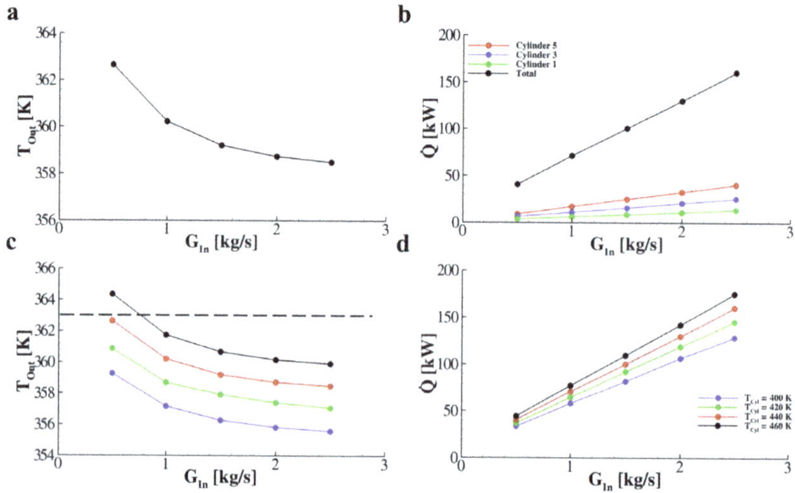

**Figure 7.** Outflow temperature and transferred heat flux for the single-inlet geometry. (**a**) Mass-averaged outflow temperature as a function of $G_{In}$ for $T_{Cyls} = 440$ K. (**b**) Cylinder specific and total transferred heat flux as a function of $G_{In}$ for $T_{Cyls} = 440$ K. (**c**) Mass-averaged outflow temperature as a function of $G_{In}$ for $T_{Cyls} = 400, 420, 440$, and 460 K. (**d**) Total transferred heat flux as a function of $G_{In}$ for $T_{Cyls} = 400, 420, 440$, and 460 K.

**Table 1.** Mass-averaged temperature, cylinder-specific and total transferred heat flux for $T_{Cyls} = 440$ K and $G_{In} = 0.5, 1.0, 1.5, 2.0$, and 2.5 kg/s obtained considering the single-inlet geometry.

| $G_{In}$ [kg/s] | $T_{Out}$ [K] | $Q_5$ [kW] | $Q_3$ [kW] | $Q_1$ [kW] | $Q_{Tot}$ [kW] |
|---|---|---|---|---|---|
| 0.5 | 362.6 | 9.6 | 6.9 | 4.4 | 40.8 |
| 1.0 | 360.2 | 17.5 | 11.4 | 6.7 | 71.4 |
| 1.5 | 359.2 | 25.4 | 16.3 | 9.0 | 100.7 |
| 2.0 | 358.7 | 32.9 | 21.3 | 11.4 | 130.3 |
| 2.5 | 358.5 | 40.8 | 25.9 | 13.9 | 160.5 |

**Table 2.** Mass-averaged outflow temperature and total transferred heat flux tabulated as a function of $G_{In}$ and $T_{Cyls}$ for the single-inlet geometry.

| $G_{In}$ [kg/s] | $T_{Cyls} = 400$ K | | $T_{Cyls} = 420$ K | | $T_{Cyls} = 440$ K | | $T_{Cyls} = 460$ K | |
|---|---|---|---|---|---|---|---|---|
| | $T_{Out}$ [K] | $Q_{Tot}$ [kW] | $T_{Out}$ [K] | $Q_{Tot}$ [kW] | $T_{Out}$ [K] | $Q_{Tot}$ [kW] | $T_{Out}$ [K] | $Q_{Tot}$ [kW] |
| 0.5 | 359.3 | 33.7 | 360.9 | 37.1 | 362.6 | 40.8 | 364.3 | 44.3 |
| 1.0 | 357.1 | 58.7 | 358.7 | 65.1 | 360.2 | 71.4 | 361.7 | 77.8 |
| 1.5 | 356.3 | 82.4 | 357.9 | 92.7 | 359.2 | 100.7 | 360.7 | 109.9 |
| 2.0 | 355.8 | 107.1 | 357.4 | 119.3 | 358.7 | 130.3 | 360.2 | 142.3 |
| 2.5 | 355.5 | 129.0 | 357.0 | 145.6 | 358.5 | 160.5 | 359.9 | 175.3 |

### 3.2. A First Cooling Improvement: The Double-Inlet Geometry

To overcome the uneven distribution of the cylinder-specific heat transfer with the single inlet geometry, a first simple geometrical modification is to double the intake manifold but with the same total inlet mass-flow rate $G_{In}$, as reported in Figure 1b. Specifically, the second inlet section is placed close to cylinder 1 to promote heat transfer for this cylinder. The non-thermal analysis is firstly considered and shown in Figure 8. The internal flow streamlines depict an even distribution of the fluid between the three cylinders. Nonetheless, it must be stressed that the up-left region close to cylinder 1 results in the less-cooled region regardless of $G_{In}$. This is qualitatively given in Figure 8a for $G_{In}$ = 0.5, 1.5, and 2.5 kg/s. The head losses computed separately for both inlets as a function of $G_{In}$ are comparable to the single inlet geometry case as shown in Figure 8c, where $G_{In}$ is the total mass flow rate through the two inlets.

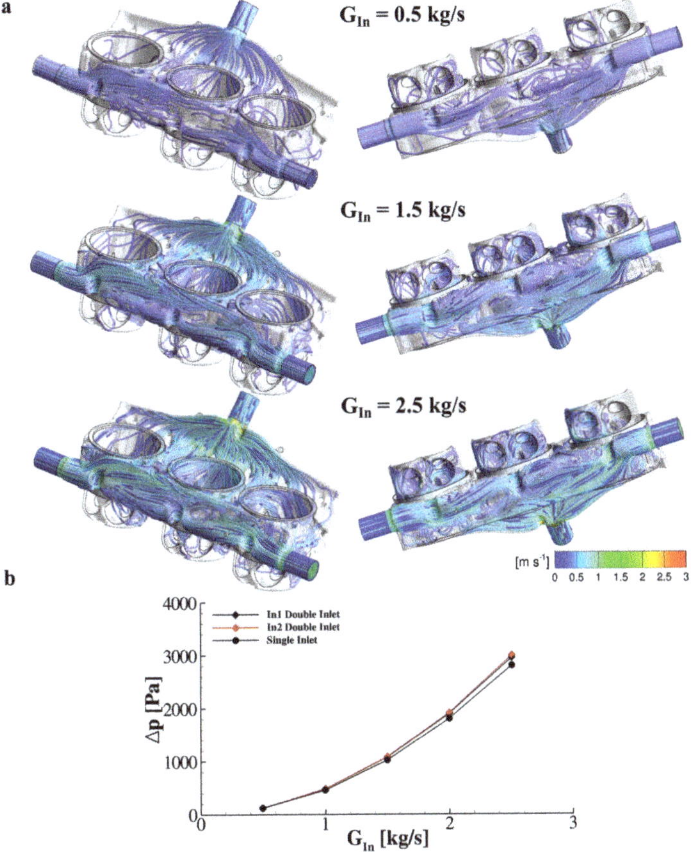


**Figure 8.** Velocity streamlines and characteristic loads for the double-inlet geometry. (**a**) Velocity magnitude colored streamlines obtained for the double-inlet geometry for $G_{In}$ = 0.5, 1.5 and 2.5 kg/s. (**b**) Distribution of the pressure loss as a function of the inlet mass-flow rate.

In order to account for the thermal effects, with the same fashion as per the single inlet geometry, the mass-averaged outflow temperature, as well as the cylinder-specific and the total transferred heat, are computed as a function of both $G_{In}$ and $T_{Cyls}$. These data are given in Table 3 for $T_{Cyls} = 440$ K. Regardless of $G_{In}$, the double inlet geometry returns outflow temperatures somewhat smaller than those of the single inlet geometry. Moreover, although an almost perfectly even distribution of transferred heat among the three cylinders, the total transferred thermal power is about 12-20% less than the previous geometry (see Table 1). This is because, even with a second inlet section, this geometry provides less cooling than the previous geometry. Indeed, it must be pointed out that the total intake mass-flow rate is the same as before but supplied by the two inlets. This reduces the amount of coolant fluid flowing into the system from each inlet, so that cylinder 5 jacket exchanges less heat with the less supplied water. As regards cylinder 3 and cylinder 1, the amount of water is about the same of cylinder 5 but not enough to efficiently cool down the overall system. For all of the investigated values of $T_{Cyls}$, the total transferred heat and the outflow temperatures are given in Table 4 which reflects the same picture drawn for the single inlet geometry.

**Table 3.** Mass-averaged temperature, cylinder-specific and total transferred heat flux for $T_{Cyls} = 440$ K and $G_{In} = 0.5, 1.0, 1.5, 2.0$, and $2.5$ kg/s when considering the double-inlet geometry.

| $G_{In}$ [kg/s] | $T_{Out}$ [K] | $Q_5$ [kW] | $Q_3$ [kW] | $Q_1$ [kW] | $Q_{Tot}$ [kW] |
|---|---|---|---|---|---|
| 0.5 | 361.2 | 6.0 | 6.1 | 5.8 | 35.8 |
| 1.0 | 358.1 | 9.7 | 9.9 | 9.4 | 58.9 |
| 1.5 | 357.2 | 13.9 | 13.5 | 13.3 | 82.6 |
| 2.0 | 356.5 | 17.6 | 17.2 | 17.3 | 104.5 |
| 2.5 | 356.3 | 21.5 | 20.9 | 21.2 | 128.5 |

**Table 4.** Mass-averaged outflow temperature and total transferred heat flux tabulated as a function of $G_{In}$ and $T_{Cyls}$ for the double inlet geometry.

| $G_{In}$ [kg/s] | $T_{Cyls} = 400$ K | | $T_{Cyls} = 420$ K | | $T_{Cyls} = 440$ K | | $T_{Cyls} = 460$ K | |
|---|---|---|---|---|---|---|---|---|
| | $T_{Out}$ [K] | $Q_{Tot}$ [kW] | $T_{Out}$ [K] | $Q_{Tot}$ [kW] | $T_{Out}$ [K] | $Q_{Tot}$ [kW] | $T_{Out}$ [K] | $Q_{Tot}$ [kW] |
| 0.5 | 357.6 | 27.7 | 359.7 | 32.6 | 361.6 | 35.8 | 362.7 | 38.9 |
| 1.0 | 355.6 | 48.3 | 356.9 | 53.6 | 358.1 | 58.9 | 359.5 | 65.0 |
| 1.5 | 354.8 | 67.6 | 356.0 | 75.1 | 357.2 | 82.6 | 358.4 | 90.3 |
| 2.0 | 354.3 | 86.2 | 355.5 | 96.0 | 356.5 | 104.5 | 357.8 | 115.7 |
| 2.5 | 353.9 | 104.5 | 355.1 | 116.5 | 356.3 | 128.5 | 357.4 | 140.5 |

*3.3. A New Single Inlet Geometry*

An interesting geometrical improvement for the cooling system is proposed in Figure 1c. This new geometry is shown for the left bank of the cooling system. Indeed, right and left banks slightly differ due to the quasi-symmetric position of the cylinder seats. However, these differences are negligible for the purposes of the present work. Specifically, the geometrical improvement is twofold. On one hand, the geometry considers some enlargements and restrictions within the bank (see the intra-cylinder jackets regions), and, on the other hand, the outlet manifold is moved from the inlet section. In this context, enlargements and restrictions are designed to guide the flow from the inlet through all of the three-cylinder jackets and then to the outflow section, while the outlet is placed between cylinder 4 and cylinder 2 to avoid any possible short-circuiting of the flow itself. The velocity magnitude colored streamlines of the simulations without thermal effects for the new geometry are depicted in Figure 9a for $G_{In}$ = 0.5, 1.5 and 2.5 kg/s. The figure shows that the fluid flow supplies, almost uniformly, the three-cylinder jackets and reaches the outlet section without being short-circuited. Indeed, the water flowing over the three cylinders results enhanced for large values of the intake mass-flow rate as shown by the increased number of streamlines sweeping over the jackets and the cylinder heads. Moreover, the increasing complexity of the flow path with respect to the previous single-inlet geometry, due to the provided enlargements and restrictions, reflects into an increase of $\Delta p$, as shown in Figure 9b.

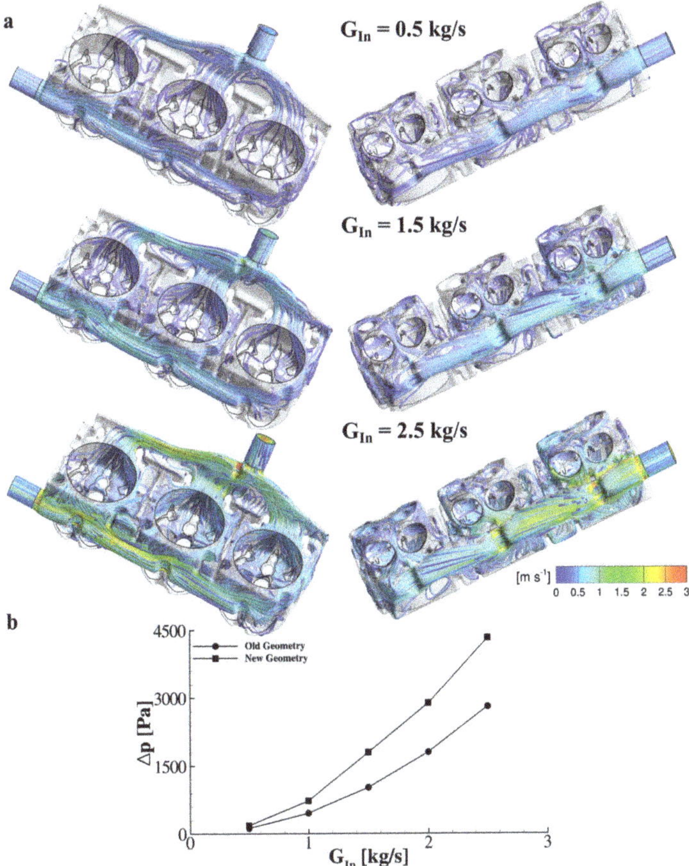


**Figure 9.** Velocity streamlines and characteristic loads for the new single-inlet geometry. (**a**) Velocity magnitude colored streamlines obtained for the new single-inlet geometry for $G_{In}$ = 0.5, 1.5 and 2.5 kg/s (**b**) Pressure loss as a function of the inlet mass-flow rate.

Thermal effects are also considered in order to analyze the cooling efficiency of this new geometry in terms of computed temperature field, mass-averaged outflow temperature and total transferred heat. As shown in Figure 10, the temperature field results quite homogeneous with respect to the old single-inlet geometry. Specifically, for $G_{In}$ = 0.5 kg/s, around the farthest cylinder jacket from the inlet, i.e., cylinder 1, a temperature between 370 and 380 K is computed for the old geometry (see Figure 6a) whereas, for the new geometry the farthest cylinder jacket from the inlet, i.e., cylinder 2, provides a temperature between 360 and 370 K. This improvement is also quantitatively assessed by the outflow mass-average temperature in Table 5. Indeed, for the $T_{Cyls}$ = 440 K case, $T_{Out}$ is higher for the new geometry with respect to the old configuration. Moreover, an improvement of about 20% for the total transferred thermal power is observed for all the investigated $G_{In}$. All of those data are summarized in Figure 11a,b for $T_{Out}$ and $\dot{Q}$, respectively. A direct comparison between the two single inlet geometries is obtained by considering $T_{Out}$ and $\dot{Q}$ as a function of $G_{In}$ and $T_{Cyls}$ and depicted in Figure 11c,d. Regardless of the cylinder jacket wall temperatures, the outflow water temperature was higher with the new configurations due to the larger amount of thermal power ($\approx$20%) transferred to the water. As per the old geometry, two thresholds are considered. The outflow temperature threshold is set to 363 K while the bandwidth for $\dot{Q}$ corresponds to 110 kW $\pm$ 10 kW. The practical values in the spanned parameter space are enlightened in Table 6.

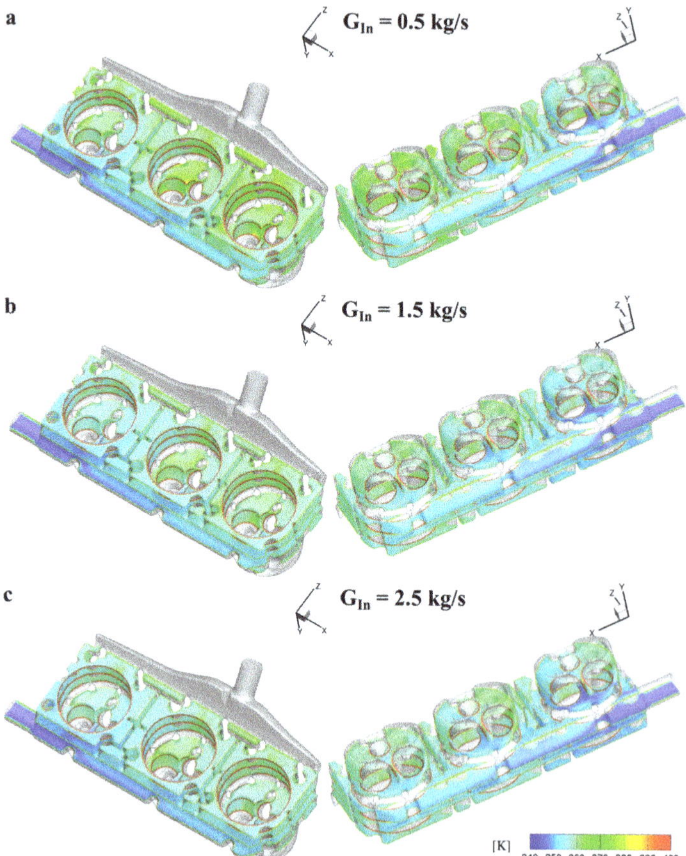


**Figure 10.** Temperature patterns for the new single-inlet geometry. Temperature contour plots obtained for $T_{Cyls}$ = 440 K and $G_{In}$ = 0.5 (**a**), 1.5 (**b**), and 2.5 (**c**) kg/s.

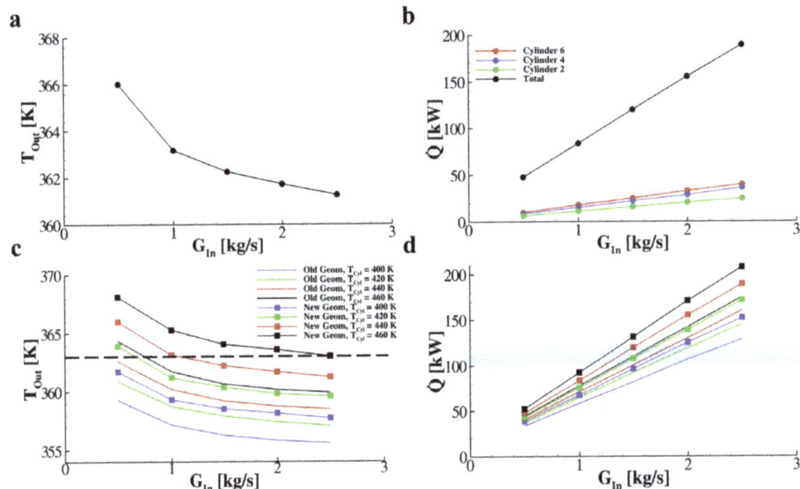

**Figure 11.** Outflow temperature and transferred heat flux for the new geometry. (**a**) Mass-averaged outflow temperature as a function of $G_{In}$ for $T_{Cyls}$ = 440 K. (**b**) Cylinder specific and total transferred heat flux as a function of inlet mass-flow rate for $T_{Cyls}$ = 440 K. (**c**) Comparison of the mass-averaged outflow temperature as a function of $G_{In}$ and $T_{Cyls}$ between the old and the new single-inlet geometry. (**d**) Total transferred heat flux as a function of the inlet mass-flow rate obtained with the old and the new geometry.

**Table 5.** Mass-averaged temperature, cylinder-specific and total transferred heat flux for $T_{Cyls}$ = 440 K and $G_{In}$ = 0.5, 1.0, 1.5, 2.0, and 2.5 kg/s when considering the new geometry.

| $G_{In}$ [kg/s] | $T_{Out}$ [K] | $Q_6$ [kW] | $Q_4$ [kW] | $Q_2$ [kW] | $Q_{Tot}$ [kW] |
|---|---|---|---|---|---|
| 0.5 | 366.0 | 10.1 | 9.0 | 6.8 | 47.9 |
| 1.0 | 363.1 | 18.3 | 15.5 | 11.4 | 83.7 |
| 1.5 | 362.2 | 25.1 | 22.5 | 16.3 | 119.9 |
| 2.0 | 361.5 | 33.2 | 28.9 | 20.8 | 155.3 |
| 2.5 | 361.2 | 39.9 | 36.4 | 25.0 | 189.4 |

**Table 6.** Mass-averaged outflow temperature and total transferred heat flux tabulated as a function of $G_{In}$ and $T_{Cyls}$ for the new geometry.

| $G_{In}$ [kg/s] | $T_{Cyls}$ = 400 K | | $T_{Cyls}$ = 420 K | | $T_{Cyls}$ = 440 K | | $T_{Cyls}$ = 460 K | |
|---|---|---|---|---|---|---|---|---|
| | $T_{Out}$ [K] | $Q_{Tot}$ [kW] | $T_{Out}$ [K] | $Q_{Tot}$ [kW] | $T_{Out}$ [K] | $Q_{Tot}$ [kW] | $T_{Out}$ [K] | $Q_{Tot}$ [kW] |
| 0.5 | 361.7 | 38.9 | 363.9 | 43.5 | 366.0 | 47.9 | 368.1 | 52.1 |
| 1.0 | 359.3 | 67.7 | 361.2 | 75.7 | 363.1 | 83.7 | 365.2 | 92.5 |
| 1.5 | 358.5 | 96.6 | 360.3 | 108.1 | 362.2 | 119.9 | 364.0 | 131.1 |
| 2.0 | 358.1 | 125.5 | 359.8 | 139.4 | 361.5 | 155.3 | 363.5 | 170.9 |
| 2.5 | 357.7 | 152.4 | 359.6 | 172.0 | 361.2 | 189.4 | 363.0 | 207.9 |

## 4. Conclusions

In this work multi-dimensional CFD is employed to enhance the performance of the cooling system of an aircraft engine. Capabilities and limitations of different geometries are highlighted, in order to give guidelines as regards an efficient design of the engine. The analysis is carried out by means of a finite-volume RANS equations solver with $k$-$\epsilon$ closure.

An initial configuration of the cooling system without any thermal effects is firstly considered, with constant temperature, density and viscosity for the water. The pressure loss of the cooling system as a function of the inlet mass-flow rate is computed. Velocity and vorticity patterns are also given to point out features of the flow field. Then, thermal effects are taken into account and the cooling performance of such a configuration is assessed as a function of both the inlet mass-flow rate and the cylinder jackets temperatures. Specifically, the outflow temperature and both the total and the cylinder-specific transferred thermal power are computed. This geometry is demonstrated to be not very efficient because the cylinder-specific transferred heat decreases as the distance between the cylinders and the intake section increases.

Two geometrical improvements are then proposed: first, the inlet sections are doubled while keeping the same total mass-flow rate; second, a new single-inlet geometry is designed by inserting restrictions and enlargements within the cooling system to constrain the coolant flow and by moving downstream the outflow duct. The double-inlet geometry exchanges less thermal power than the initial single inlet geometry ($\approx$12-20%) with an almost perfectly even cylinder-specific transferred heat distribution. The new single inlet geometry improves the overall abilities of the cooling system of about 20% (referring to the total transferred thermal power) while keeping the three cylinders only slightly unevenly cooled. Moreover, the new single-inlet geometry presents, as expected, an increase of the characteristic loads due to the restrictions and enlargements within the flow field. However, since this head loss represents only about 10% of the total head loss of the whole cooling circuit, the new single-inlet geometry may be a reasonable choice to optimize the efficiency of the cooling system.

**Author Contributions:** Data curation, A.V. and V.M.; funding acquisition, V.M.; investigation, A. C. and M.F.; methodology, A.C. and M.F.; project administration, A.V., P.C. and V.M.; writing, review and editing, A.V. and V.M. All authors have read and agreed to the published version of the manuscript.

**Funding:** Project SIMPA "Sistemi Innovativi per Motori a Pistoni Aeronautici", Asse I, Priorità di Investimento 1.b, Azione 1.1.3 LDR granted by MISE "Ministero dello Sviluppo Economico" and MUR "Ministero dell'Università e della Ricerca".

**Acknowledgments:** This work is financially supported by MISE "Ministero dello Sviluppo Economico" and MUR "Ministero dell'Università e della Ricerca" under project SIMPA "Sistemi Innovativi per Motori a Pistoni Aeronautici", Asse I, Priorità di Investimento 1.b, Azione 1.1.3 LDR.

**Conflicts of Interest:** The authors declare no conflict of interest.

## References

1. Jacobs, T.J. Internal Combustion Engines, Developments in. In *Fossil Energy*; Malhotra, R., Ed.; Encyclopedia of Sustainability Science and Technology Series; Springer, New York, NY, USA, 2020; pp. 133–184, ISBN 978-1-4939-9762-6._430. [CrossRef]
2. Ryuichi, M.; Toshihiko, I.; Hiroyuki, F.; Yasutoshi, Y. Internal combustion engine cooling apparatus. *App. Therm. Eng.* **1996**, *16*, 101–132. [CrossRef]
3. Griffiths, J.; Barnard, J. *Flame and Combustion*; Routledge: Abingdon, UK, 2019; ISBN 9780203755976. [CrossRef]
4. Zhukov, V.; Melnik, O.; Logunov, N.; Chernyi, S. Regulation and control in cooling systems of internal combustion engines. *E3S Web Conf.* **2019**, *02015*, 135. [CrossRef]
5. Farokhi, S. *Future Propulsion Systems and Energy Sources in Sustainable Aviation*; John Wiley & Sons: Hoboken, NJ, USA, 2020. [CrossRef]

6. Vasilyev, A.V.; Lartsev, A.; Fedyanov, E. Evaluation of Possible Limits of Forcing of High-Capacity Air-Cooled Engines. In Proceedings of the International Conference on Industrial Engineering, Sochi, Russia, 25–29 March 2019; Springer: Berlin, Germany, 2019._18. [CrossRef]

7. Bleier, F.; Bauer, H. Prediction of Heat Transfer in a Jet Cooled Aircraft Engine Compressor Cone Based on Statistical Methods. *Aerospace* **2018**, *5*, 51. [CrossRef]

8. Jafari, S.; Nikolaidis, T.; Thermal Management Systems for Civil Aircraft Engines: Review, Challenges and Exploring the Future. *Appl. Sci.* **2018**, *8*, 11. [CrossRef]

9. Wang, D.; Naterer, G.; Wang, G.; Adaptive response surface method for thermal optimization: application to aircraft engine cooling system. In Proceedings of the 8th AIAA/ASME Joint Thermophysics and Heat Transfer Conference, St. Louis, MO, USA, 24–26 June 2002. [CrossRef]

10. Jaffe, R.; Taylor, W. *The Physics of Energy*; Cambridge University Press: Cambridge, UK, 2018.

11. Jaffe, R.; Taylor, W. *Internal Combustion Engines*; Cambridge University Press: Cambridge, UK, 2019; pp. 203–218. [CrossRef]

12. Aschemann, H.; Prabel, R.; Gross, C.; Schindele, D. Flatness-based control for an internal combustion engine cooling system. In Proceedings of the 2011 IEEE International Conference on Mechatronics, Istanbul, Turkey, 13–15 April 2011; pp. 140–145. [CrossRef]

13. Petrov, A.P.; Bannikov, S.N. Active Shutters of the Internal Combustion Engine Cooling System of a Passenger Car. In Proceedings of the Higher Educational Institutions, Palma, Spain, 1–3 July 2019; pp. 44–51. [CrossRef]

14. Razuvaev, A.; Slobodina, E. The operating conditions of the internal combustion engine with high temperature cooling *J. Phys. Conf. Ser.* **2020**, *1441*, 120–126. [CrossRef]

15. Keskinen, K.; Nuutinen, M.; Kaario, O.; Vuorinen, V.; Koch, Jann, W.; Yuri, M.; Larmi, M.; Boulouchos, K. Hybrid LES/RANS with wall treatment in tangential and impinging flow configurations. *Int. J. Heat Fluid Flow* **2017**, *65*, 141–158. [CrossRef]

16. Krastev, V.K.; Silvestri, L.; Bella, G. Effects of turbulence modeling and grid quality on the zonal URANS/LES simulation of static and reciprocating engine-like geometries. *SAE Int. J. Eng.* **2018**, *11*, 669–686. [CrossRef]

17. Krastev, V.K.; Di Ilio, G.; Falcucci, G.; Bella, G. Notes on the hybrid URANS/LES turbulence modeling for Internal Combustion Engines simulation. *Energy Procedia* **2018**, *148*, 1098–1104, DOI 10.1016/j.egypro.2018.08.047. [CrossRef]

18. Chorin, A.J. Numerical solution of the Navier-Stokes equations. *Math. Comput.* **1968**, *22*, 745–762. [CrossRef]

19. Patankar, S.V.; Spalding, D.B. A calculation procedure for heat, mass and momentum transfer in three-dimensional parabolic flows. *Int. J. Heat Mass Transfer* **1972**, *15*, 54–73. [CrossRef]

20. Anderson, W.K.; Bonhaus, D.L. An implicit upwind algorithm for computing turbulent flows on unstructured grids. *Comput. Fluids* **1994**, *23*, 1–21. [CrossRef]

21. Barth, T.; Jespersen, D. The design and application of upwind schemes on unstructured meshes. In Proceedings of the 27th Aerospace Sciences Meeting, Reno, NV, USA, 9–12 January 1989; pp. 366–390.

22. Hutchinson, B.R.; Raithby, G.D. A multigrid method based on the additive correction strategy. *Numer. Heat Transf. Part A Appl.* **1986**, *9*, 511–537.

23. Launder, B.E.; Spalding, D.B. The numerical computation of turbulent flows. *Comput. Methods Appl. Mech. Eng.* **1974**, *3*, 269–289. [CrossRef]

24. Likhachev, E.R. Dependence of Water Viscosity on Temperature and Pressure. *Tech. Phys.* **2003**, *48*, 4. [CrossRef]

MDPI
St. Alban-Anlage 66
4052 Basel
Switzerland
Tel. +41 61 683 77 34
Fax +41 61 302 89 18
www.mdpi.com

*Fluids* Editorial Office
E-mail: fluids@mdpi.com
www.mdpi.com/journal/fluids